CLAY MINERALS: THEIR STRUCTURE, BEHAVIOUR AND USE

CLAY MINERALS: THEIR STRUCTURE, BEHAVIOUR AND USE

PROCEEDINGS OF
A ROYAL SOCIETY DISCUSSION MEETING
HELD ON 9 AND 10 NOVEMBER 1983

ORGANIZED AND EDITED BY
SIR LESLIE FOWDEN, F.R.S., R. M. BARRER F.R.S.,
AND P. B. TINKER

LONDON
THE ROYAL SOCIETY
1984

Printed in Great Britain for the Royal Society
at the
University Press, Cambridge

ISBN 0 85403 232 0

First published in *Philosophical Transactions of the Royal Society of London*,
series A, volume 311 (no. 1517), pages 219–432

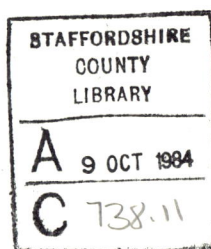

Published by the Royal Society
6 Carlton House Terrace, London SW1Y 5AG

PREFACE

Clay minerals are ubiquitous in Nature. They occur in discrete deposits, in soils and in all products of erosion and breakdown of rocks. Their special properties result from their colloidal size, layer-like structure and tendency to be electrically charged. These properties cause them to dominate the properties of soils and sediments in which they form a significant part, and to be used in very large amounts as valuable raw materials for the ceramic and many other industries.

Progress in understanding clays thus arises from a wide variety of industries and disciplines. X-ray diffraction methods have been pre-eminent for half a century, but new and sophisticated applications are allowing much greater detail to be detected, and other new techniques are being used to good effect. In particular it is clear that many natural clays are interlayered, or in other ways more complex than the standard 'type' clays.

The meeting held at the Royal Society in November 1983 allowed all these aspects to be compared and assessed. The organizers hope that this resulting volume will provide a full statement of the 'state of the art' of this multi-faceted subject.

March 1984   P. B. TINKER

CONTENTS

CONTENTS

Phil. Trans. R. Soc. Lond. A **311**, 221–240 (1984)   [221]
Printed in Great Britain

Crystal structures of clay minerals and related phyllosilicates

By G. Brown

Rothamsted Experimental Station, Harpenden, Hertfordshire AL5 2JQ, U.K.

The paper reviews the structures and crystal chemistry of clay minerals, including interstratified minerals and related phyllosilicates. Brief accounts are presented of the disordered structures encountered in the different mineral groups and a conceptual structural model for interstratified minerals is outlined.

Introduction

Clays are by definition fine-grained solids and many of their properties derive from this fact. They are hydrous silicates with layer structures and they belong to the larger group of phyllosilicates of which micas are probably the most widely known. Layer silicates are so named because the ions (or atoms) in their structures are arranged in sets of parallel planes which are strongly bonded together to form layers. Their morphology, usually thin platy crystals, reflects the underlying atomic arrangement. Clay minerals have been defined (Bailey 1980*a*) as follows: 'Clay minerals belong to the family of phyllosilicates and contain continuous two-dimensional tetrahedral sheets of composition T_2O_5 (T = Si, Al, Fe^{3+}, ...) with tetrahedra linked by sharing three corners of each with a further corner pointing in any direction. The tetrahedral sheets are linked in the unit structure to octahedral sheets and to groups of coordinated cations or individual cations.'

The essential features of phyllosilicate structures are the continuous two-dimensional tetrahedral T_2O_5 sheets, the basal oxygens of which form a hexagonal (in ideal configuration) mesh pattern (figure 1*a*). The apical oxygen at the fourth corner of the tetrahedra, usually directed normal (or nearly normal) to the sheet, forms part of an immediately adjacent octahedral sheet in which octahedra are linked by sharing edges (figure 1*b*). The junction plane between tetrahedral and octahedral sheets consists of the shared apical oxygens of the tetrahedra and unshared OH groups that lie in projection at the centre of each sixfold ring of tetrahedra. Usually coordinating cations in the octahedral sheet are Al, Mg, Fe^{3+} and Fe^{2+} but other cations such as Li, Cr, Mn, Ni, Cu, Zn occur in some species. There are two kinds of octahedral sheets. If all the octahedra are occupied the sheet is called trioctahedral; if two-thirds of the octahedra are occupied and the third octahedron is vacant, the sheet is called dioctahedral.

The unit formed by linking one octahedral sheet to one tetrahedral sheet is named a 1:1 layer and the exposed surface of the octahedral sheet consists of OH groups (figure 1*c*). A similar linkage can occur on the other side of an octahedral sheet to form a 2:1 layer, both surfaces of which consist of the hexagonal mesh of basal oxygens. Structures exist in which the 1:1 or 2:1 layers are not electrically neutral. Charge balance is maintained in such structures by interlayer material which may be individual cations as in the mica group, hydrated cations as in vermiculites and smectites, or octahedrally coordinated hydroxide groups or linked sheets of the latter, as in the chlorite minerals. The composite unit layer plus interlayer forms a structure unit, e.g. a mica unit of structure: a mica unit. The terms plane, sheet, layer and

[1]

(a)

(b)

(c)

tetrahedral
octahedral
tetrahedral

1:1 layer   2:1 layer

• tetrahedral cation   ○ oxygen
○ octahedral cation   ⊖ oxygen + hydroxyl (in projection)
                         ◎ hydroxyl group

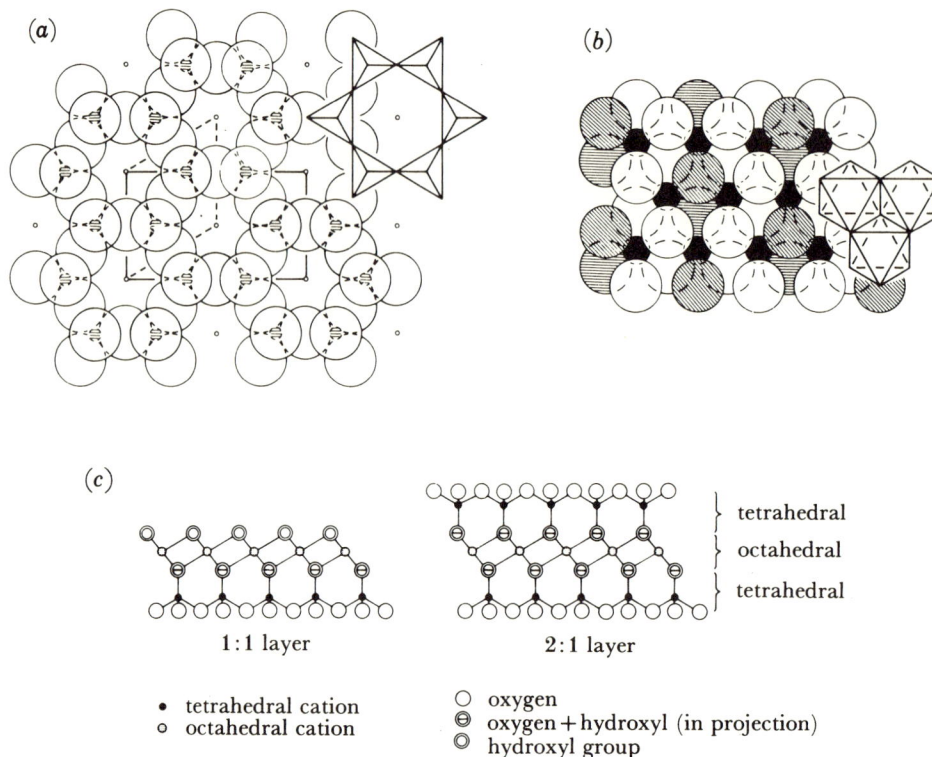

FIGURE 1. (a) Plan of ideally hexagonal tetrahedral sheet; alternative hexagonal P- (dashed) and orthogonal C-cells (full lines) are shown. (b) Octahedral sheet with inner hydroxyls of 2:1 layer shaded. (c) 1:1 and 2:1 layers (from Bailey 1980 b).

TABLE 1. CLASSIFICATION OF CLAY MINERALS AND RELATED PHYLLOSILICATES

structure type	group	charge†
1:1 layer	serpentine–kaolin	ca. 0
2:1 layer	talc–pyrophyllite	ca. 0
	smectite	ca. 0.2–0.6
	vermiculite	ca. 0.6–0.9
	mica	ca. 1.0
	brittle mica	ca. 2.0
	chlorite	variable
2:1 inverted ribbons	sepiolite–palygorskite	variable

† Negative charge per formula unit layer, $O_5(OH)_4$ for 1:1 layer minerals and $O_{10}(OH)_2$ for mica, brittle mica, vermiculite and smectite.

structure unit have precise meanings and define increasingly thicker parts of the layered arrangement.

CLASSIFICATION

Phyllosilicates can be classified on the basis of layer type (2:1 or 1:1), charge on the layers, and the type of material occurring between the layers. Such a classification is set out in table 1. Further subdivision can be made based on the nature of the octahedral sheets into dioctahedral and trioctahedral subgroups, into polytypes (or polytypoids) on the basis of the way the structure units are stacked to form the crystals, and also on the basis of chemical composition into mineral species. The structures of the major groups are shown in a schematic form in figure 2.

[2]

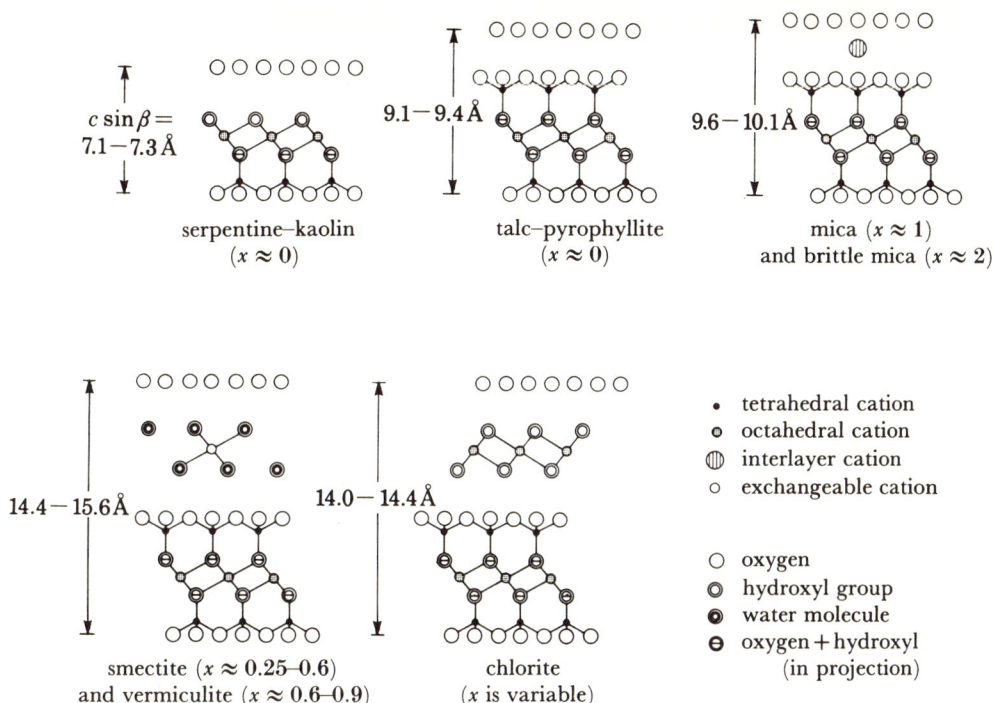

FIGURE 2. Structures of major clay minerals groups (from Bailey 1980 b).

REAL STRUCTURES

The discussion so far has been limited to idealized geometrical arrangements: hexagonal arrangements of tetrahedra in the tetrahedral sheets and regular octahedra in octahedral sheets. For tetrahedral sheets to link with octahedral sheets in a common junction plane the sheets must have similar lateral dimensions.

Dimensions of free unlinked octahedral and tetrahedral sheets can be estimated. For octahedral sheets the cell parameters of brucite, $Mg(OH)_2$, and of gibbsite and bayerite, $Al(OH)_3$, lead to b-dimensions for the conventional C-face-centred orthogonal cell of 9.4 Å and 8.65 Å† respectively. Unconstrained tetrahedral sheets do not exist in minerals but an estimate of the lateral dimensions they would have can be obtained assuming undistorted tetrahedra and T–O bond lengths found in silicates. For ideal hexagonal geometry, $b = (4\sqrt{2})(T–O)$. Structure determinations of many silicates have shown that Si–O $= 1.618 \pm 0.01$ Å and tetrahedral Al–O $= 1.748 \pm 0.01$ Å. For a tetrahedral sheet containing Si only, the lateral dimension of an ideal hexagonal array is estimated to be 9.15 ± 0.06 Å. Substitution of Al increases this and, if a linear relation between substitution of Al for Si and average T–O bond length is assumed, the corresponding b-dimension for a tetrahedral sheet of average composition $(Si_{1-x}Al_xO_5)$ would be close to $9.15 + 0.74x$. Thus the estimated b-dimension of the tetrahedral sheets in muscovite $(Si_{0.75}Al_{0.25})$ composition would be about 9.335 Å. Structural adjustments are therefore necessary to allow tetrahedral and octahedral sheets to link together. These adjustments involve distortions of both tetrahedral sheets and octahedral sheets from their ideal geometry.

Lateral dimensions of tetrahedral sheets can be decreased by twisting adjacent tetrahedra

† 1 Å $= 10^{-10}$ m $= 0.1$ nm.

[3]

in alternate directions through an angle α in the plane of the sheet. The resulting b-dimension is given by $b = b_{\text{ideal}} \cos \alpha$. The amount of lateral decrease ranges from about 1 % for $\alpha = 8.1°$ to a maximum of 13.4 % for the maximum geometrically possible value of $\alpha = 30°$. Additional adjustment can be obtained by thinning or thickening the tetrahedral sheets by changing $O_{\text{basal}}\text{-}T\text{-}O_{\text{apical}}$ angles. An increase in this angle thickens the sheet and a decrease makes the sheet thinner. Further adjustments of tetrahedral sheets can be made by tilting of tetrahedra.

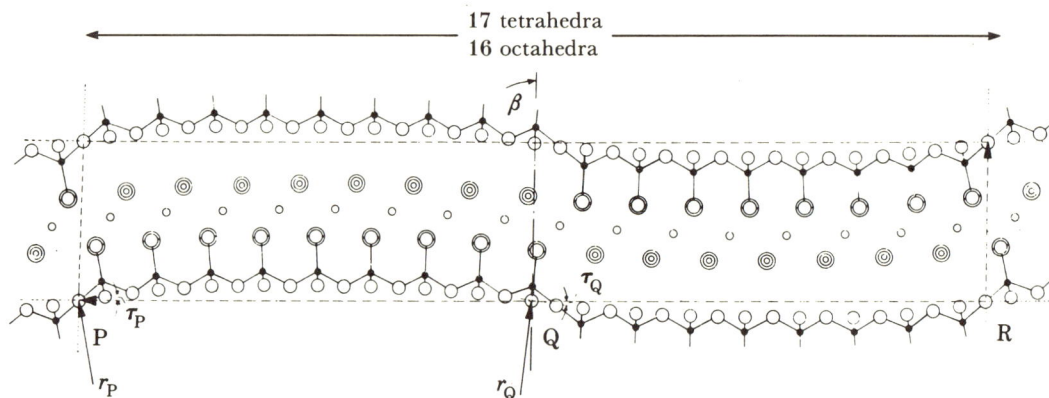

FIGURE 3. Antigorite alternating wave structure (from Bailey 1980 b).

Octahedral dimensions are altered mainly by changes in thickness. Even when as in brucite and gibbsite, they are not linked to tetrahedral sheets, octahedral sheets are enlarged because of cation repulsion. In dioctahedral structures this leads to two types of octahedra, one larger and unoccupied and two smaller containing cations. The amount of octahedral flattening can be related to the angle between the normal to the sheet and a body diagonal. For an ideal octahedron this angle $\psi = 54.73°$ and the octahedral thickness is $2(\text{M–O}) \cos \psi$, where M–O is the octahedral site to anion distances averaged over all the octahedra including those that are unoccupied. The lateral dimension of an octahedral sheet is given by

$$b = (3\sqrt{3})\,(\sin \psi)(\text{M–O}).$$

In layer silicates ψ usually lies between 58° and 60°.

In summary, by means of the distortions from ideal geometry, tetrahedral and octahedral sheets of a wide range of composition can link together and form 1:1 and 2:1 layers.

SERPENTINE FORMULAE

Many aspects of the crystal chemistry of minerals are conveniently summarized by structural formulae. In these the species and proportion of cations in differently coordinated sites are grouped together and the formula is expressed in terms of the anion content of the unit cell (or other defined structure unit).

The anions in phyllosilicates are predominantly O and OH but F may replace some or all of the OH in some species. Comparison of the number and kinds of ions in successive planes in the structure unit of pyrophyllite and muscovite, both minerals having 2:1 layers, shows the information summarized in structural formulae. Both minerals have the same anionic framework. In pyrophyllite the succession of planes of atoms in the structure unit is

negative charge: 6O:12; 4O+2OH:10; 4O+2OH:10; 6O:12;
positive charge: 4Si:16; 4Al:12; 4Si:16; 6O:12.

[4]

Charge on layer is: positive, 44; negative, 44. The electrical charge on the 2:1 layer is balanced, no interlayer material is required, and the structural formula of pyrophyllite can be written

$$(Al_4)(Si_8)O_{20}(OH)_4.$$

The corresponding succession of planes in the 2:1 layer of muscovite is

negative charge: $6O:12$; $4O+2OH:10$; $4O+2OH:10$; $6O:12$;
positive charge: $3Si+Al:15$; $Al:12$; $3Si+1Al:15$.

Charge on layer is: positive, 44; negative, 42. The 2:1 layer unit has a net negative charge which is balanced by 2 K^+ ions as interlayer material and the structural formula is

$$K_2(Al_4)(Si_6Al_2)O_{20}(OH)_4.$$

The formulae given above are expressed on the basis of an $O_{20}(OH)_4$ anion unit, the anion content of a 2:1 structure unit. The volume, V, of the structure unit in layer silicates is given by

$$V = \text{repeat period normal to the plane of the layers} \times ac$$
(the area of the base of the C-face centred orthogonal cell).

For minerals with 2:1 type layers, other than chlorites, the anionic content of the ideal structure unit is $O_{20}(OH)_4$; for minerals with 1:1 layers the content is $O_{10}(OH)_8$. Because these structure units are face-centred, structural formulae are frequently expressed on the basis of an $O_{10}(OH)_2$ formula unit for 2:1 layer minerals and $O_5(OH)_4$ for 1:1 minerals.

Serpentine minerals

The serpentine minerals have 1:1 trioctahedral layers and their structural formulae are given in table 2. Their crystal forms range from platy (lizardite, amesite) to cylindrical rolls (chrysotile). Antigorite with a periodic alternating wave structure and greenalite and caryopilite with saucer-shaped domains are intermediate. These different structures have been attributed to the degree of misfit between the dimensions of the tetrahedral and octahedral sheets.

In chrysotile, magnesian lizardite and antigorite there is little or no tetrahedral substitution so the Si-rich tetrahedral sheet is smaller than the Mg-rich octahedral sheet. The linking of the tetrahedral and octahedral sheets in chrysotile is achieved by tilting the tetrahedra which causes the 1:1 layers to form cylindrical rolls with the larger octahedral sheet on the outside. In antigorite (figure 3) the tendency to curl is periodically interrupted by inverting the tetrahedral sheet and producing an alternating wave structure which allows 17 tetrahedra to link with 16 octahedra. This leads to smaller Mg/Si and OH/Si ratios than those of the ideal formula, $Mg_3Si_2O_5(OH)_4$. In the region of inversion fourfold and eightfold rings with half the tetrahedra facing in opposite directions replace the normal sixfold rings (figure 4). Recently Guggenheim et al. (1982) have indicated that in greenalite and caryopilite a different pattern of tetrahedral inversion relieves the strain of the lateral misfit. They have postulated structures consisting of islands of six-membered rings of tetrahedra surrounded by three- and four-membered rings of inverted tetrahedra. A deficiency of octahedral cations relative to the ideal formula results and this is substantiated by chemical analysis. Structural formulae of nepouites (the Ni-rich analogues of lizardite) show similar deficiency of octahedral cations (Brindley & Wan 1975) and their powder diffraction patterns are said to be highly disordered.

[5]

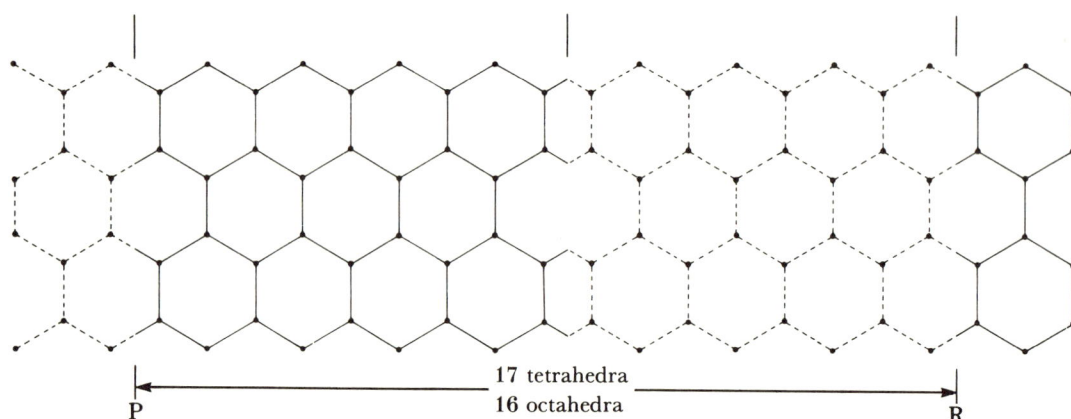

FIGURE 4. Fourfold and eightfold rings of tetrahedra in the region where tetrahedra are inverted in antigorite: sixfold rings of tetrahedra point up (full lines) and down (dashed lines) in alternating waves (from Bailey 1980 *b*).

TABLE 2. STRUCTURAL FORMULAE TYPICAL OF SOME SERPENTINE–KAOLIN GROUP MINERALS

	cations		anions
	octahedral	tetrahedral	
trioctahedral sub-group			
chrysotile	Mg_3	Si_2	$O_5(OH)_4$
lizardite	$(Mg_{2.8}Al_{0.2})$	$(Si_{1.8}Al_{0.2})$	$O_5(OH)_4$
berthierine	$[(Fe^{2+}, Mn^{2+}, Mg)_{3-x}(Fe^{3+}, Al)_x]$	$(Si_{2-x}Al_x)$	$O_5(OH)_4$
amesite	$[(Mg, Fe^{2+})_2 Al_{1.0}]$	$(Si_1 Al_1)$	$O_5(OH)_4$
cronstedtite	$(R^{2+}_{3-x}, Fe^{3+}_x)$	(Si_{2-x}, Fe^{3+}_x)	$O_5(OH)_4$
antigorite†	Mg_3	Si_2	$O_5(OH)_4$
greenalite†	Fe^{2+}_3	Si_2	$O_5(OH)_4$
dioctahedral sub-group			
kaolinite⎫ dickite ⎬ nacrite ⎭	Al_2	Si_2	$O_5(OH)_4$
halloysite – 7 Å	Al_2	Si_2	$O_5(OH)_4$ c $\frac{1}{2}H_2O$
halloysite – 10 Å	Al_2	Si_2	$O_5(OH)_4$ c $2H_2O$

† These formulae are only approximate, see text.

In the structures adopted by the planar serpentine minerals the positions of adjacent layers are governed by the pairing of the basal oxygens in the surface of the tetrahedral sheet of one layer with hydroxyls on the surface of the adjacent layer to form hydrogen bonds (figure 5). There are two possible sites for octahedral cations in a given 1:1 layer (figure 6) and there are three relative positions of adjacent 1:1 layers that allow the O–OH pairing required for interlayer hydrogen bonds (Bailey 1969). Bailey (1969) has shown that if only one of the three relative positions of adjacent layers is allowed in the same crystal there are twelve trioctahedral polytypes, i.e. structures consisting of layers of the same composition differing only in the stacking sequence. There is a tendency for certain compositions to adopt specific stacking sequences; these structures are more correctly called polytypoids rather than polytypes. Different polytypes lead to different patterns of superimposition of ions and hence to the balance of attractive and repulsive forces in the structure, and it has been found that structures that have the most energetically favourable arrangements are more abundant naturally.

[6]

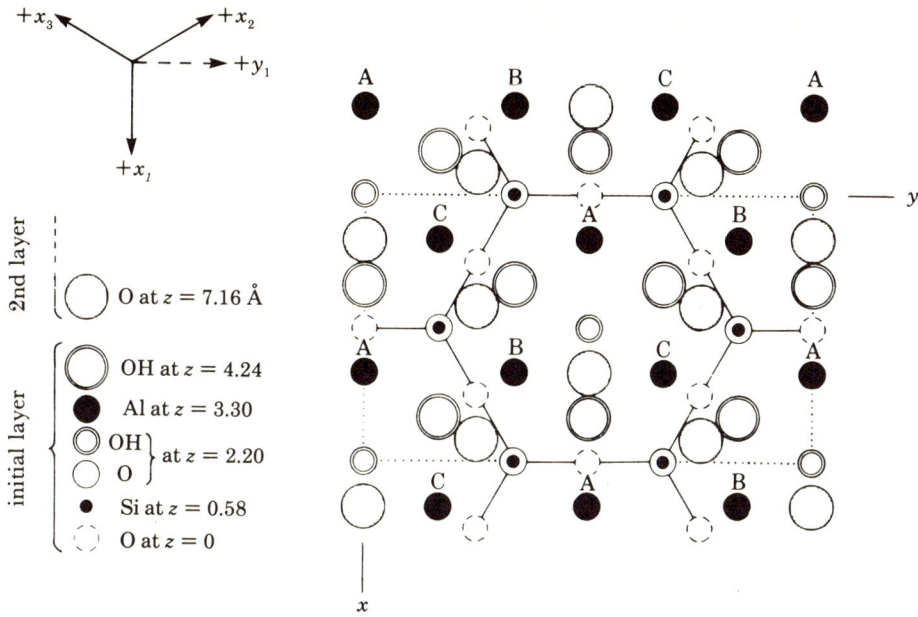

FIGURE 5. Pairing of basal tetrahedral oxygens (stippled) of upper 1:1 layer with surface hydroxyls (large double circles) of layer below (from Bailey 1980b).

(a)

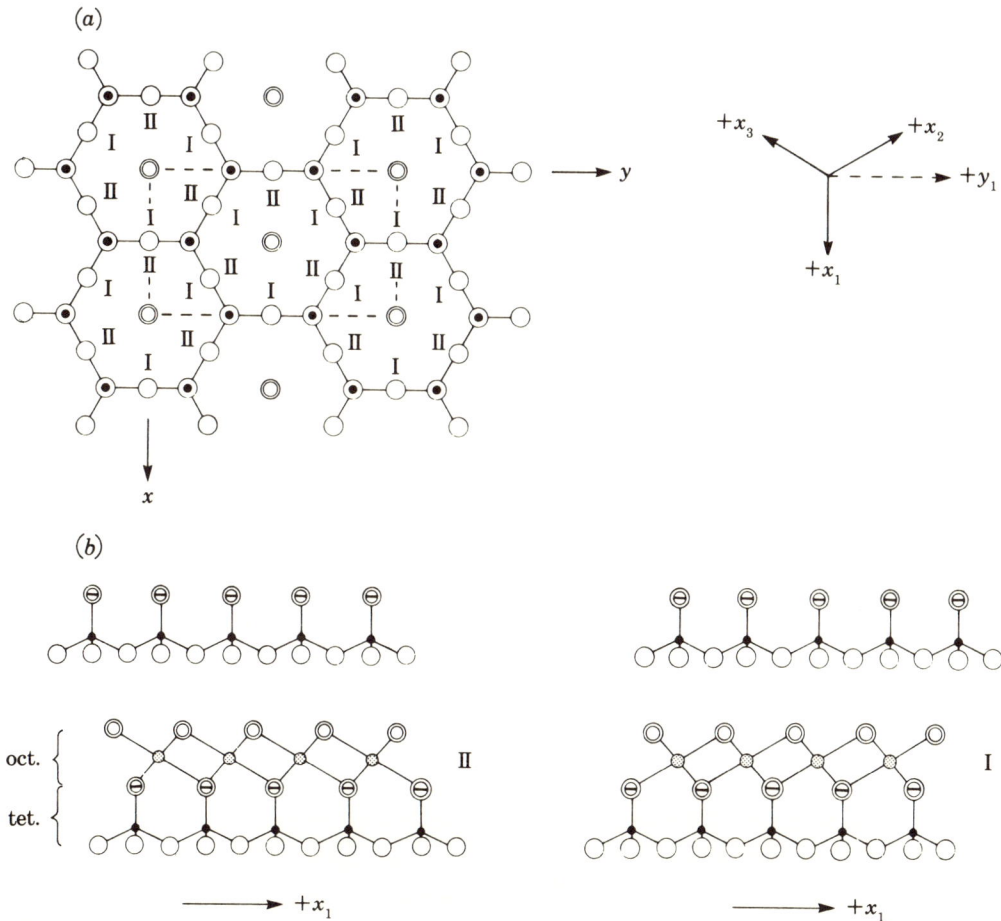

(b)

FIGURE 6. (a) Sets I and II of octahedral cation positions above a tetrahedral sheet of a 1:1 layer relative to a fixed set of hexagonal axes. (b) Octahedral sheets in different orientations due to occupancy of set I or set II octahedral cation sites (from Bailey 1980b).

[7]

Partly disordered layer stacking is common in serpentine minerals that arise from $\frac{1}{3}b$ displacements at interlayer junctions and disorder in the sets of octahedral cations sites that are occupied.

KAOLIN MINERALS

The kaolin minerals consist of 1:1 dioctahedral layers and their structural formulae are given in table 2. There are three polytypes kaolinite, dickite and nacrite. Only two of the three octahedral sites are occupied in any 1:1 layer and the larger vacant octahedral site is surrounded by six smaller occupied sites (figure 7). Zvyagin (1962) showed that there were 52 possible dioctahedral structures with periodicities between one and six layers. From analysis of attractive and repulsive forces between adjacent layers Newnham (1961) showed that dickite and nacrite should be the most stable 2-layer forms and kaolinite the most stable 1-layer form. Bailey (1963) has shown that the structures of kaolinite, dickite and nacrite differ in the arrangement of vacant and occupied octahedral cation sites in successive layers.

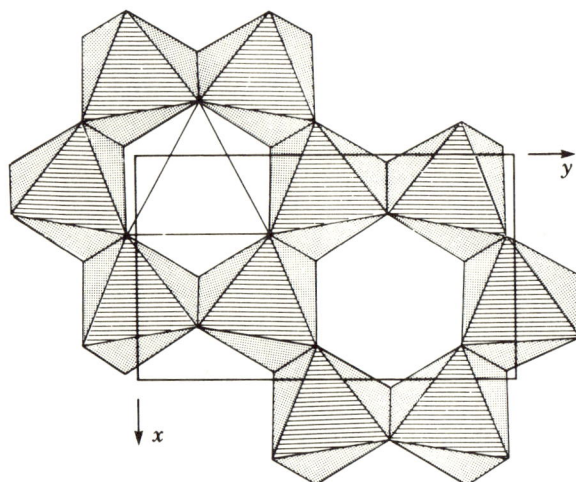

FIGURE 7. Occupied octahedra (shaded) surrounding larger vacant octahedral site in dickite structure (from Bailey 1980 b).

Partly disordered kaolin minerals are common. Careful analysis of powder diffraction patterns by Plançon & Tchoubar (1975, 1976, 1977) showed that faults involving displacement of octahedral cation vacancies and stacking faults by linear displacements of layers by $\frac{1}{3}b$ and by arbitrary distances were involved. Their results showed that the most important type of defect was that arising from variations in occupancy of the three types of octahedral cation sites. Faults arising from linear displacements of adjacent layers occurred but were of less importance.

Halloysite minerals show morphologies that range from the most common tubular form to pseudo-spherical forms. In addition the 10 Å form has been reported to exist as in a platy form. Halloysite consists of 1:1 dioctahedral layers that are stacked turbostratically. It exists in two forms, a less-hydrated 7 Å form in which there is little or no water between the layers, and a 10 Å-hydrated halloysite which has a single sheet of water molecules between the layers. In partly dehydrated halloysite some of the interlayers contain water, others do not, and hydrated and dehydrated interlayers appear to be in a random sequence.

TALC AND PYROPHYLLITE

The minerals of this group have neutral 2:1 type layers. Structural formulae are given in table 3. Talc and the disordered fine grained varieties kerolite (Mg-rich) and pimelite (Ni-rich) have trioctahedral layers; pyrophyllite has dioctahedral layers.

The electrically neutral layers of pyrophyllite and talc are held together by van der Waals bonding reinforced by a small ionic attraction. Layer stacking is restricted neither by the need to superimpose rings of tetrahedral basal oxygens around interlayer cations nor by arrangements required for hydrogen bonding. In ordered structures oxygens in adjacent layers do not adopt the closest possible packing arrangement. In both talc and pyrophyllite four of the six basal oxygens of one layer superimpose between two oxygens of the hexagonal ring of basal oxygens in the next layer. This arrangement leads to a minimum Si–Si repulsion between adjacent layers.

TABLE 3. STRUCTURAL FORMULAE OF TALC–PYROPHYLLITE GROUP MINERALS

	cations		anions
	octahedral	tetrahedral	
trioctahedral			
talc	Mg_3	Si_4	$O_{10}(OH)_2$
kerolite	Mg_3	Si_4	$O_{10}(OH)_2$ c $1H_2O$
dioctahedral			
pyrophyllite	Al_2	Si_4	$O_{10}(OH)_2$

Partial disorder in layer stacking is common in pyrophyllite and talc. Layer displacements of $\frac{1}{3}b$ lead to identical positions of the atoms in the surfaces of opposing layers and hence displacements occur easily.

Kerolite is a very fine-grained material with composition identical to talc except that more water is found by analysis. Layer stacking appears to be completely turbostratic. Thus the oxygens in the surface of adjacent layers do not interweave as they do in talc and as a result the layer spacing is slightly larger than talc. It is also possible that occasional water interlayers or occluded water molecules are situated between the talc-like layers.

MICAS

The 2:1 layers in micas are negatively charged and this charge is balanced by large univalent cations, usually potassium. As in the other mineral groups discussed previously there are trioctahedral and dioctahedral subgroups. Table 4 summarizes the structural formulae of some common micas. The layer charges arise by (i) substitution of Al for Si in tetrahedral sites, as in muscovite, and (ii) substitution of univalent or divalent cations for the divalent or trivalent octahedral cations in trioctahedral and dioctahedral micas respectively, as in celadonite.

The stacking of 2:1 layers of micas is controlled by the 'keying' of the large interlayer cation into the ideally hexagonal rings of surface oxygens in adjacent layers. Frequently because of tetrahedral rotation the surface oxygen arrangement is ditrigonal. Smith & Yoder (1956) derived the possible stacking sequences that apply to both trioctahedral and dioctahedral subgroups. Based on the assumption that the same interlayer stacking angle is found between all the layers in a given crystal, they showed that there are only six possible polytypes. They used

[9]

a convenient and now widely accepted nomenclature for these in terms of the symmetry of the polytype and the number of mica units in the repeat unit. These polytypes are $1M$ (one-layer monoclinic), $2M_1$ and $2M_2$ (both two-layer monoclinic structures), $2Or$ (two-layer orthorhombic), $3T$ (three-layer trigonal) and $6H$ (six-layer hexagonal). The $1M$, $2M_1$ and $3T$ structures are most abundant because in these the surface oxygen arrangement is ditrigonal (owing to tetrahedral twist), and the coordination of the large interlayer cation is octahedral with respect to its nearest oxygen neighbours. In the $2M_2$, $2Or$ and $6H$ structures, the six oxygens around the interlayer cation form a trigonal prism in which three surface oxygens of one layer superimpose on three oxygens of the adjacent layer, presumably a less energetically favourable situation.

TABLE 4. STRUCTURAL FORMULAE OF SOME MICA GROUP MINERALS INCLUDING CLAY MICAS

	interlayer	cations octahedral	tetrahedral	anions
		trioctahedral		
phlogopite	K	Mg_3	(Si_3Al)	$O_{10}(OH)_2$
biotite	K	$(MgFe^{2+})_3$	(Si_3Al)	$O_{10}(OH)_2$
		dioctahedral		
muscovite	K	Al_2	(Si_3Al)	$O_{10}(OH)_2$
paragonite	K	Al_2	(Si_3Al)	$O_{10}(OH)_2$
phengite	K	$(Al_{1.5}Mg_{0.5})$	$(Si_{3.5}Al_{0.5})$	$O_{10}(OH)_2$
celadonite	K	$(AlFe^{3+})_1(MgFe^{2+})_1$	Si_4	$O_{10}(OH)_2$
glauconite	$K_{0.85}$	$(R^{3+}_{1.4}R^{2+}_{0.6})$	$(Si_{3.75}Al_{0.25})$	$O_{10}(OH)_2$
illite	$K_{0.8}$	$(R^{3+}_{1.65}R^{2+}_{0.35})$	$(Si_{3.55}Al_{0.45})$	$O_{10}(OH)_2$

If only an occasional mistake occurs in layer stacking angles, twinned crystals result. If the stacking angle varies regularly or randomly, disordered crystals are formed. In addition to variation in layer stacking angles, disorder may arise from variations in the distances between different 2:1 layers. Kodama *et al.* (1971) from a careful analysis of line profiles of basal reflections concluded that variation in layer to layer separation was related to the number of interlayer cations per formula unit, ideally 1. In the samples they studied these ranged from 0.913 to 0.837 per formula unit.

Micas that occur widely in fine grained form are illite, glauconite and celadonite. Celadonite appears to have an ordered $1M$ structure; illite and glauconite are $1M$ disordered. Illite and glauconite units commonly occur interstratified with smectite layers and it is difficult to detect with certainty the presence of less than 10% swelling interlayers. Even so, whereas celadonite appears to have close to one large cation per formula unit, glauconites and illites that show no evidence of interstratification rarely contain as many as 0.8 large cations per formula unit. Another feature of illite (and glauconite) is that chemical analyses show more structural water per formula unit than well-crystallized micas (Hower & Mowatt 1966; Čimbálníková 1971). The modern consensus is that this water is located between occasional 2:1 layers either as water molecules or lenses of water trapped between layers.

BRITTLE MICAS

Brittle micas are distinguished from true micas in having a negative charge of 2 per formula unit. Table 5 gives ideal structural formula for clintonite (trioctahedral) and margarite (dioctahedral). The layer charge may originate entirely in the tetrahedral sheet, or partly in the tetrahedral sheet and partly in the octahedral sheet. Clintonite with $SiAl_3$ in the tetrahedral sheet has the largest tetrahedral twist, $\alpha = 23°$, found in layer silicates which decreases the lateral dimensions of the sheet by about 10%. Margarite, the dioctahedral variety, is of interest because it has been shown that in one specimen the Al content of the two tetrahedral sheets in the same 2:1 layer differ by about 10% (Guggenheim & Bailey 1975).

TABLE 5. STRUCTURAL FORMULAE OF BRITTLE MICA GROUP MINERALS

	interlayer	cations octahedral	tetrahedral	anions
		trioctahedral		
clintonite	Ca	(Mg_2Al)	$(SiAl_3)$	$O_{10}(OH)_2$
		dioctahedral		
margarite	Ca	Al_2	(Si_2Al_2)	$O_{10}(OH)_2$

TABLE 6. STRUCTURAL FORMULAE OF SOME CHLORITE GROUP MINERALS

	hydroxide sheet cations	anions	2:1 layer cations octahedral	tetrahedral	anions
			trioctahedral		
	$(R^{2+}_{1.4}R^{3+}_{1.6})$	$(OH)_6$	R^{2+}_3	$(Si_{2.4}Al_{1.6})$	$O_{10}(OH)_2$
	$(R^{2+}_{1.1}R^{3+}_{1.5}\square_{0.4})$	$(OH)_6$	R^{2+}_3	$(Si_{3.3}Al_{0.7})$	$O_{10}(OH)_2$
			di-trioctahedral		
sudoite	$(R^{2+}_{2.1}R^{3+}_{0.9})$	$(OH)_6$	$Al_{2.1}$	$(Si_{2.8}Al_{1.2})$	$O_{10}(OH)_2$
cookeite	(Al_2Li)	$(OH)_6$	Al_2	(Si_3Al)	$O_{10}(OH)_2$
			di-dioctahedral		
donbassite	$(Al_{2.2}Mg_{0.1})$	$(OH)_6$	Al_2	$(Si_{3.2}Al_{0.8})$	$O_{10}(OH)_2$

CHLORITES

The chlorite structure consists of negatively charged 2:1 layers that are regularly interleaved with positively charged sheets of cations octahedrally coordinated to hydroxyls. In addition to the electrostatic attraction between the negatively charged 2:1 layer and the positively-charged hydroxide sheet, the hydroxyls of the hydroxide sheet form hydrogen bonds with the surface oxygens of the 2:1 layers on either side.

There are two octahedrally coordinated sheets in the chlorite structure, one in the 2:1 layer and another in the hydroxide interlayer sheet. These are either both trioctahedral (trioctahedral chlorites), dioctahedral in the 2:1 layer and trioctahedral in the hydroxide sheet (di-trioctahedral) or dioctahedral in the 2:1 layer and in the hydroxide sheet (di-dioctahedral). Structural formulae for these different chlorites are given in table 6. Tri-dioctahedral varieties may be possible but have not yet been recognized naturally.

Chlorites are usually either regular one-layer structures or more commonly irregular structures with $\frac{1}{3}b$ displacements. Bailey & Brown (1962) have shown that there are twelve possible different regular chlorite polytypes designated I or II, according to which of the two sets of octahedral cation positions are occupied in the interlayer sheet, *a* or *b*, according to the position of the interlayer sheet relative to the plane of surface oxygens of an underlying 2:1 layer, and 1–6, depending on the position of the second 2:1 layer above the interlayer sheet (figure 8).

Many chlorites have what has been termed semi-random stacking sequences. Their X-ray diffraction patterns indicate that adjacent 14 Å thick chlorite units (2:1 layer plus hydroxide sheet) are irregularly stacked but the layers and interlayers maintain hydrogen bonding and only occur in positions that are related by $\pm \frac{1}{3}b$ shifts along the three pseudo-hexagonal Y-axes. These semi-random structures allow irregular stacking sequences of either the three a or three b positions of the interlayer sheet on the original 2:1 layer and of the 2,4,6 ('even') or 1,3,5 ('odd') positions of the following 2:1 layer on the hydroxide sheet. Because of equivalence of some of the structures, there are found to be six semi-random structures. Four of these are based on a monoclinic-shaped cell with $\beta = 97°$, Ia-even, Ib-even, IIa-odd, IIb-even, and two

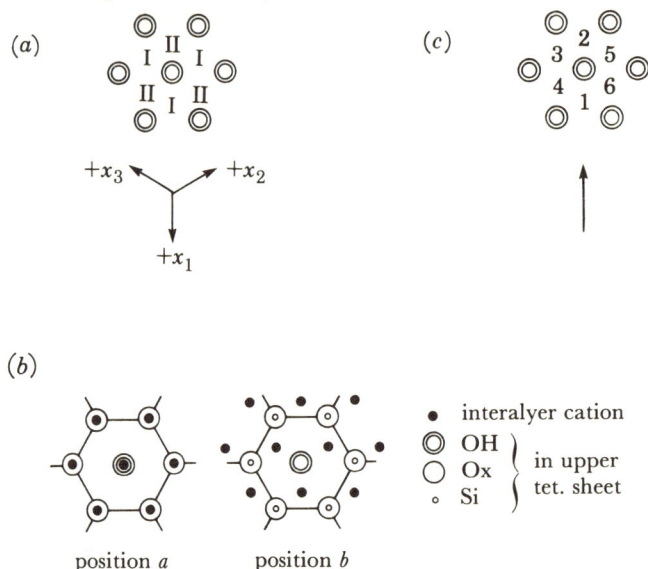

FIGURE 8. (a) Choice of octahedral cation sets I and II in interlayer hydroxyl sheet in chlorite. (b) Two different positions a and b of cations of interlayer hydroxyl sheet over upper tetrahedral sheet of underlying 2:1 layer. (c) Six positions for centre of hexagonal ring in 2:1 layer above interlayer relative to upper hydroxyl plane of interlayer sheet (from Bailey 1980b).

are based on an orthohombic-shaped cell $\beta = 90°$; Ib-odd and IIa-even. These six semi-random sequences can be recognized from X-ray powder diffraction patterns. Four have been found to occur naturally, IIb-even (80%), Ib-odd, $\beta = 90°$ (12%), Ib-even, $\beta = 97°$ (4%), Ia-even, $\beta = 97°$ (3%). The relative abundance of these structures can be explained on the basis of differences in attractive and repulsive forces of which repulsion of the superposed interlayer cations and tetrahedral cations is probably the most important in decreasing the stability of Ia and IIa structural units relative to Ib and IIb. Bailey (1975) has given a detailed analysis of the structural factors affecting the relative stabilities of chlorite structures.

VERMICULITES

Vermiculites and smectites are minerals whose 2:1 layers have a smaller negative charge than micas. As indicated in table 1, the layer charge on 2:1 layers in vermiculites ranges from about 0.6–0.9 per formula unit; in smectites the layer charge lies in the range 0.2–0.6. Vermiculites are trioctahedral; dioctahedral varieties may exist but have not been found free of other minerals in nature. A typical structural formula for trioctahedral vermiculite is

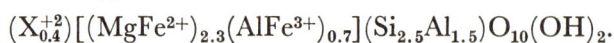

$$(X_{0.4}^{+2})[(MgFe^{2+})_{2.3}(AlFe^{3+})_{0.7}](Si_{2.5}Al_{1.5})O_{10}(OH)_2.$$

[12]

Because vermiculites and smectites have smaller layer charge than micas the attractive forces between the 2:1 layers and the interlayer cation are less. The hydration energy of the interlayer cation may then be sufficient to overcome the attractive forces of the layer to the cations and allow water to hydrate the interlayer cation which causes swelling normal to the plane of the layers. The ability of vermiculites and smectites to swell in water allows cation exchange between the interlayer cation and cations in an external solution. Both groups of minerals can also sorb organic cations by cation exchange and other organic molecules by solvation of the interlayer cations. In fact a widely used diagnostic test for identifying vermiculites and smectites is based on the amount of swelling when ethylene glycol or glycerol is sorbed between the 2:1 layers. Generally, vermiculites swell less than smectites because the interlayer cation to 2:1 layer attractive force is greater.

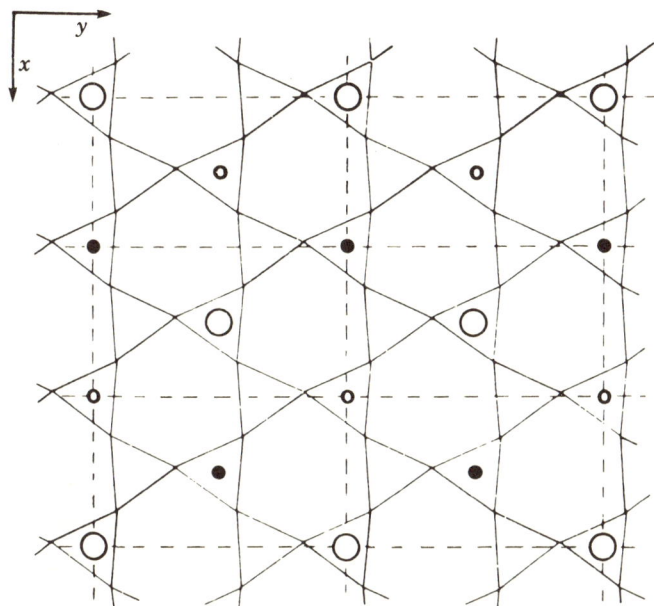

FIGURE 9. Superlattice of dimensions $3a \times b$ caused by short-range order in distribution of exchangeable inter-layer cations (large circles) in local domains in vermiculite. Smaller solid or open circles are possible sets of cation positions in adjacent domains (from Bailey 1980 b).

The crystal structure of the 14.36 Å phase of Mg-vermiculite (Mathieson & Walker 1954; Mathieson 1958; Shirozu & Bailey 1966) shows that the exchangeable Mg^{2+} cations lie in a plane midway between adjacent 2:1 layers and have a plane of water molecules on either side. The interlayer material is similar to the interlayer sheet in chlorite but with about $\frac{1}{9}$ of the cation sites occupied and $\frac{2}{3}$ of the hydroxyl sites of chlorite occupied by water molecules in vermiculite. The tetrahedral cation sites in vermiculite are ordered, the T_1 site having more Al than the T_2 site and the interlayer cations are found in those sites which lie between the Al-rich tetrahedra in adjacent layers. Although all the chlorite stacking sequences should be possible, only the I a chlorite structure has been found in vermiculites. I a chlorites are less stable than II b and I b types because of repulsion between superposed interlayer and tetrahedral cations. This cation repulsion is less in vermiculite because there are fewer interlayer cations and their situation in the sites that give local charge balance may be the factor that causes the I a type structure to be favoured.

[13]

The interlayer cations and water molecules in vermiculite appear on average to be randomly distributed over the possible sites. Alcover *et al.* (1973) have shown short-range ordering of interlayer cations over a face-centred superlattice with dimensions $3a \times b$ (figure 9). This ordering is two-dimensional only, so that cation distribution in one interlayer is not related to that in the next and may even change in different regions of the same interlayer.

The structural similarities of chlorite and vermiculite give rise to materials of intermediate type which may have hydrated cations and islands of hydroxide material in the interlayer. Minerals of this kind are common and appear to be formed by the alteration of micas in moderately acid environments, and give rise to materials that have been referred to as dioctahedral vermiculite, soil chlorites or smectite intergrades commonly found in soils (Brown 1953; Rich 1968).

SMECTITES

As discussed in the section on vermiculites, smectites have 2:1 layers with layer charge 0.2–0.6 per $O_{10}(OH)_2$ formula unit and have the ability to swell by absorption of water and other substances between the 2:1 layers. The interlayer cations are usually easily exchangeable with these in external solutions.

TABLE 7. IDEAL STRUCTURAL FORMULAE OF SOME SMECTITE GROUP MINERALS

	interlayer†	cations octahedral	tetrahedral	anions
		trioctahedral sub-group		
saponite	$X^{+1}_{0.5}$	Mg_3	$(Si_{3.5}Al_{0.5})$	$O_{10}(OH)_2$
hectorite	$X^{+1}_{0.3}$	$(Mg_{2.7}Li_{0.3})$	Si_4	$O_{10}(OH)_2$
		dioctahedral sub-group		
montmorillonite	$X^{+1}_{0.35}$	$(Al_{1.65}Mg_{0.35})$	Si_4	$O_{10}(OH)_2$
beidellite	$X^{+1}_{0.4}$	Al_2	$(Si_{3.6}Al_{0.4})$	$O_{10}(OH)_2$
nontronite	$X^{+1}_{0.4}$	Fe^{3+}_2	$[Si_{3.6}(Al, Fe^{3+})_{0.4}]$	$O_{10}(OH)_2$

† X denotes an exchangeable cation.

Both trioctahedral and dioctahedral smectites occur with a wide range of compositions. Table 7 gives structural formulae for some of the more common varieties. The charge on the layers arises from tetrahedral substitution, octahedral substitution or a balance of both. In the dioctahedral subgroup ideal montmorillonite has octahedral substitution only; beidellite and nontronite have tetrahedral substitution. Many dioctahedral smectites show both octahedral and tetrahedral substitutions. In saponites, trioctahedral Li-poor smectites, tetrahedral substitution predominates whereas in hectorite, a Li-rich trioctahedral smectite, the charge arises from replacement of some divalent Mg^{2+} octahedral cations by Li ions. Stevensite is a trioctahedral smectite with less than the ideal octahedral occupancy of three cations and little or no tetrahedral Al for Si substitution. The deficiency of octahedral cations gives rise to the negative charge on the layers.

Layer stacking in most smectites is random, there being little or no relation in the *ab*-plane in the positions of adjacent layers. Some saponites and beidellites, notably those with higher charge on the 2:1 layers, show some ordering in layer stacking when the appropriate interlayer cations are present and when in appropriate hydration states.

[14]

Crystalline swelling of smectites in water is caused by hydration of the interlayer cation. The amount of swelling depends mainly on relative humidity and the nature of the interlayer cation. When Na is the interlayer cation, hydrates with one, two and three layers of water molecules in the interlamellar space are formed and at large relative humidities 2:1 layers may disperse. With Ca interlayer cations, swelling appears to be limited to the three water-layer hydrate in which the 2:1 layers are separated by about 9 Å. The two-layer hydrate is stable over a wide range of relative humidities (*ca.* 0.98–0.3, Ormerod & Newman 1983), and the one-layer hydrate is found at small relative humidities.

SEPIOLITE AND PALYGORSKITE

Both sepiolite and palygorskite have a lath-like or fibrous morphology that reflects their crystal structures (figures 10 and 11). Both contain continuous planes of oxygen about 6.5 Å apart that are the basal oxygens of dominantly Si tetrahedra. Apical oxygens point up and

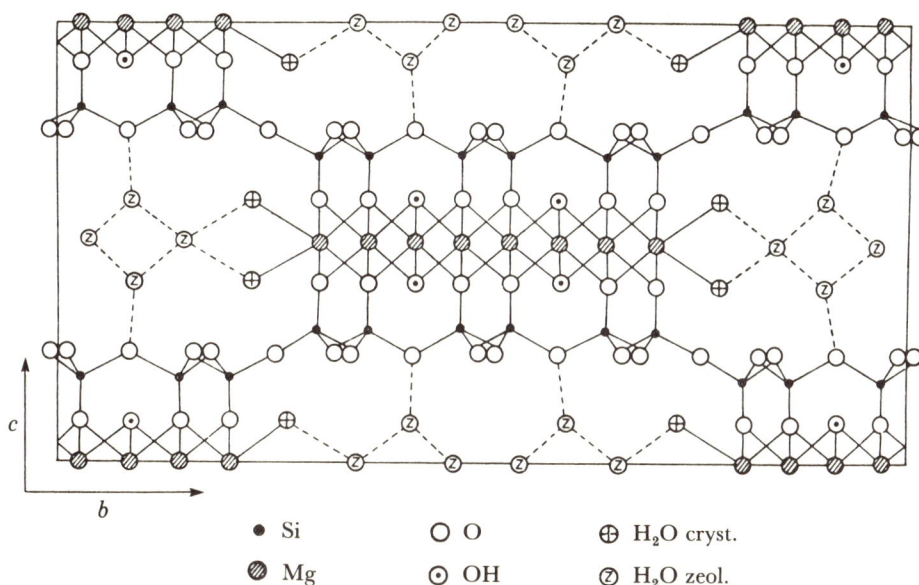

FIGURE 10. Structure of sepiolite projected along the fibre axis (Bailey 1980 *b*).

down from the basal oxygen planes so that tetrahedra pointing in the same direction form ribbons in the 5.3 Å repeat fibre direction. The tetrahedra in the ribbons with apices facing up and down are linked via octahedral cations thus forming a 2:1 structure continuous in the fibre direction but limited in extent to the direction perpendicular to the ribbons. The ribbons have a width of three linked pyroxene chains in sepiolite and two pyroxene chains in palygorskite. Thus the basal oxygens of the tetrahedra form continuous two-dimensional planes but with apices that point either up or down in adjacent ribbons. In the ideally rectangular channels between the ribbons there are exchangeable Ca and Mg ions and and variable amounts of zeolitic water. Water molecules denoted (OH$_2$) in the structural formulae on table 8 are coordinated to octahedral Mg or Al at the edges of the ribbons and are more strongly held than the zeolitic water in the channels. Evidence has been presented (Yücel *et al.* 1981) that in less-well-crystallized sepiolites the silicate ribbons are not parallel but form a 'herringbone'

[15]

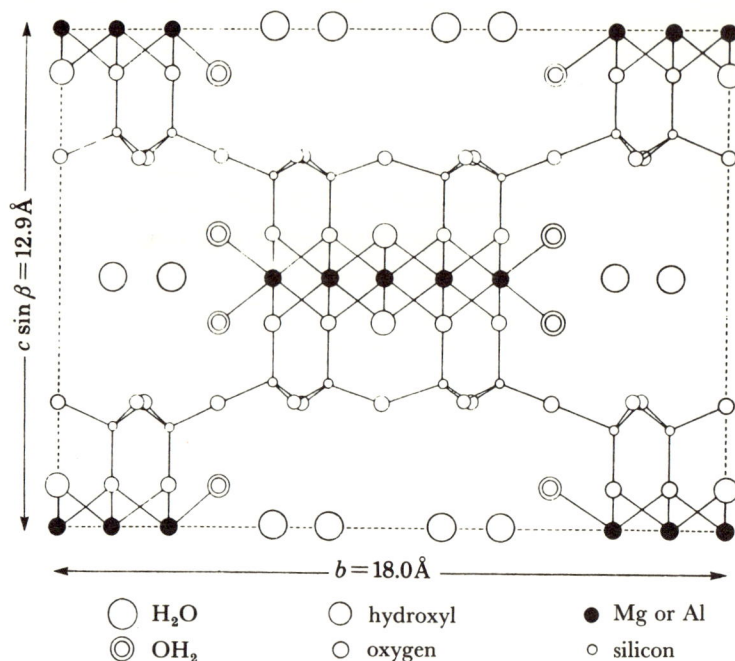

FIGURE 11. Structure of palygorskite projected along fibre axis (from Bailey 1980 b).

TABLE 8. STRUCTURAL FORMULAE OF SEPIOLITE–PALYGORSKITE GROUP MINERALS

	2:1 ribbons			channels	
	cations		anions		
	octahedral	tetrahedral		exchangeable cations	
		trioctahedral			
sepiolite (ideal)	Mg_8	Si_{12}	$O_{30}(OH)_4(OH_2)_4$		
	$(Mg_{7.3}Al_{0.5})$	$(Si_{11.8}Al_{0.2})$	$O_{30}(OH)_4(OH_2)_4$	$X^{+1}_{0.1}$	$0.8H_2O$
		dioctahedral			
palygorskite (ideal)	Mg_5	Si_8	$O_{20}(OH)_2(OH_2)_4$		
	(Mg_2Al_2)	$(Si_{7.8}Al_{0.2})$	$O_{20}(OH)_2(OH_2)_4$	$X^{+1}_{0.2}$	$0.4H_2O$

pattern by small rotations of the ribbons in alternate directions about the fibre axis at the corners of the ribbons.

INTERSTRATIFIED MINERALS

Interstratification is the term used to describe phyllosilicate materials in which structural units of different kinds occur in the stacking sequence normal to the plane of the layers.

Interstratified materials arise in two ways. Some clay minerals (notably smectites, vermiculites and halloysite) can sorb inorganic (e.g. water and cations) or organic (cations, alcohols, etc.) material between the layers. This process may not take place equally in all the interlayers at the same time so more than one kind of interlayer separation in the same crystallite may result. Second, because the 2:1 and 1:1 layers are strongly bonded internally but relatively weakly bonded to each other and because the surface planes of the different kinds of layers and the hydroxide sheet of chlorites are geometrically similar, layers with different internal arrangements can stack together and articulate well at their interfaces.

[16]

Interstratification may be random, in which no discernible pattern exists in the stacking sequence, ordered, with regular periodic stacking schemes, and partly ordered, in which there is a tendency for ordering to exist.

Regularly ordered sequences have been found for two-component systems only. Those which were judged (Bailey 1982) to be sufficiently regular to be allocated mineral names, plus hydrobiotite (Brindley et al. 1983), are listed in table 9. In these the stacking sequence of structure units is ABABAB ... and the repeat distance normal to the layers is the sum of that of the components. Another two-component system, named tarasovite (Lazarenko & Korolev 1970) and described by them as an interstratification of mica and rectorite (itself a regular 1:1 interstratification of mica and smectite units) has been included although it does not quite meet Bailey's (1982) criteria. In this mineral ideally the layer sequence would be ... AAABAAABA ... in which A represents a mica unit and B represents a smectite unit. Brindley & Suzuki (1983) have examined this material and report that there is a small excess of mica units over and above those required for the ideal sequence. These they believe to be randomly interstratified among the AAAB units.

TABLE 9. REGULARLY INTERSTRATIFIED MINERALS

mineral	component layers
	1:1 regular
aliettite	talc–saponite
corrensite	trioctahedral chlorite–trioctahedral smectite
	trioctahedral chlorite–trioctahedral vermiculite
hydrobiotite	biotite–trioctahedral vermiculite
kulkeite	trioctahedral chlorite–talc
rectorite	dioctahedral mica–dioctahedral smectite
tosudite	chlorite–smectite (dioctahedral)
	2:1 regular
tarasovite	mica–rectorite (3:1 mica–smectite)

Less regularly ordered sequences than those in table 9 are very common ranging from partly ordered to random. The following two-component systems have been reported: illite–smectite, glauconite–smectite, mica–vermiculite, mica–dioctahedral chlorite, smectite–chlorite, chlorite–vermiculite, kaolinite–smectite, and mica–chlorite. Three component systems, illite–chlorite–smectite and illite–smectite–vermiculite are also found (Reynolds 1980). In addition Eberl et al. (1982) have described a kerolite–stevensite interstratification.

The need for precise use of the terms outlined earlier – 'plane', 'sheet', 'layer', 'structure unit' and 'interlayer' – for the various components of layer silicate structures is specially evident in discussion of interstratified minerals and in hypotheses about their structure and paragenesis.

A CONCEPTUAL MODEL FOR INTERSTRATIFIED MINERALS

Consider interstratifications of mica units (M) and smectite units (S). The mica units consist of 2:1 layers with a negative charge of about 1 equivalent per formula unit which is balanced by an interlayer K ion. The smectite units are 2:1 layers having a negative charge of, say, about 0.3 equivalents per formula unit balanced by a solvated interlayer exchangeable cations. The basic assumption is that local charge balance is maintained across the interlayer between

[17]

adjacent 2:1 layers and that the charges on the surfaces of the 2:1 layers on either side of an interface are equal. In pure mica and smectite, the stacking sequences are ... MMMM ... and ... SSSSS ... respectively. In both mica and smectite structures (and in all non-interstratified structures) there is one kind of interlayer and one kind of interface and interlayer. Under the assumptions made above the mica 2:1 and smectite 2:1 layers must be homogeneous or nearly so, i.e. have about the same negative charge on the silicate oxygen surfaces on each side of the 2:1 layer that face each other across every interlayer. We can symbolize these by ↔↔ for mica and — — for smectite 2:1 layers, where an arrowhead represents a high-charge half-layer and no arrowhead, a low-charge half-layer, and the distance between the symbols indicates the separation of adjacent 2:1 layers.

Next consider the regularly interstratified mineral rectorite in which the unit sequence is MSMSMSMS .... . To obtain two kinds of interlayer regions the material must contain polar 2:1 layers (i.e. layers with asymmetric charge distribution) that have high charge (mica-like) on one side of each 2:1 layer and low charge (smectite-like) on the other side. Layers of this type, symbolized ← can be stacked →← →← to provide an alternation of high-charged non-expanding K-containing interlayers with low-charged expanding interlayers that contain exchangeable cations and water. The rectorite structure would then have one kind of layer, two kinds of interfaces and two kinds of interlayers. An ideal tarasovite, MMMSMMMS ..., which consists of 2:1 layers in which three non-expanding interlayers are regularly followed by an expanding interlayer can be symbolized

$$→↔↔←\quad→↔↔←$$

It has two kinds of interfaces in the proportion of three non-expanding to one expanding, and two kinds of 2:1 layers, one non-polar and the other polar, and polar and non-polar layers that are equally abundant. The sequence can be readily adjusted to include the small extra proportion of mica units found by Brindley & Suzuki (1983) in their tarasovite specimen by the addition of occasional extra non-polar high charge 2:1 layers, ↔, into the middle of the SMMMS packets. Such a sequence would be

$$→↔↔←\quad→↔↔↔←\quad→↔↔←$$

The reverse of this mechanism would allow for smaller proportions of non-expanding interlayers in tarasovite. The process of addition of extra non-polar high charge 2:1 layers (i.e. more than 50% mica units) could also apply to the MSMS ordered rectorite structure, but removal of such layers from the ideal rectorite (50% M 50% S units) is not possible. The fact that partly ordered MS structures appear to be found only when more than 50% of the interlayers do not expand may be related to this observation.

It is of interest to note that the regular sequence ... SMMS ... has not yet been reported and that Reynolds & Hower (1970) found that only models based on ... SMSM ... ('allevardite-type ordering') and ... SMMMS ... ordering ('Kalkberg type ordering') were required to match the X-ray patterns observed for partly-ordered natural illite–smectites.

If random interstratification of mica and smectite units is considered, two kinds of interfaces and interlayers are required. One of these is high charge and non-expanding, the other low charge and expanding. In a randomly interstratified two-component sequence, three kinds of 2:1 layers are necessary, high charge and low charge non-polar 2:1 layers, ↔ and — respectively, and a polar high–low charge 2:1 layer, ←. A short illustrative sequence is

$$—\quad—\quad→↔←\quad—\quad→↔←\quad—\quad→←\quad→↔↔←.$$

For a three-component random interstratification, for example, a mica–vermiculite–smectite interstratification, six kinds of 2:1 layers are necessary; three non-polar and three polar. This provides three kinds of interfaces; high-charge non-expanding, medium- and low-charge each with different abilities to expand.

These considerations are relevant to the description of interstratified clay minerals. What do we imply when a mineral is described as a random interstratification of illite and smectite or a partly-ordered IS-type material? Unless care is exercised in terminology, the nature, properties and proportions of the different kinds of layers, interlayers and interfaces can be obscured.

Understanding the genetic changes involved when clays, rich in expanding interlayers, are buried, or clays rich in collapsed layers are altered by leaching requires careful consideration of the structural modifications in the 2:1 layers and of the resulting patterns of stacking sequences. The structural changes in 2:1 layers and their observed sequences (see, for example, Hower *et al.* 1976; Weaver & Beck 1971) seem to require the production of different kinds of 2:1 layers and their arrangement in stacking sequences similar to those described above.

The author thanks the Mineralogical Society for permission to reproduce the figures in this paper which are taken from *Crystal structures of clay minerals and their X-ray identification* (Brindley & Brown 1980). Chapter 1 (Bailey 1980*b*), chapter 2 (Brindley 1980) and chapter 4 (Reynolds 1980) in the same monograph were also invaluable in compiling this brief account of the structures of clay minerals.

References

Alcover, J. F., Gatineau, L. & Méring, J. 1973 *Clays Clay Miner.* **21**, 131–136.
Bailey, S. W. 1963 *Am. Miner.* **48**, 1196–1209.
Bailey, S. W. 1969 *Clays Clay Miner.* **17**, 355–371.
Bailey, S. W. 1975 Chlorites. In *Soil constituents* (ed. J. E. Gieseking), vol. 2 (*Inorganic constituents*), ch. 7. Berlin, Heidelberg, New York: Springer-Verlag.
Bailey, S. W. 1980*a* *Am. Miner.* **65**, 1–7.
Bailey, S. W. 1980*b* Structures of layer silicates. In *Crystal structures of clay minerals and their X-ray identification* (ed. G. W. Brindley & G. Brown), ch. 1. London: Mineralogical Society.
Bailey, S. W. 1982 *Clay Miner.* **17**, 243–248.
Bailey, S. W. & Brown, B. E. 1962 *Am. Miner.* **47**, 819–850.
Brindley, G. W. 1980 Order–disorder in clay mineral structures. In *Crystal structures of clay minerals and their X-ray identification* (ed. G. W. Brindley & G. Brown), ch. 2. London: Mineralogical Society.
Brindley, G. W. & Brown, G. 1980 (eds) *Crystal structure of clay minerals and their X-ray identification*. London: Mineralogical Society.
Brindley, G. W. & Suzuki, T. 1983 *Clay Miner.* **18**, 89–94.
Brindley, G. W. & Wan, H. M. 1975 *Am. Miner.* **60**, 863–871.
Brindley, G. W., Zalba, P. E. & Bethke, C. M. 1983 *Am. Miner.* **68**, 420–425.
Brown, G. 1953 *Clay Miner. Bull.* **2**, 64–70.
Cimbálniková, A. 1971 *Am. Miner.* **56**, 1385–1392.
Eberl, D. D., Jones, B. F. & Khoury, H. N. 1982 *Clays Clay Miner.* **30**, 321–326.
Guggenheim, S. & Bailey, S. W. 1975 *Am. Miner.* **60**, 1023–1029.
Guggenheim, S., Bailey, S. W., Eggleton, R. A. & Wilkes, P. 1982 *Can. Miner.* **20**, 1–18.
Hower, J., Eslinger, E. V., Hower, M. E. & Perry, E. A. 1976 *Bull. geol. Soc. Am.* **87**, 725–736.
Hower, J. & Mowatt, T. C. 1966 *Am. Miner.* **51**, 825–854.
Kodama, H., Gatineau, L. & Mering, J. 1971 *Clays Clay Miner.* **19**, 405–414.
Lazarenko, E. K. & Korolev, Yu. M. 1970 *Zapiski Vses. Mineralog. Obshch.* **99**, 214–224.
Mathieson, A. McL. 1958 *Am. Miner.* **43**, 216–227.
Mathieson, A. McL. & Walker, G. F. 1954 *Am. Miner.* **39**, 231–255.
Newnham, R. 1961 *Mineralog. Mag.* **32**, 683–704.
Ormerod, E. C. & Newman, A. C. D. 1983 *Clays Clay Miner.* **18**, 289–299.

Plançon, A. & Tchoubar, C. 1975 *J. appl. Crystallogr.* **8**, 582–588.
Plançon, A. & Tchoubar, C. 1976 *J. appl. Crystallogr.* **9**, 279–285.
Plançon, A. & Tchoubar, C. 1977 *Clays Clay Miner.* **25**, 436–450.
Reynolds, R. C. 1980 Interstratified clay minerals. In *Crystal structures of clay minerals and their identification* (ed. G. W. Brindley & G. Brown), ch. 4. London: Mineralogical Society.
Reynolds, R. C. & Hower, J. 1970 *Clays Clay Miner.* **18**, 25–36.
Rich, C. 1968 *Clays Clay Miner.* **16**, 15–30.
Shirozu, H. & Bailey, S. W. 1965 *Am. Miner.* **50**, 868–885.
Smith, J. V. & Yoder, H. S. 1956 *Mineralog. Mag.* **31**, 209–235.
Weaver, C. E. & Beck, K. C. 1971 *Geol. Soc. Am.* special paper 134.
Yücel, A., Rautureau, M., Tchoubar, D. & Tchoubar, C. 1981 *J. appl. Crystallogr.* **14**, 451–454.
Zvyagin, B. B. 1962 *Soviet Phys. Crystallogr.* **7**, 38–51.

Discussion

P. NADEAU (*The Macaulay Institute for Soil Research, Craigiebuckler, Aberdeen, U.K.*). Concerning the proposed interstratified mineral tarasovite, I should ask the speaker to consider a clay material composed of 40 Å thick mica particles, corresponding to 4 silicate layers coordinated by 3 planes of K^+ ions. How would the interfaces of those particles behave, as pyrophyllite, or as smectite, i.e. non-swelling or swelling? If the interfaces behave as smectite, which I suggest they do, then a sedimented aggregate of such particles would be indistinguishable from a so called ideal tarasovite structure by X-ray diffraction, that is to say 3 mica layers followed by a smectite layer.

G. BROWN. Dr Nadeau's comments re-emphasize the need for precise use of terminology in discussions of the structural components of clays.

If it were possible to cleave a mica crystal, e.g. muscovite, into crystallites 40 Å thick, i.e. crystallites composed of four 2:1 layers each bearing a negative charge of about 1 equivalent per $O_{10}(OH)_2$ formula unit, the charge being symmetrically disposed about the plane of the octahedrally coordinated cations, with K^+ ions balancing the negative charge on the 2:1 layers, including the newly formed external surfaces, the resulting crystallites could be referred to as 40 Å thick mica particles. The units referred to by Dr Nadeau are not of this kind.

The existence of material consisting *entirely* of crystallites 40 Å thick, presumably obtained by dispersion in an aqueous environment, virtually presupposes that the original sample was ideal tarasovite, which consists of a stack of 2:1 layers in which the succession of interlayers is a regular sequence of three high-charge non-expanding interlayers followed by one low-charge expanding interlayer.

Phil. Trans. R. Soc. Lond. A **311**, 241–257 (1984)   [241]
Printed in Great Britain

Clay mineral formation and transformation in rocks and soils

By D. D. Eberl

U.S. Geological Survey, Federal Center, Mail Stop 420, Denver, Colorado 80225, U.S.A.

Three mechanisms for clay mineral formation (inheritance, neoformation, and transformation) operating in three geological environments (weathering, sedimentary, and diagenetic-hydrothermal) yield nine possibilities for the origin of clay minerals in nature. Several of these possibilities are discussed in terms of the rock cycle. The mineralogy of clays neoformed in the weathering environment is a function of solution chemistry, with the most dilute solutions favouring formation of the least soluble clays. After erosion and transportation, these clays may be deposited on the ocean floor in a lateral sequence that depends on floccule size. Clays undergo little reaction in the ocean, except for ion exchange and the neoformation of smectite; therefore, most clays found on the ocean floor are inherited from adjacent continents. Upon burial and heating, however, dioctahedral smectite reacts in the diagenetic environment to yield mixed-layer illite–smectite, and finally illite. With uplift and weathering, the cycle begins again.

Introduction

Three questions may arise when clay is observed in nature. What is the clay? How did it form? Under what conditions did it form? The answers to these questions are the substance of the science of clay mineral petrology.

An answer to the first question requires an analysis of the clay's chemical composition and crystal structure, topics discussed by Brown (this symposium) and in books by Grim (1968) and Brindley & Brown (1980).

The second question concerns mechanisms of clay formation. This paper will consider three mechanisms, generally based on ideas of Esquevin (1958) and Millot (1970). These mechanisms are: (i) inheritance, (ii) neoformation and (iii) transformation. Origin by inheritance simply means that a clay mineral found in a natural deposit originated from reactions that occurred in another area during a previous stage in the rock cycle, and that the clay is stable enough to remain inert in its present environment. Its stability may result either from slow reaction rates or from being in chemical equilibrium. Origin by neoformation means that the clay has precipitated from solution or has formed from reaction of amorphous material. Origin by transformation requires that the clay has kept some of its inherited structure intact while undergoing chemical reaction. This reaction may take two forms: (i) ion exchange, in which loosely bound ions are exchanged with those of the environment; and (ii) layer transformation, in which the arrangements of tightly bound octahedral, tetrahedral, or fixed interlayer cations are modified. Layer transformation will be emphasized in this discussion because the results of this type of reaction are better preserved in the geological record. From a geological perspective, it is important to determine which mechanism gave rise to clays in a natural deposit. Clays that have inherited their crystal structures are indicators of provenance, and provide information about environmental conditions in the sediment source area. Neoformed clays have precipitated in response to *in situ* conditions, past or present. Transformed clays carry both types of information, having inherited characteristics from the source area and having reacted in response to *in situ* changes in environment.

The third question of interest to the clay petrologist concerns the clay's environment of formation. This environment can be described in terms of temperature, pressure, chemical composition and reaction time. In order to generalize this discussion, however, these variables are condensed into three geological situations, the conditions within each of which vary over a limited range. The situations, again based on the ideas of Esquevin (1958) and Millot (1970), are: (i) the weathering environment, (ii) the sedimentary environment, and (iii) the diagenetic–hydrothermal environment. The weathering environment is the upper zone of the Earth's crust that is at or near the atmospheric interface, where temperature and pressure vary over the relatively narrow range of Earth surface conditions. Reaction times are therefore relatively short, usually of the order of thousands of years, because the upper layers of a soil undergo continuous erosion, and solution composition is variable, depending mainly on original rock type, rainfall, evaporation and drainage. The sedimentary environment most often is found near or below sea level or lake level, in depressed areas of the crust, and refers to the zone near the sediment–water interface. In the most common sedimentary environment for clays, the

increasing energy required →

increasing temperature ↓

mechanism / environment	inheritance	neoformation	layer transformation
sedimentary			
weathering			
diagenetic–hydrothermal			

FIGURE 1. Nine possibilities for the formation of clay minerals in Nature (after Esquevin 1958 and Millot 1970).

ocean floor, temperatures are generally lower and restricted to a narrower range than those found in the weathering environment, pressures may range to more than 1 kilobar† in the deepest parts of the ocean, and the composition is that of sea water or related pore water. Reaction time, generally measured in millions of years, depends on rates of sedimentation and subsidence, and on rates of sea floor subduction, processes that move clays into higher temperature environments. The diagenetic–hydrothermal environment includes all zones that have been in contact with hot water. Clays in this situation may experience a wide range of environmental conditions.

The three mechanisms for mineral formation operating in three geological environments yield nine possibilities for the evolution of clay minerals (figure 1). Although exceptions are numerous, a pattern exists relating mechanism to environment. Generally, inheritance dominates in the sedimentary environment where reaction rates are slow, whereas layer transformation, a mechanism that can require large inputs of energy, becomes prevalent in the higher temperature diagenetic–hydrothermal environment. Between these extremes is the weathering environment in which examples of all three mechanisms are common.

Several of the nine possibilities given in figure 1 will be discussed in terms of a simplified rock

† 1 bar $= 10^5$ Pa.

cycle (figure 2). Clays neoformed from crystalline rock in the weathering environment will be traced as they are transported into the sedimentary environment, buried and heated in the diagenetic–hydrothermal environment, and eventually recrystallized during metamorphism. With uplift and weathering, the cycle begins again.

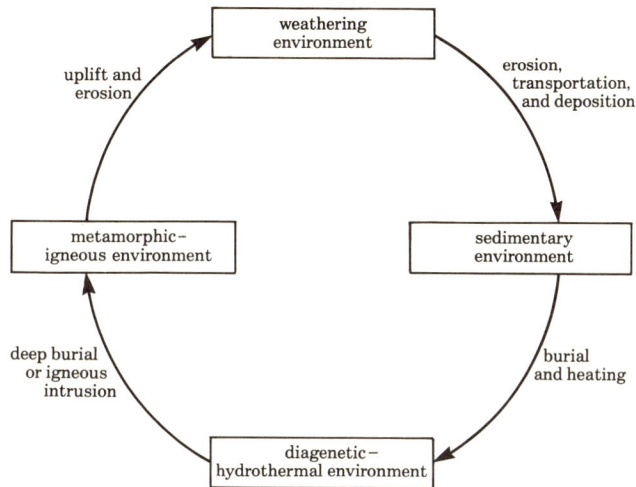

FIGURE 2. A simplified version of the rock cycle.

NEOFORMATION IN THE WEATHERING ENVIRONMENT

Let the cycle as depicted in figure 2 begin with a crystalline rock. The type of clay mineral neoformed from this rock in the weathering environment may change as a function of climate, drainage, original rock type, vegetation and weathering time. Bates (1962) studied weathering in the Hawaiian Islands, an area in which rock type (basalt) is constant, and clay neoformation could be observed mainly as a function of rainfall and drainage. Bates found that smectite forms on the dry, leeward sides of the islands, where rainfall is 0–50 cm a^{-1}, whereas gibbsite forms on the windward sides and in the mountains, where rainfall can be more than 1000 cm a^{-1}. Halloysite predominates in zones with intermediate rainfall. Bates's study illustrates the simple principle that clay minerals composed of the more soluble elements (for example, smectite) are formed in environments where these ions can accumulate (for example, in a dry climate or in a poorly drained soil), whereas clays composed of the least soluble elements (for example, gibbsite) form under severe leaching conditions (for example, on a hilltop in the wet tropics) where only sparingly soluble elements such as aluminium and ferric iron can remain. Kaolinite minerals form in intermediate zones where silicon, as well as aluminium, can be retained.

In a complementary study, Clemency (1975) examined a deeply weathered outcrop near São Paulo, Brazil, a situation in which climate, vegetation, and time were constant, and mineralogy could be studied as a function of original rock type. The outcrop consisted of reddish kaolinite, formed by weathering of a precursor schist or gneiss, and two greenish dykes, composed of ferric iron-rich smectite, formed by weathering of amphibolite. Once again, clay mineralogy could be attributed to differences in leaching intensity because the smectite-rich dyke rock was much less permeable than the kaolinite-rich host rock, thereby enabling soluble ions to accumulate in the dyke.

These patterns of neoformation also were found by Barshad (1966) in a study of weathering in the foothills of the Sierra Nevada mountains in California. The weathering products of acid and basic igneous rocks were studied as a function of rainfall (figure 3). The same pattern emerged: smectite forms in the dry climate, gibbsite in the wet, and kaolinite in the inter- medite. Patterns for other clays shown in figure 3 may arise from inheritance or transformation mechanisms as well as from neoformation.

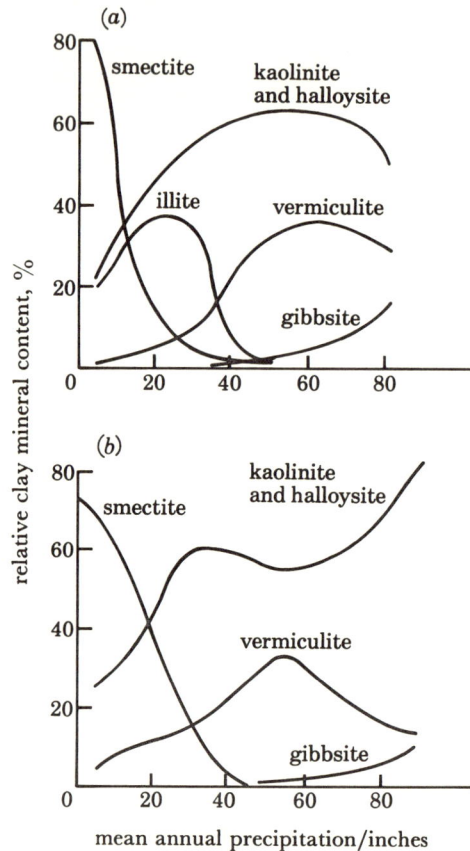

FIGURE 3. The effect of precipitation on the frequency distribution of the clay minerals in the surface layer of residual soils from acid (a) and basic (b) igneous rocks in California (from Barshad 1966). 1 inch = 2.54×10^{-2} m.

The above examples illustrate the importance of soil-solution chemistry in determining the mineralogy of a neoformed clay, and that the variables of climate, drainage, rock type, and vegetation are important insofar as they affect solution chemistry. A clay mineral may be neo- formed when its solubility in a soil solution has been exceeded, given favourable reaction rates. This principle can be represented quantitatively in stability diagrams (for example, figure 4); these diagrams are constructed from thermochemical data as outlined by Garrels & Christ (1965) and Drever (1982). If a soil-solution composition lies within one of the mineral stability fields, as shown in figure 4, then that mineral should be neoformed at equilibrium. Geological evidence qualitatively supports these thermodynamic predictions. Moving from the smectite field in figure 4 to the lower left, and keeping pH constant, one passes through stability fields for smectite, kaolinite, and gibbsite as solution composition becomes more dilute, thereby duplicating mineralogical patterns found with increasing rainfall in Hawaii and in the Sierra Nevada foothills, and with increasing permeability in the outcrop near São Paulo.

These patterns for clay mineral neoformation have been used in relation to geology, for example, to study variations in palaeoclimate. Thompson *et al.* (1982) traced four major changes in precipitation in the northern Rocky Mountains during the Tertiary era by studying fauna, sedimentary structures, sediment accumulation, and clay mineralogy in continental

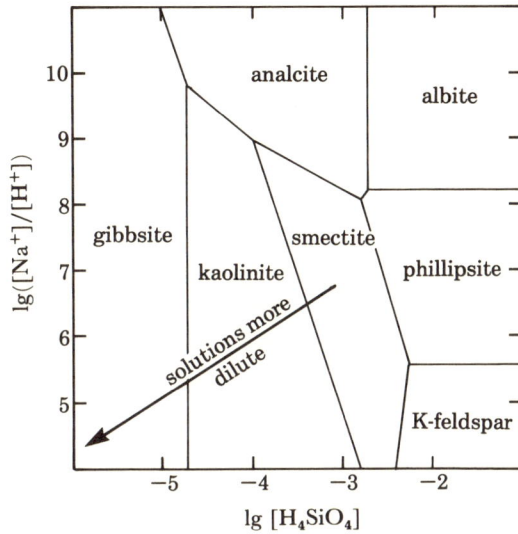

FIGURE 4. An example of a stability diagram for the system $Na_2O–K_2O–Al_2O_3–SiO_2–H_2O$ drawn for $lg ([K^+]/[H^+]) = 4$. The bracketed species are activities (from Drever 1982).

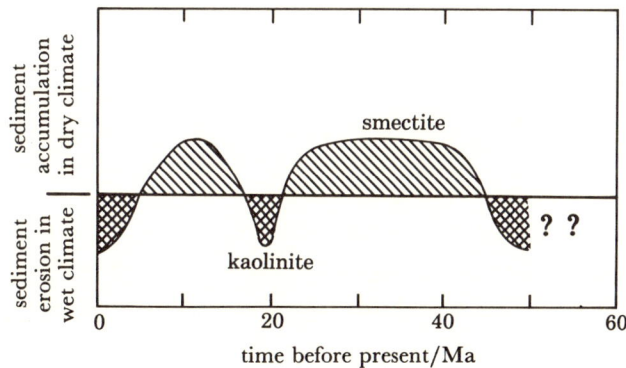

FIGURE 5. Erosion and deposition patterns through time in basins in Montana and Idaho. The vertical axis is in arbitrary units (after Thompson *et al.* 1982).

clastics found in intermontane basins. They found that wet climates, which were synchronous with times of sediment removal from the basins, are characterized by red, kaolinite-rich palaeosols that underlay unconformities (figure 5). These saprolitic rocks are similar to many modern soils formed in the wet tropics. Dry periods, which corresponded to times of sediment accumulation in the basins, are characterized by sediment types (for example, evaporites) and by fossils (for example, camels) formed in arid to semiarid climates, and by a clay mineralogy that is exclusively smectite.

Exceptions to the trends depicted in figures 3 and 4 are found in rocks and soils that contain

[25]

significant amounts of organic matter and living organisms. Here chelating agents such as organic acids may render aluminium and iron more mobile than the more soluble elements (Huang & Keller 1970), and crystallization processes may be extremely slow. Weathering under these conditions may yield non-crystalline or poorly ordered materials and organomineral complexes (Wilson & Jones 1983).

Neoformation has been highlighted in this discussion of the weathering environment, but inheritance and transformation also are important mechanisms functioning in a soil. Inheritance dominates where weathering is mainly mechanical, such as under climatic conditions found in polar regions. Layer transformation, such as leaching of interlayer potassium from illite to form soil vermiculite (Weaver 1958), or precipitation of aluminium hydroxide interlayers in smectite to form soil chlorite (Rich 1968), is important in zones of intermediate weathering intensity, whereas neoformation prevails under more severe chemical weathering conditions. Thus, mechanisms of clay formation in the weathering environment are related approximately to latitude, with inheritance abundant near the poles, and neoformation dominant in the wet tropics. Of course there are many exceptions to these trends. For example, Reynolds (1971) described neoformation of gibbsite above the snowline in the northern Cascade Mountains of Washington State. The rocks in this area are exposed to high insolation and melt-water runoff during spring and summer and therefore undergo intense leaching during part of the year. Weathering of other rocks in this region leads to a neoformation of smectite and to a transformation of phlogopite into vermiculite, thereby demonstrating the importance of local chemical environment to clay mineral formation.

INHERITANCE IN THE SEDIMENTARY ENVIRONMENT

In the second stage of the cycle (figure 2), clays are eroded and transported to the ocean floor. Transportation into regions adjacent to continents is mainly by water, and sedimentation rates may be large. Clays are transported into regions far from continental shelves and slopes mainly by wind, and sedimentation rates may be very small. Minimal changes in composition are undergone by clays during transportation and deposition, and are primarily confined to ion-exchange reactions. On entering the ocean, the dominant exchangeable ions on clays from the Amazon River system, for example, change from Ca^{2+} (80%) to Na^+ and Mg^{2+} (about 38% each), with a significant increase in K^+ (9%) (Sayles & Margelsdorf 1979). This K^+ may be fixed in soil vermiculite and thereby transform it back into illiite (Weaver 1959).

By far the most important origin for clays on the ocean floor is inheritance from adjacent continents. Pelagic clays generally correspond both in mineralogy and in relative abundance to clays being supplied by rivers that are eroding adjacent terrestrial zones (Weaver 1959; Biscaye 1965; Griffin et al. 1968; Lisitzin 1972). The latitudinal distribution found for clays in the weathering environment leads to a bipolar distribution for marine clays. Chlorite and illite, minerals generally inherited in soils forming on diagenetic or low-grade metamorphic rocks, are common in marine sediments that adjoin land areas with cold climates, where mechanical weathering dominates, whereas kaolinite and gibbsite, clays neoformed in soil undergoing intense chemical leaching, are common in marine sediments bordering equatorial, humid zones (for example, figure 6). Potassium–argon age dates of oceanic illite confirm that it has been derived by inheritance from the continents rather than by neoformation or transformation reactions occurring on the ocean floor, and that it has not undergone extensive alteration (Hurley *et al.* 1963; Lisitzin 1972).

The distribution of smectite on the ocean floor is not as regular as that found for the other minerals, and may result from neoformation of smectite in the marine environment and in diverse continental environments. Griffin *et al.* (1968), for example, found that there is more smectite in the southern oceans because of greater volcanic activity, and Hein *et al.* (1979) suggest that smectite forms at a uniform rate over much of the north Pacific deep sea floor.

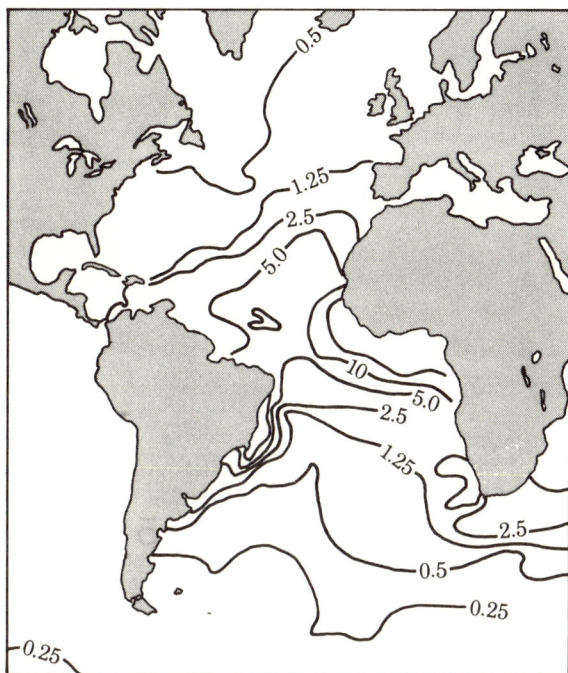

FIGURE 6. Kaolinite–chlorite ratio for the under 2 μm size fraction of deep sea sediments (from Biscaye 1965).

Glauconite also is neoformed (and transformed) in the ocean on continental shelves (Odin & Matter 1981). Neoformation of oceanic smectite indicates that smectite could be thermo-dynamically stable in seawater, although metastability resulting from slow reaction is also a strong possibility (Eberl & Hower 1976). The apparent stability of most clays in the ocean, however, results from slow reaction rates. It is unlikely that clay minerals formed in diverse terrestrial environments would be in equilibrium with the same seawater solution.

When a mixture of waterborne clay minerals enters the ocean, differential settling is expected, thereby giving rise to a systematic distribution of inherited clay minerals with distance from shore. The experiments of Whitehouse *et al.* (1960) give the following general sequence for settling rate under quiet conditions at 26 °C in seawater with approximately normal composition:

$$\text{illite } (15.8 \text{ m d}^{-1}) > \text{chlorite} > \text{kaolinite } (11.8 \text{ m d}^{-1})$$
$$\gg \text{montmorillonite } (1.3 \text{ m d}^{-1}).$$

Settling rates for clays in saline water were found to depend on floccule size rather than on grain size. Relative settling rates were not greatly changed by changes in temperature, organic matter, or salinity, with the exception that decreasing salinity caused the settling rate for

[27]

chlorite to fall slightly below that for kaolinite. The experimental pattern has been observed in nature, for example, in recent sediments of the Niger delta (Porrenga 1966). Smectite in these sediments increases, relative to kaolinite, with water depth and distance from shore. This simple pattern for lateral clay mineral distribution may be altered by having several source rivers for a basin, and by complicating wind, density, and, or, sea current patterns. Conversely, these currents can be traced by studying the distribution of inherited clays (for example, Sawhney & Frink 1978).

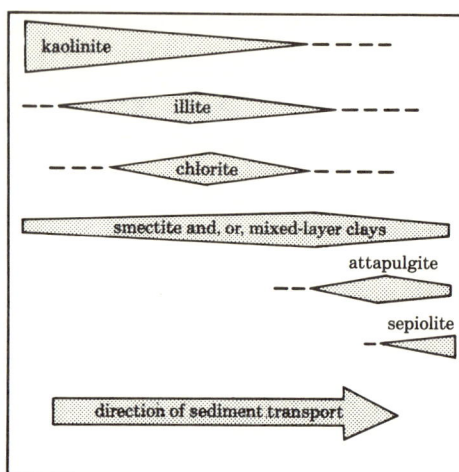

FIGURE 7. Generalized lateral variations in clay mineral assemblages with direction of transport (from Parham 1966).

Parham (1966), and many workers before him, studied lateral variations of clay minerals in sediments of various ages and lithologies and found a systematic pattern with distance from shore (figure 7). These trends do not agree with those of Whitehouse et al. (1960), and may reflect the effect of burial diagenesis, in the case of illite, and neoformation, in the case of chlorite, attapulgite, and sepiolite, on trends that originated from differential settling. Exceptions to the trends shown in figure 7 are legion, but studies of such trends have proven to be useful: maps of the relative amounts of kaolinite, illite, and chlorite have been used to define the margins and centres of ancient sedimentary basins in the search for oil.

TRANSFORMATION AND NEOFORMATION IN THE DIAGENETIC–HYDROTHERMAL ENVIRONMENT

Clays deposited in a sedimentary environment such as a subsiding basin, are buried, heated, and eventually enter the diagenetic environment (figure 2). An important reaction occurring in shales formed from muds that originally contained smectite is the gradual conversion of smectite into mixed-layer illite–smectite (I–S) with increasing burial depth and increasing temperature (figure 8a). Perry & Hower (1970) found that this reaction begins at about 60 °C in U.S. Gulf Coast sediments. Initially, the distribution of illite and smectite layers in I–S is random, but it becomes ordered when about 65 % of the layers are illite. Neoformation of chlorite accompanies this reaction (figure 8b) as does dissolution of detrital potassium feldspar (figure 8c). Generally, the overall reaction is (Hower et al. 1976):

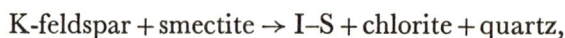

$$\text{K-feldspar} + \text{smectite} \rightarrow \text{I–S} + \text{chlorite} + \text{quartz},$$

[28]

although the stoichiometry of this reaction is as yet unclear (Boles & Franks 1979). The extent of this reaction in shales is most likely a function of time, temperature and porewater composition (Hower 1981).

The reaction of smectite to I–S is considered to proceed by transformation, although this mechanism has not been proven. Chemical data from U.S. Gulf Coast burial sequences indicate that Al^{3+} substitutes for Si^{4+} in the clays' tetrahedral sheets, thereby increasing the negative charge on smectite interlayers (Weaver & Beck 1971; Hower et al. 1976). When a critical

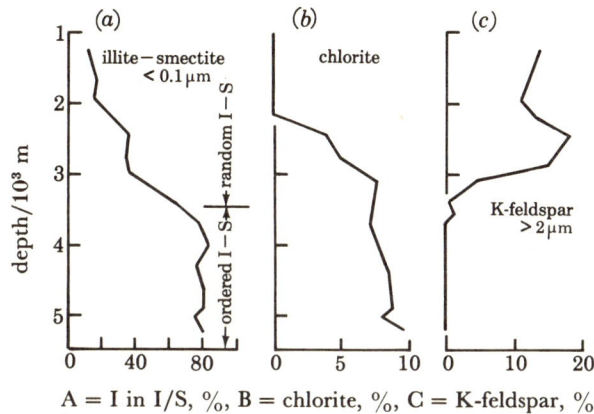

A = I in I/S, %, B = chlorite, %, C = K-feldspar, %.

FIGURE 8. Depth-dependent changes in composition of illite–smectite (a), percentage chlorite (b), and percentage K-feldspar (c) from the Matagorda well (data from Hower et al. 1976; figure from Hower 1981).

layer charge is reached, interlayer potassium dehydrates, thereby transforming expanded smectite interlayers into nonexpanding illite interlayers. The layer charge required for interlayer potassium dehydration is about −0.75 equivalents per half unit cell (Eberl 1980). Interlayer cations with hydration energies greater than that of potassium require greater layer charges for dehydration. Potassium is greatly preferred over these other cations in interlayers that have reached a charge of −0.75, because dehydrated potassium has a much smaller ionic radius than hydrated cations and, therefore, according to Coulomb's law, is adsorbed with greater energy. Thus, burial diagenesis tends to concentrate potassium in illite because K^+ is the first ion of abundance to dehydrate in response to increasing layer charge.

This transformation of smectite into illite may affect the generation of oil in sedimentary basins. The tops of most oil bearing horizons in the U.S. Gulf Coast Tertiary occur in the depth interval where the reaction takes place (figure 9). Although this correlation may be fortuitous, there are several reasons why this reaction could help produce oil: fine-grained smectite layers initially may protect organic matter from oxidation and then catalyse its transformation into petroleum (Johns 1979); water released by the fixation of interlayer potassium may aid in flushing hydrocarbons out of source rocks and into reservoirs (Weaver 1960; Perry & Hower 1972); and pore space resulting from the collapse of smectite layers could provide pathways through source rocks for petroleum migration.

Trends in the reaction found with increasing temperature for clays in shales (figure 8) are not necessarily applicable to rocks with different chemical compositions. For example, hydrothermal alteration of basalts at ocean ridges produces saponites, interstratified chlorite–smectites, and chlorite with increasing temperature (Kristmannsdottir 1979), and figure 10 shows a comparison between temperature-dependent mineral assemblages found in shales, sandstones, and volcanic rocks (Hoffman & Hower 1979). In addition, laboratory studies of hydrothermal systems of more diverse compositions have shown that the conversion of smectite into I–S by fixation of interlayer potassium is but one example of a more general type of reaction, in which smectite may be converted into a variety of mixed-layer clays by reaction with interlayer cations other than potassium (Eberl 1978). For example, montmorillonites that

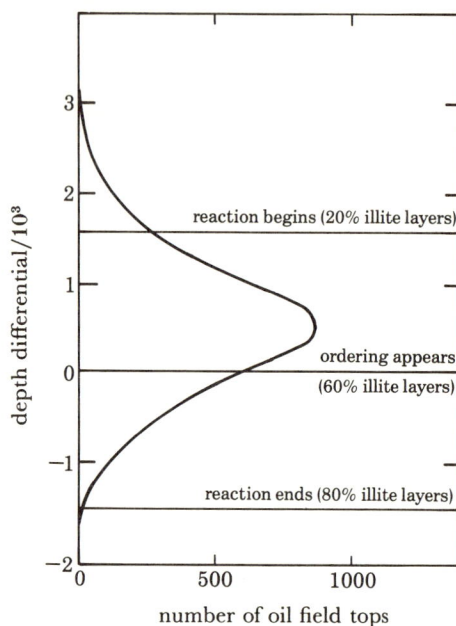

FIGURE 9. Relation between the beginning and end of the reaction of smectite to illite–smectite and the number of producing oil fields 'tops' in the U.S. Gulf Coast Tertiary (after Weaver 1979). The zero line refers to the depth at which the 1.7 nm X-ray peak can no longer be observed. This depth usually corresponds to the first appearance of ordered I–S.

contain potassium yield the following series on hydrothermal treatment with increasing reaction time and temperature: K-montmorillonite → I–S (random)→ I–S (ordered) → illite; whereas montmorillonites with lithium as the interlayer cation yield: Li-montmorillonite → Li-tosudite → cookeite (?). These differences in reaction products probably reflect differences in an interplay between growth of layer charge and interlayer cation hydration energies. Other interlayer cations give other reaction series; therefore, the origin of several geologically diverse clay minerals can be traced to hydrothermal reaction of precursor montmorillonites having different interlayer compositions.

   Hydrothermal experiments also have shown that trioctahedral smectites react very differently than do dioctahedral smectites commonly found in burial diagenetic sequences. Trioctahedral smectites such as saponite and stevensite can be much less reactive than their dioctahedral counterparts. They can remain highly expanded even after hydrothermal treatment at 400 °C for 200 d, although eventually they may react to form mixed-layer talc–smectites or chlorite–smectites (Whitney 1983). Evidently, these trioctahedral systems contain insufficient aluminium to build the layer charge needed to form a mixed-layer mica–smectite structure (Eberl *et al.* 1978). Thus, aluminium, one of the least reactive elements in the weathering environment, plays a central role in reactions that occur in the diagenetic–hydrothermal environment.

Synthesis experiments also show that clays in hydrothermal systems frequently react according to Ostwald's step rule. This rule states that the first phase to form in a system held at constant temperature, pressure and composition may not be the thermodynamically stable phase; and that the initial phase may disappear with time yielding new phases as equilibrium is gradually approached in a series of steps. I–S formed in U.S. Gulf Coast sediments (figure 8a) may be a natural example of this rule, as indicated by hydrothermal experiments on similar systems (Eberl & Hower 1976). A clear example of Ostwald's rule in a synthetic system was

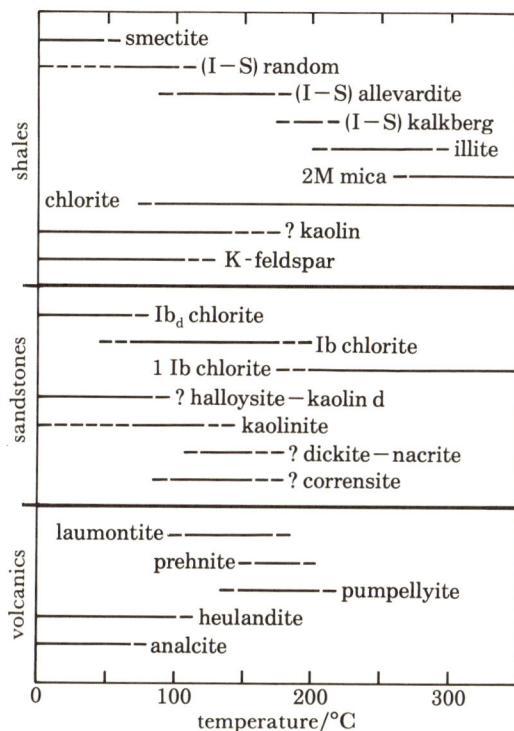

FIGURE 10. Comparison of temperature-dependent mineral assemblages in shales, sandstones and volcanic rocks. Temperatures do not represent equilibrium, but are applicable (with caution) to lower Tertiary and Mesozoic pre-greenschist facies rocks (from Hoffman & Hower 1979).

demonstrated by Whitney (1979) and Whitney & Eberl (1982). Starting with a gel of composition similar to that of talc, the first phase to form at all temperatures in these experiments was talc. At 450 °C and below, however, the talc structure became more disordered with increasing reaction time, eventually yielding kerolite. Kerolite then dissolved, forming trioctahedral smectite (stevensite), and then this smectite reacted to form corrensite (figure 11). At 500 °C and above, only talc was synthesized throughout the time of the experiments. These experiments show that metastable clays may persist for many months even in relatively high temperature hydrothermal systems. The experiments suggest that may clays in nature also could be metastable, forming in response to Ostwald's step rule rather than in response to equilibrium conditions. It remains to be demonstrated whether or not reaction pathways followed in an Ostwald sequence can be described by irreversible thermodynamics (Helgeson 1968; Helgeson et al. 1969), by nucleation kinetics (Fyfe et al. 1958), or by some other model.

[31]

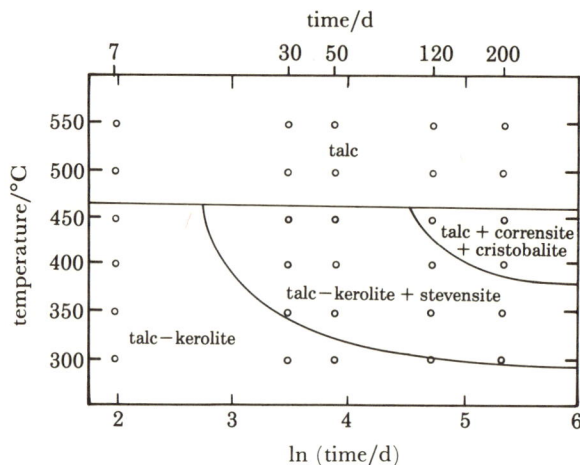

FIGURE 11. Synthesis diagram for the reaction of gel having the Mg/Si ratio of talc in an hydrothermal system held at 1 kilobar pressure. The circles are hydrothermal runs (from Whitney 1979).

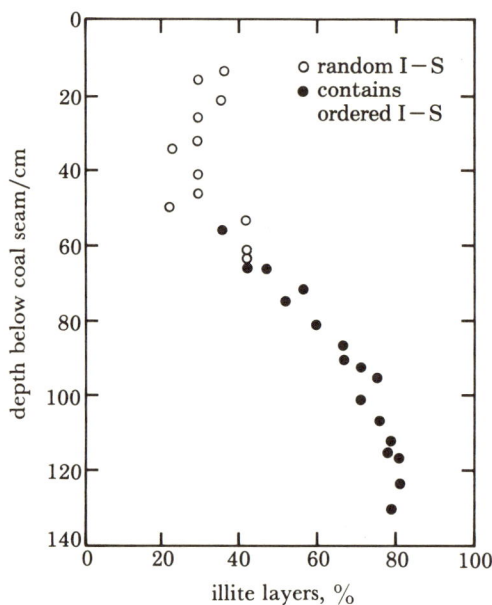

FIGURE 12. Relation between depth below coal, percentage illite layers in illite–smectite and type of interlayering for the under 2 μm size fraction of an Illinois underclay (after Rimmer & Eberl 1982).

TRANSFORMATION IN THE WEATHERING ENVIRONMENT

Clays transformed or neoformed in the diagenetic–hydrothermal environment may be buried deeper, or subjected to igneous intrusion, and eventually may recrystallize to coarser grained micas and chlorites as they enter the metamorphic environment at the greenschist facies (figures 2 and 10). Rather than trace this cycle any further, let shales containing illite be exposed to the weathering environment once again by uplift and erosion.

An interesting weathering pattern develops for such shales that have undergone acid leaching. The profile in figure 12 shows a gradual increase in percentage illite in I–S with depth in an Illinois underclay. (An underclay is a non-bedded, light coloured argillaceous rock that is commonly found beneath beds of coal.) Over a depth of 140 cm, the Illinois clays appear to duplicate trends observed for I–S over depth intervals of thousands of metres in burial diagenetic sequences (figure 8a). The clays in the underclay, however, have responded to a chemical rather than to a temperature gradient, and the reaction is different. Illite layers were transformed to smectite by a reaction in response to removal of interlayer potassium by acid leaching. Transformation was greatest at the top of the profile where leaching was most intense (Rimmer & Eberl 1982). The leaching sequence is: illite → I–S (ordered) → I–S (random) with increasing intensity. This series encompasses the plastic underclays. With more intense leaching, these transformed minerals will disappear, giving rise to neoformed kaolinite and aluminium hydroxide-bearing underclays (Huddle & Patterson 1961). Thus, all three mechanisms for clay formation may operate simultaneously, a situation commonly found in nature. In fact, all three mechanisms may operate to yield a single species of mixed-layer clay. Data from Jones & Weir (1983) suggest that clay minerals from alkaline, saline Lake Abert may originate from an interlayering of inherited montmorillonite, neoformed kerolite and stevensite, and transformed or neoformed illite and chlorite. An intimate mixture of these layers may diffract X-rays as a single mixed-layer clay mineral through the process of interparticle diffraction (Nadeau et al. 1984).

CONCLUSION

Clay minerals are nearly ubiquitous in the Earth's upper crust, and they offer a unique record of Earth processes and Earth history. Thus, clay mineral petrology forms an important branch of the science of geology. Clay minerals may be very responsive to environmental changes; yet they are not so responsive that all previous history is lost. For example, clays may react almost instantaneously to environmental changes through ion exchange and neoformation reactions; they may respond more slowly by layer transformation and by neoformation in an Ostwald step series; or they may remain unreactive. Different clays in a given deposit may react in a different manner and at different rates, thereby offering the potential for a unique solution to the problem of determining environmental conditions, past and present.

In order to read this complex record, it first is necessary to identify the clays precisely and then to determine their mechanisms of formation. Evidence for the occurrence of a transformation reaction would be the discovery of serial relationships in which one clay is seen to alter gradually into another clay over a temperature, pressure, compositional or time gradient (for example, figures 8a and 12). The sudden disappearance of one clay and the appearance of another, with no transitional phases, suggests neoformation (for example, figure 8b), or a change in source area. Radiometric age dates and textural information obtained from scanning electron microscopy are useful in distinguishing neoformation from inheritance (for example, Hower et al. 1963; Wilson & Pittman 1977).

The formation of clay minerals has been presented as a cyclic process (figure 2), but clay minerals in shales have changed through geological time. Young shales and recent sediments are relatively enriched in kaolinite and smectite-rich mixed-layer clay, whereas older, pre-Mesozoic shales are depleted in these phases and are enriched in chlorite and illite (Weaver 1967; Hower et al. 1976). It has been suggested that these differences in mineralogy are due to differences in chemical composition between recent and ancient weathering environments (Weaver 1967), or to a greater degree of diagenesis for older rocks (Grim 1968; Weaver & Beck 1971; Hower et al. 1976). The latter explanation is favoured by remarkable parallels found between trends in mineralogy and composition of shales with increasing geological age, and of shales subjected to increasing diagenetic grade (Garrels & Mackenzie 1975; Hower et al. 1976).

[33]

Finally, it is emphasized that clay mineral petrology is a young science that is still inexact and incomplete. For example, mixed-layer clays are among the most abundant of the clay minerals, but only recently has it been possible to characterize them with confidence (Reynolds & Hower 1970; Reynolds 1980; Srodon 1980). Mixed-layer clays may prove to be sensitive indicators of Earth history once their mechanisms and environments of formation are understood.

The author thanks C. Clemency, C. Farmer, D. Horton, W. Keller, B. Kimball, P. Nadeau, J. Russell, L. Schultz, J. Srodon, C. Weaver, A. Weir, and G. Whitney for their reviews of this manuscript. This paper is dedicated to the memory of G. W. Brindley and John Hower.

REFERENCES

Barshad, I. 1966 The effect of variation in precipitation on the nature of clay mineral formation in soils from acid and basic igneous rocks. *Proc. int. Clay Conf.* (Jerusalem) **1**, 167–173.
Bates, T. 1962 Halloysite and gibbsite formation in Hawaii. *Clays Clay Miner.* **9**, 315–328.
Biscaye, P. E. 1965 Mineralogy and sedimentation of recent deep-sea clay in the Atlantic Ocean and adjacent seas and oceans. *Bull. geol. Soc. Am.* **76**, 803–832.
Boles, J. R. & Franks, S. G. 1979 Clay diagenesis in Wilcox sandstones of southwest Texas: Implications of smectite diagenesis on sandstone cementation. *J. sedim. Petrol.* **49**, 55–70.
Brindley, G. W. & Brown, G. 1980 *Crystal structures of clay minerals and their X-ray identification.* London: Mineralogical Society.
Clemency, C. V. 1975 Simultaneous weathering of a granitic gneiss and an intrusive amphibolite dike near São Paulo, Brazil, and the origin of clay minerals. *Proc. int. Clay Conf.* (Mexico), 157–172.
Drever, J. I. 1982 *The geochemistry of natural waters.* Englewood Cliffs, N.J.: Prentice-Hall.
Eberl, D. 1978 Reaction series for dioctahedral smectites. *Clays Clay Miner.* **26**, 327–340.
Eberl, D. D. 1980 Alkali cation selectivity and fixation by clay minerals. *Clays Clay Miner.* **28**, 161–172.
Eberl, D. & Hower, J. 1976 Kinetics of illite formation. *Bull. geol. Soc. Am.* **87**, 1326–1330.
Eberl, D., Whitney, G. & Khoury, H. 1978 Hydrothermal reactivity of smectite. *Am. Miner.* **63**, 401–409.
Esquevin, J. 1958 Les silicates de zinc. Etude de produits de synthèse et des minéraux naturels. *Theèse Sci. Paris.*
Fyfe, W. S., Turner, F. J. & Verhoogen, J. 1958 *Metamorphic reactions and metamorphic facies.* Geological Society of America Memoir 73.
Garrels, R. M. & Christ, C. L. 1965 *Solutions, minerals and equilibria.* San Francisco: Freeman Cooper & Company.
Garrels, R. M. & Mackenzie, F. T. 1975 Chemical history of the oceans deduced from post-depositional changes in sedimentary rocks. In *Studies in paleo-oceanography* (ed. W. W. Hay). Soc. Econ. Paleontologists & Mineralologists Spec. Pub. **20**, 193–204.
Griffin, J. J. Windom, H. & Goldberg, E. D. 1968 The distribution of clay minerals in the World Ocean. *Deep-Sea Res.* **15**, 433–459.
Grim, R. E. 1968 *Clay mineralogy.* New York: McGraw-Hill Book Co.
Hein, J. R. Yeh, H.-W. & Alexander, E. 1979 Origin of iron-rich montmorillonite from the manganese nodule belt of the north equatorial Pacific. *Clays Clay Miner.* **27**, 185–194.
Helgeson, H. C. 1968 Evaluation of irreversible reactions in geochemical processes involving minerals and aqueous solutions. I. Thermodynamic relations. *Geochim. cosmochim. Acta* **32**, 853–877.
Helgeson, H. C., Garrels, R. M. & Mackenzie, F. T. 1969 Evaluation of irreversible reactions in geochemical processes involving minerals and aqueous solution. II. Applications. *Geochim. cosmochim. Acta* **33**, 455–481.
Hoffman, J. & Hower, J. 1979 Clay mineral assemblages as low grade metamorphic geothermometers: application to the thrust faulted disturbed belt of Montana, U.S.A. In *Aspects of diagenesis* (ed. P. A. Scholle & P. R. Schluger). Soc. Econ. Paleontologists & Mineralogists Spec. Pub. **26**, 55–79.
Hower, J. 1981 Shale diagenesis. In *Clays and the resource geologist* (ed. F. J. Longstaffe). *Short course handbook* **7**, 60–80. Toronto: Mineralogical Association of Canada.
Hower, J., Hurley, P. M., Pinson, W. H. & Fairbairn, H. W. 1963 The dependence of K-Ar age on the mineralogy of various particle size ranges in a shale. *Geochim. cosmochim. Acta* **27**, 405–410.
Hower, J., Eslinger, E. V., Hower, M. E. & Perry, E. A. 1976 Mechanism of burial metamorphism of argillaceous sediment. I. Mineralogical and chemical evidence. *Bull. geol. Soc. Am.* **87**, 725–737.
Huang, W. H. & Keller, W. D. 1970 Dissolution of rock forming minerals in organic acids. Simulated first stage weathering of fresh mineral surfaces. *Am. Miner.* **55**, 2076–2094.
Huddle, J. W. & Patterson, S. H. 1961 Origin of Pennsylvanian underclay and related seat rocks. *Bull. geol. Soc. Am.* **72**, 1643–1660.
Hurley, P. M., Heezen, B. C., Pinson, W. H. & Fairbairn, H. W. 1963 K-Ar age values in pelagic sediments of the North Atlantic. *Geochim. cosmochim. Acta* **27**, 393–399.

Johns, W. D. 1979 Clay mineral catalysis and petroleum generation. *Am. Rev. Earth Planet Sci.* **7**, 183–198.

Jones, B. F. & Weir, A. H. 1983 Clay minerals of Lake Abert, an alkaline, saline lake. *Clays Clay Miner.* **31**, 161–172.

Kristmannsdottir, H. 1979 Alteration of basaltic rocks by hydrothermal activity at 100–300 °C. *Proc. int. Clay Conf.* (Oxford), 359–367.

Lisitzin, A. P. 1972 In *Sedimentation in the world ocean* (ed. K. S. Rodolfo). Soc. Econ. Paleontologists & Mineralogists Spec. Pub. **17**.

Millot, G. 1970 *Geology of clays* (trans. W. R. Farrand & H. Paquet). New York: Springer-Verlag.

Nadeau, P. H., Tait, J. M., McHardy, W. J. & Wilson, M. J. 1984 Interstratified XRD characteristics of physical mixtures of elementary clay particles (In the press.)

Odin, G. S. & Matter, A. 1981 De glauconiarum origine. *Sedimentology* **28**, 611–641.

Parham, W. E. 1966 Lateral variations of clay mineral assemblages in modern and ancient sediments. *Proc. int. Clay Conf.* (Jerusalem) **1**, 135–145.

Perry, E. A. & Hower, J. 1970 Burial diagenesis in Gulf Coast pelitic sediments. *Clays Clay Miner.* **18**, 165–177.

Perry, E. A. & Hower, J. 1972 Late-stage dehydration in deeply buried pelitic sediments. *Bull. Am. Ass. Petrol. Geol.* **56**, 2013–2021.

Porrenga, D. H. 1966 Clay minerals in recent sediments of the Niger delta. *Clays Clay Miner.* **14**, 221–233.

Reynolds, R. C. 1971 Clay mineral formation in an alpine environment. *Clays Clay Miner.* **19**, 361–374.

Reynolds, R. C. 1980 Interstratified clay minerals. In *Crystal structures of clay minerals and their X-ray identification* (ed. G. W. Brindley & G. Brown), pp. 249–303. London: Mineralogical Society.

Reynolds, R. C. & Hower, J. 1970 The nature of interlayering in mixed-layer illite–montmorillonite. *Clays Clay Miner.* **18**, 25–36.

Rich, C. I. 1968 Hydroxyl interlayers in expansible layer silicates. *Clays Clay Miner.* **16**, 15–30.

Rimmer, S. M. & Eberl, D. D. 1982 Origin of an underclay as revealed by vertical variations in mineralogy and chemistry. *Clays Clay Miner.* **30**, 422–430.

Sawhney, B. L. & Frink, C. R. 1978 Clay minerals as indicators of sediment source in tidal estuaries of Long Island Sound. *Clays Clay Miner.* **26**, 227–230.

Sayles, F. L. & Mangelsdorf, P. C. Jr. 1979 Cation-exchange characteristics of Amazon River suspended sediment and its reaction with seawater. *Geochim. cosmochim. Acta* **43**, 767–779.

Środoń, J. 1980 Precise identification of illite–smectite interstratifications by X-ray powder diffraction. *Clays Clay Miner.* **28**, 401–411.

Thompson, G. R., Fields, R. W. & Alt, D. 1982 Land-based evidence for Tertiary climatic variations: Northern Rockies. *Geology* **10**, 413–417.

Weaver, C. E. 1958 The effects and geologic significance of potassium 'fixation' by expandable clay minerals derived from muscovite, biotite, chlorite, and volcanic material. *Am. Miner.* **43**, 839–861.

Weaver, C. E. 1959 The clay petrology of sediments. *Clays Clay Miner.* **6**, 154–187.

Weaver, C. E. 1960 Possible uses of clay minerals in the search for oil. *Bull. Am. Ass. Petrol. Geol.* **44**, 1505–1518.

Weaver, C. E. 1967 Potassium, illite, and the ocean. *Geochim. cosmochim. Acta* **31**, 2181–2196.

Weaver, C. E. 1979 Geothermal alteration of clay minerals and shales: Diagenesis. Technical Report UC-70. Office of Nuclear Waste Isolation, Battelle.

Weaver, C. E. & Beck, K. C. 1971 Clay water diagenesis during burial: How mud becomes gneiss. *Geol. Soc. Am. Spec. Paper* **134**, 1–96.

Whitehouse, U. G., Jeffrey, L. M. & Debbrecht, J. D. 1960 Differential settling tendencies of clay minerals in saline waters. *Clays Clay Miner.* **7**, 1–79.

Whitney, G. 1979 The paragenesis of synthetic phyllosilicates on the talc–phlogopite join. Ph.D. Thesis. University of Illinois, Urbana.

Whitney, G. & Eberl, D. D. 1982 Mineral paragenesis in a talc–water experimental hydrothermal system. *Am. Miner.* **67**, 944–949.

Whitney, G. 1983 Hydrothermal reactivity of saponite. *Clays Clay Miner.* **31**, 1–8.

Wilson, M. D. & Pittman, E. D. 1977 Authigenic clays in sandstones: recognition and influence on reservoir properties and paleoenvironmental analysis. *J. sedim. Petrol.* **47**, 3–31.

Wilson, M. J. & Jones, D. 1983 Lichen weathering of minerals: implications for pedogenesis. *Spec. Publ. Geol. Soc. London* **11**, 5–12.

Discussion

V. C. FARMER. (*Macaulay Institute for Soil Research, Craigiebuckler, Aberdeen AB9 2QJ, U.K.*). It seems unlikely that the diagenetic alteration of montmorillonite to illite is a transformation which preserves the layer structure of the parent montmorillonite, because (see Tchoubar, this meeting) (i) successive layers in montmorillonite are randomly oriented and randomly displaced relative to each other, whereas in illite the hexagonal holes in the tetrahedral layer are directly superimposed in successive layers; (ii) The structures of the octahedral sheets differ, as in illite only *cis* sites are occupied, whereas in montmorillonite, a *cis* and *trans* site are occupied by octahedral cations.

Thus the formation of illite from smectite involves major restructuring, and is best considered a neoformation, consistent with the larger and more regular crystals of illite and interstratified illite–smectites, compared with montmorillonite. The widespread occurrence of regularly inter-stratified illite–smectites can be understood if the restructuring largely involves adjacent smectite sheets. In the process, tetrahedral aluminium will be concentrated in the inner tetra-hedral sheets, adjacent to potassium ions, and the outer tetrahedral sheet will be depleted in aluminium, compared with the original smectite. Thus a third mica layer is unlikely to build onto the initial two, so interstratification results.

D. D. EBERL. Whether illite forms from smectite during burial diagenesis by neoformation or by transformation is an open question. There is, however, no problem in making holes in adjacent 2:1 layers line up. K-smectite can be converted readily into an illite-like structure in the laboratory by dry heating or by wetting and drying. Also, the collapse of an expanded Na-vermiculite structure to an illite-like structure can be made to occur spontaneously in suspension by adding potassium. Rearrangements in the octahedral sheet also could occur during transformation. In fact, magnesium needed to form chlorite (figure 8*b*) may come from smectite's octahedral sheet.

Regularly shaped illite crystals have been observed in sandstones, thereby suggesting neo-formation. But transmission electron microscopy studies of shales that initially contain detrital smectite suggest that both neoformation and transformation take place.

I agree with Dr Farmer's view of the illite crystals as having asymmetrical layer charges, but illite packets must be thicker than two 2:1 layers, otherwise there could never be a 20% ex-pandable clay (see figure 8*a*). Only 50% expandable K-rectorite and clays of higher expand-abilities could exist. When a two-layer thick illite crystal is converted into a three-layer thick illite, does the two-layer crystal completely dissolve and reprecipitate as a thicker crystal, or does it grow by increasing layer charge on its edges? This question could be resolved by careful K–Ar studies of an appropriate illite–smectite sequence. In the dissolution–reprecipitation mechanism, all radiogenic argon should be lost as a gas at each increase in illite content.

R. M. BARRER, F.R.S. (*Chemistry Department, Imperial College, London SW7, U.K.*). You mentioned the hydrothermal régime for clay mineral synthesis as beginning above about 60 °C. This low temperature is not strictly in the hydrothermal region which one usually associates with temperatures above 100 °C. Can you comment further on the temperature ranges over which smectites and kandites have been synthesized or have formed in Nature? Are there approximate upper limits?

D. D. EBERL. The conversion of montmorillonite into illite is detectable in U.S. Gulf Coast sediments when the burial temperature has reached approximately 60 °C. This temperature is in the realm of 'diagenesis'. Some authors distinguish between 'early' and 'late' diagenesis. Early diagenesis refers to changes that occur in the first few hundred metres of sediment where elevated temperatures are not encountered. In the present paper, these reactions are considered to occur in the sedimentary environment. Late stage diagenesis is synonymous with burial metamorphism. Precise temperature boundaries between 'early diagenesis', 'late diagenesis', 'hydrothermal', and 'metamorphic' are arbitrary.

Smectite has been synthesized at temperatures ranging from below 25 °C to 850 °C. In fact, an expanding phase has been synthesized at 1000 °C at 5 kilobar pressure. High pressures favour the development of expandable phases in systems of appropriate chemistries, thereby suggesting (from the Le Chatelier principle) that interlayer water is denser than pore water in hydrothermal systems. Smectite synthesis at low temperatures is favoured by the presence of magnesium. It is difficult to synthesize aluminium-rich smectites such as beidellite from gel at temperatures less than 200 °C.

Kaolinite, like beidellite, is difficult to synthesize at low temperatures. We were able to obtain a good yield of kaolinite starting with gel at 150 °C after 80 d, but were unsuccessful at 100 °C. The problem appears to be kinetic. Using organic catalysts, however, kaolinite has been synthesized at 25° C. The upper stability limit of kaolinite is about 405 °C where it reacts to form pyrophyllite. It can react to form pyrophyllite at a lower temperature if the Si/Al atomic ratio of the system is greater than one.

Phil. Trans. R. Soc. Lond. A **311**, 259–269 (1984)   [259]
Printed in Great Britain

X-ray studies of defects in clays

By C. Tchoubar

*Laboratoire de Cristallographie (Equipe Associée au C.N.R.S. no. 841),
Université d'Orléans, 45046 Orléans, Cedex, France*

Defects are any departures from the ideal well-ordered structure. Defects in clays can be classified in three categories: those affecting the layers themselves (*cis*- or *trans*-vacancies, rotation of tetrahedra, localization of the isomorphous substitutions, etc.); those specific to the interlamellar space (position of the cations and, for example, of the water molecules); and stacking faults (including a change in the nature of stacked layers). X-ray studies of clays usually involve the analysis of powder diagrams generally perturbed by a partial orientation of the particles in the powder. The paper will present a general approach to the determination of defects in clays with an indirect method of analysis. The intensities and shapes of the diffraction bands are calculated for model structures and are fitted to the experimental pattern. In such an approach the powder orientation, the shape and sizes of the coherent domains, the various backgrounds and perturbations due to the apparatus are taken into account.

1. Introduction

Defects can be defined as any departures from the ideal well-ordered structure. It is well known that the presence of structural faults in the clays modifies some of their properties and, consequently, their behaviour and use.

Defects in clays can be classified in three categories.

(1) Those affecting the layers themselves, such as: (i) localization of the octahedral vacancies in *cis*- or *trans*-positions; (ii) rotation of tetrahedra leading to a more or less important deformation of the 'hexagonal' cavities. This rotation depends, partly, on the interlamellar cation size.

(2) Those specific to the interlamellar space and corresponding with a modification of the positions of either cations or intercalated molecules or both. This kind of defect appears, for instance, when the number of energetically equivalent sites is higher than the number of cations or molecules in these sites.

(3) Stacking faults that concern either only the relative position of layers (translational or rotational stacking faults) or the kind of successive layers (mixed layer structures).

Lamellar silicates with irregular and disordered structures present a particle size range between 5 nm and 10 μm and only powders can be used for the X-ray diffraction studies.

To recognize the kind, the proportion and the distribution rule of structural defects in clays, the powder pattern analysis is always made with an indirect approach that consists of predicting the effect on the diffraction phenomenon of one given kind of fault. Brindley & Robinson (1946) have been the first to use such a way of research to interpret patterns of disordered kaolinites: to explain the simultaneous presence of *hkl* reflections (with $k = 3n$, n integer) and of (*hk*) bands (with $k \neq 3n$), these authors had suggested the existence of translational $\pm \frac{1}{3}\boldsymbol{b}$ stacking faults. Since these years, much work has been devoted to qualitative and then quantitative interpretation of the powder patterns of phyllosilicates that contain a large amount of structural defects. A good survey of these studies is given in the book edited by Brindley & Brown (1980). We shall just point out some fundamental works: those of Mering who was the first to explain, by means of a mathematical formalism, the main features of powder patterns of lamellar systems that contain different kinds of defects (Mering 1949). In the same way, MacEwan (MacEwen *et al.* 1961), Drits (Drits & Sakharov 1976) and Reynolds (1980) have given principles of calculation and interpretation of the 00*l* intensity distribution for the interstratified clay minerals.

All these studies were made on the basis of an indirect analysis of patterns consisting of devising a structural model adapted to each clay mineral studied. The model takes into account the presence of different kinds of defects with their abundances and the rule of their distribution. By starting from such a model, a synthetic pattern is calculated. Then, by modifying the values of characteristic parameters of the defects, the theoretical diagram is fitted with the experimental one, the agreement being obtained simultaneously for profiles and for intensities of the reflections.

Now such a method based on the model is fully developed, mainly in the Soviet and French laboratories of clay crystallography.

The following will be devoted to the description of modelling possibilities and to the power of this method for the determination of defects in clays.

2. EXPERIMENTAL CONDITIONS AND PRINCIPLE OF CALCULATION

An effective modelling method requires, first of all, the recording of patterns in very rigorous experimental conditions in which all the functions that can perturb the diffraction phenomenon are controlled. Specifically, it is necessary to use monochromatic radiation, to have good precision in the intensity measurements and good angular resolution, to control the width of the beam and slits, the thickness of samples and the orientation of particles in the powders. Most of the disturbing functions can be minimized so as to give effects in patterns several times lower than those generated by defects. When such a function cannot be neglected, it is necessary to characterize its influence by introducing it to calculations. This is the case, for instance, with the residual orientation of particles in powder. The orientation distribution is experimentally determined from the 00*l* reflections (Taylor & Norrish 1966; De Courville *et al.* 1979). Calculations show that orientation influences relative intensities of reflections as well as their profiles (Plançon & Tchoubar 1977; Plançon 1980). Figure 1 illustrates this phenomenon for a partly disordered kaolinite (Plançon 1980). Curves correspond to the (02, 11) and (20, 13) domains. For random orientation, the calculated diagram is represented by the dashed line, while the solid line is obtained with the particle orientation given in figure 2.

As for the mathematical formalism that describes the diffraction phenomenon, it varies with authors. Nevertheless it seems that the most powerful is the matrix notation. It was introduced for lamellar systems by Hendricks & Teller in 1942. Since then, several papers have developed analogous formalisms adapted to the complexity of the structural model considered or to the kind of reflections studied (Kakinoki & Komura 1952; Drits & Sakharov 1976; Plançon & Tchoubar 1976, 1977; Plançon 1981; Sakharov *et al.* 1982*a, b*).

In the general case, for a powder with particle orientation, various thicknesses of coherent domains, where each stack contains, simultaneously or not, (i) layers of different kinds, and (ii) different translations or rotations between these layers (with or without correlations between the defects), the intensity diffracted at the 2θ angle by an *hk* rod is given by:

$$I_{hk}(s) = \frac{1}{s\Omega\sigma} \sum_M \alpha(M) \int \bar{N}(\phi) \text{ trace Re } \{[F][W][R(M)]\} \, T(X) \, \mathrm{d}\phi,$$

[40]

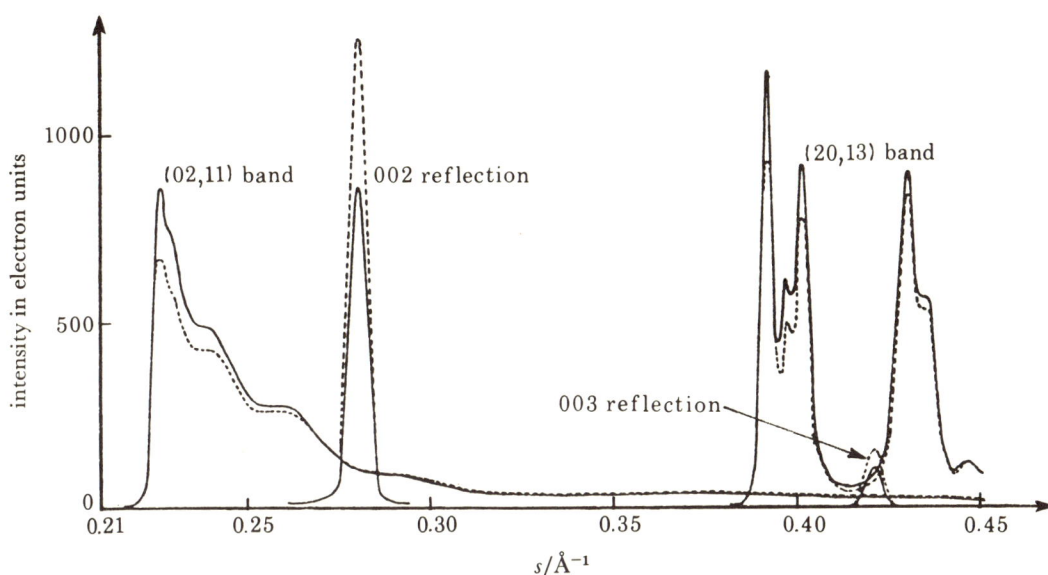

FIGURE 1. Effect of particle orientation on the X-ray intensities in the case of symmetric transmission; dashed line: sample with random particle orientation; solid line: sample with particle orientation corresponding to figure 2.

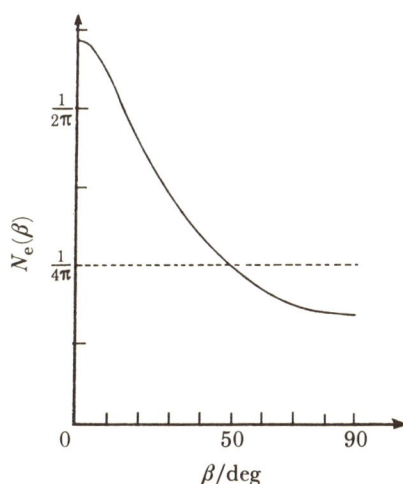

FIGURE 2. Orientation function that defines the orientation of particles in the powder.

where $s = 2 \sin \theta / \lambda$ is the modulus of the scattering vector; Ω is the area of the (a, b) unit cell; σ is the area of the coherent interferential domain in the (a, b) plane; $\alpha(M)$ is the statistical weight distribution of the stackings with M layers; ϕ is the angle between the scattering vector and a plane normal to the hk rod. Summation ϕ expresses a powder pattern calculation and the integration includes the effect usually described by the Lorentz factor; $\bar{N}(\phi)$ is the proportion of particles in powder which participate in diffraction at the chosen 2θ angle. $\bar{N}(\phi)$ is obtained from the experimental orientation function (De Courville $et\ al.$ 1979; Plançon & Tchoubar 1977; Plançon 1980); $T(X)$ is a function characteristic of the shape and sizes of the coherent interferential domain. It has been introduced in the powder integration by Brindley & Mering

[41]

(1951); $[F]$, $[W]$, $[R(M)]$ are three square matrices describing the structural model itself and Re means real part of their product; $[F]$ is the matrix characterizing the layer kinds; each ij matrix element is equal to $F_i^* F_j$, product of the structure factor F_j of the j-type layer by the conjugate structure factor F_i^* of the i-type layer; $[W]$ is the matrix that takes into account the abundance of each layer kind; $[R(M)]$ is the matrix that expresses the interference phenomenon between the waves scattered by the different layers of a stacking. This matrix takes into account the relative positions of layers, the kind and rate of stacking faults and their distribution.

The complexity of these matrices varies with the choice of the structural model, but the $I_{hk}(s)$ relation permits calculation of all kinds of powder patterns observed with clay minerals, even if more or less long range correlations exist either between layers or between defects or both (Plançon 1981).

The principle of defect determination by fitting of theoretical and experimental patterns leads immediately to a question: is it possible to obtain practically the same calculated diagram, starting from several absolutely different models?

For the purpose of answering the question, a systematic analysis, based on the crystallo-chemical viewpoint of the relationship between structural characteristics and diffraction patterns, has been made for the dioctahedral phyllosilicates (Drits et al. 1983). This study has shown that each kind of fault and structural characteristic leads to specific features in definite parts of the patterns and is insensitive in others.

A general approach to defect and fine structure determination for a given clay should include:

(1) successive consideration of all models which are crystallochemically possible;

(2) calculation, in all accessible domains of the reciprocal space, of intensity distribution and profile variation, obtained by changing only one parameter at a time that defines one type of structural feature (e.g. cation distribution in individual layers, nature of stacking faults, etc.);

(3) Systematic analysis of the calculated diffraction patterns to establish the diffraction criteria which will help to interpret the experimental data;

(4) Agreement of experimental and calculated diffraction patterns, to obtain the values of all parameters characteristic of the defects and fine structure of the studied mineral.

Such an approach will be illustrated by some examples.

3. EXAMPLES OF X-RAY STUDIES OF DEFECTS IN CLAYS

(a) Specific modifications in intensity distribution correlated with the kind of stacking faults

The more frequently imagined and crystallochemically possible stacking faults are:

(i) arbitrary translations or rotations of layers in their plane (probability of observing such a fault between two adjacent layers is p_A).

(ii) faults which we will name 'partly defined faults', in contrast to the arbitrary ones: they correspond to $\pm 120°$ or $n\,60°$ rotations of successive layers or to $-\frac{1}{3}na \pm \frac{1}{3}mb$ translations (with $n = 0$ or 1, $m = 0$ or 1).

If the structural model contains only arbitrary stacking faults, calculations show, as is well known, an evolution of the pattern to a superposition of two-dimensional (hk) bands when p_A increases. In contrast, the presence of only 'partly defined' faults does not modify identically all the (hk) domains and moreover each kind of these faults can be identified.

[42]

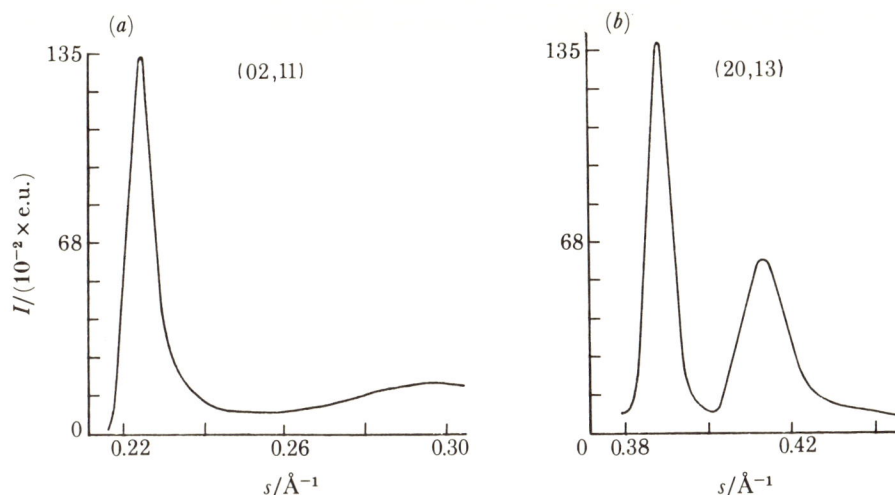

FIGURE 3. Calculated profiles for a *cis*-vacant dioctahedral smectite containing $\pm n\,60°$ rotational stacking faults ($p_r = 0.67$). (*a*) (02, 11) band; (*b*) (20, 13) band.

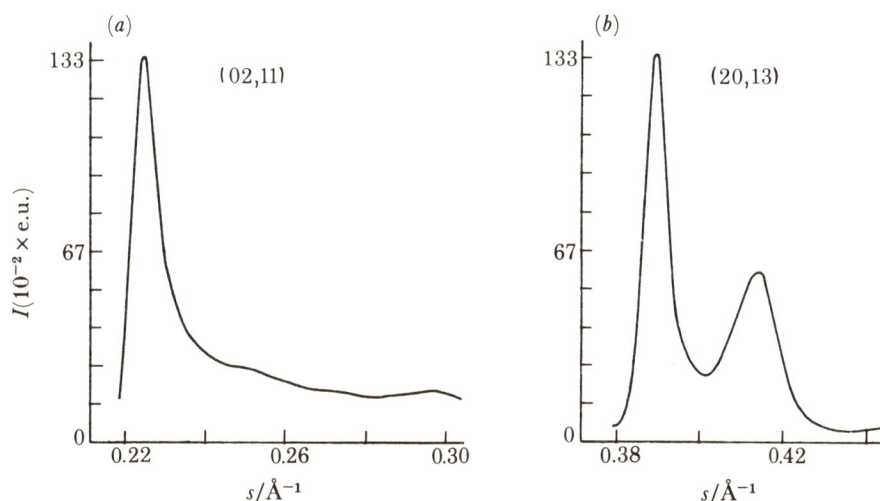

FIGURE 4. Calculated profiles for a *cis*-vacant dioctahedral smectite containing $-\frac{1}{3}na \pm \frac{1}{3}mb$ ($n = 0$ or 1, $m = 0$ or 1) translational stacking faults ($p_t = 0.67$). (*a*) (02, 11) band; (*b*) (20, 13) band.

For instance it is also well known that $\pm 120°$ faults can be easily distinguished from the $\pm \frac{1}{3}b$ translations on the basis of the (20, 13) domain which is strongly modified only in the case of rotational faults.

The distinction is not so evident in the case of $n\,60°$ and $(-\frac{1}{3}na \pm \frac{1}{3}mb)$ faults. Figures 3 and 4 correspond to a dioctahedral smectite with *cis*-vacant positions and containing a probability p_R or p_T, equal to 0.67 of observing a stacking fault between adjacent layers (Drits *et al.* 1984). A superficial qualitative examination seems to reveal a strict analogy between the patterns calculated in the (02, 11) and (20, 13) domains respectively: for the two types of faults, there exists only one modulation at $s \approx 0.415$ Å^{-1}† on the two (20, 13) bands, and the two (02, 11) bands present a strong tendency to a two-dimensional structure. But in fact a quantitative comparison of the two patterns shows that the (20, 13) band corresponding to $n\,60°$ faults has,

† 1 Å $= 10^{-10}$ m $= 0.1$ nm.

[43]

at $s \approx 0.40$ Å^{-1}, a minimum more clearly pronounced. In the same way, for the translational faults, the (02, 11) band presents an almost regular decrease of the intensity between $s \approx 0.23$ Å^{-1} and $s \approx 0.30$ Å^{-1}, while in the case of n 60° faults a minimum at $s \approx 0.25$ Å^{-1} is visible as well as a modulation at $s \approx 0.30$ Å^{-1}.

(b) Determination of defects inside the layers

Clays defects inside layers usually concern modifications of a part of the atomic positions in the unit cell or modifications of the kind of cations. Because of the weakness of the X-ray atomic scattering power, it is often difficult to detect these categories of defects in the pattern of a natural untreated clay. Such a sample generally contains a large amount of stacking faults which gives a more important intensity modification than defects inside the layers. To reveal these defects it is necessary to minimize the effects of stacking faults by reorganizing the stack. Two ways permit one to obtain better ordering in the layer stackings: by K-saturation and several wetting and drying cycles (Mamy & Gaultier 1976) or, more simply, by Cs-saturation (Besson et al. 1983).

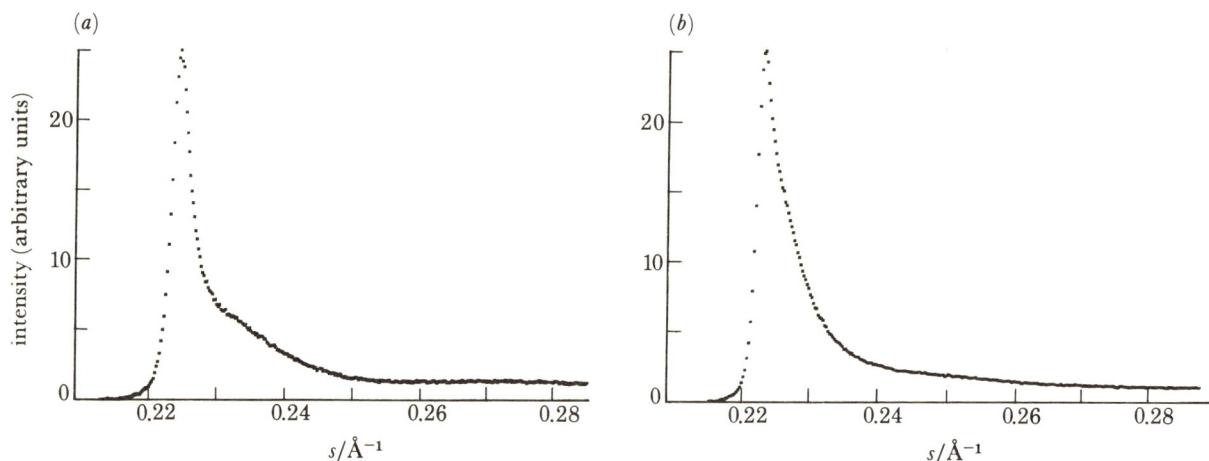

FIGURE 5. Experimental (02, 11) band profiles of anhydrous Na-beidellites. (a) Black Jack Mine beidellite; (b) Rupsroth beidellite.

If we plan to determine, for example, the nature of octahedral vacancies in a dioctahedral smectite, the analysis of the Na-saturated mineral does not lead to a conclusion. Figure 5 shows the experimental (02, 11) bands of the Black Jack Mine Na-beidellite (figure 5a) and of the Rupsroth Na-beidellite (figure 5b) in the anhydrous state: the (02, 11) bands are very similar (slight differences are localized in a part of the pattern sensitive to the nature of stacking faults) and do not permit any statement about octahedral vacancy positions. But calculations of the diffraction (02, 11) domain for a model of anhydrous Cs-beidellite leads to very different patterns whether vacancies are in *trans*-positions (figure 6a) or in *cis*-positions (figure 6b), or are statistically distributed between all octahedral sites (figure 6c). Comparing these theoretical patterns with the experimental ones obtained from Black Jack Mine and Rupsroth anhydrous Cs-beidellites (figures 7a and b), it can be concluded, without any ambiguity and even on the basis of only qualitative examination, that the Black Jack Mine beidellite has *trans*-vacant octahedral sheets while these sheets are *cis*-vacant in the Rupsroth beidellite (Besson 1980; Besson et al. 1983). A quantitative study, with agreement of intensities and profiles between

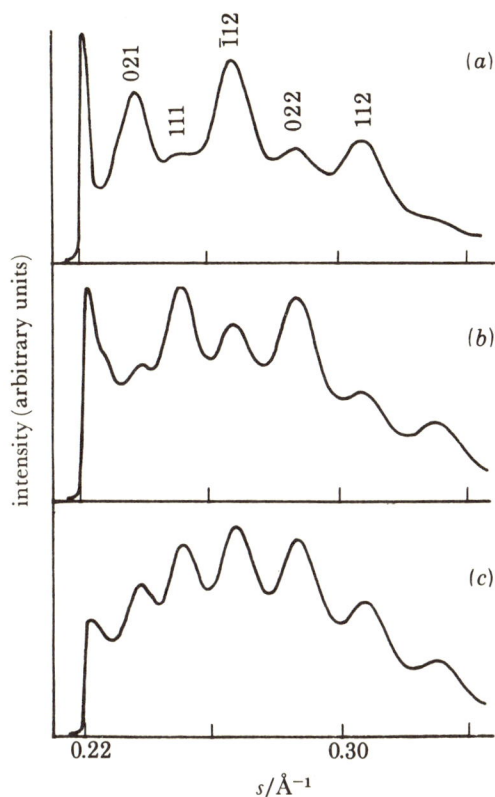

FIGURE 6. Calculated (02, 11) band profiles of anhydrous Cs-beidellite with three different positions of the octahedral vacancies. (a) Vacant *trans*-positions; (b) vacant *cis*-positions; (c) statistical distribution of vacancies between all the octahedral sites.

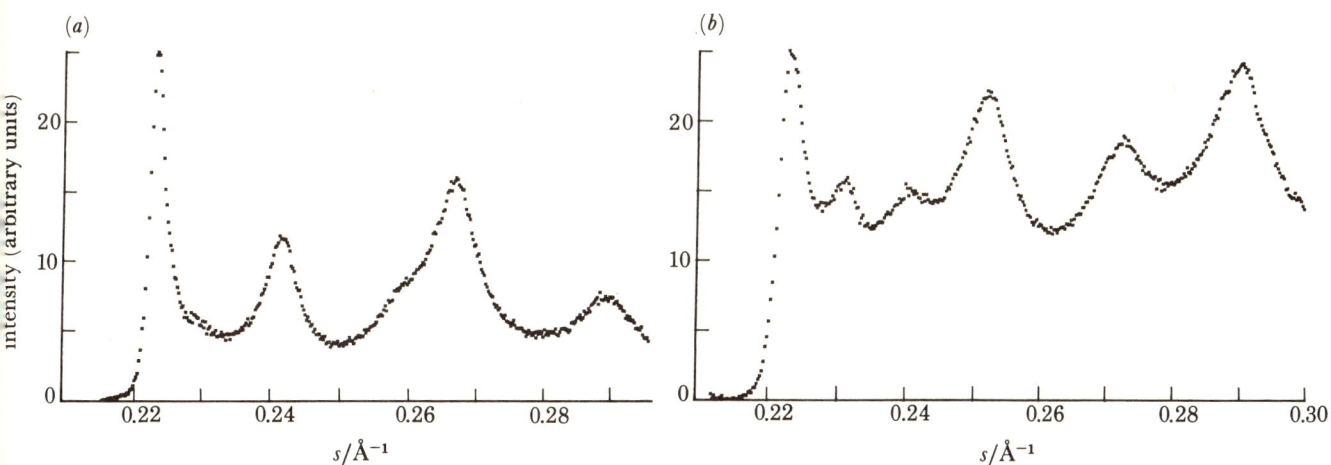

FIGURE 7. Experimental (02, 11) band profiles of anhydrous Cs-beidellites. (a) Black Jack Mine beidellite; (b) Rupsroth beidellite.

[45]

calculated and experimental patterns, shows that the presence of faults in the octahedral vacancies positions cannot be detected if the proportion of defects is less than 10 % (because of the existence of residual stacking faults).

(c) Distribution of atoms and molecules in the interlamellar spaces

This problem will be illustrated by work of Ben Brahim *et al.* (1984) concerning the X-ray study of the two-water-layer homogeneous hydration state of the Rupsroth Na-beidellite.

Three different categories of sites are proposed in the literature for water molecules in the two layer homogeneous state of hydrated clays. The water position projections on the (a, b) plane are given in figure 8a: we distinguish (i) sites projected near the basal oxygens of the layer (A_1 sites), (ii) those projected close to the tetrahedral cations (B sites), and (iii) those neighbouring to the centre of the 'hexagonal' cavities (C sites).

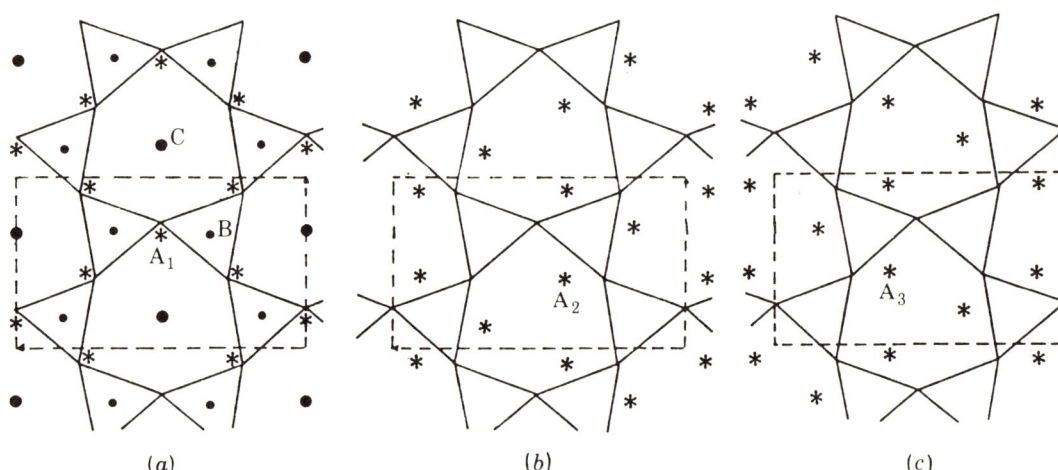

FIGURE 8. Projection, on the (a, b) plane, of possible water molecules sites in a two-water-layer hydration state of smectite. (a) First possibility with three types (A_1, B, C) of water positions. (b) and (c). Enantiomorphic distributions (A_2 and A_3) of the A sites, obtained by changing y-coordinates of A_1 sites into $y \pm \frac{1}{3}$.

Seeing that the two B and C site categories can be deduced from one another by changing their y-coordinates into $y \pm \frac{1}{3}$, the authors have imagined a simultaneous identical modification of the y-coordinates of the A_1 sites, leading to two enantiomorphic A_2 and A_3 site distributions (see figure 8b and c).

Theoretical pattern calculations have led to the following conclusions.

(1) The (20, 13) domain does not permit one to distinguish A_1, A_2, A_3 sites from one another, or B sites from C sites. In this diffraction domain, profiles and intensities of reflections are sentitive only to the ratio $(A_1 + A_2 + A_3)/(B + C)$, $(A_1 + A_2 + A_3)$ and $(B + C)$ being the total quantity of water molecules in the A and $(B + C)$ sites.

(2) The (02, 11) domain varies with the B/C ratio, but it is also absolutely insensitive to the sharing of water molecules between the different A site types.

(3) On the other hand the (04, 22) domain shows important differences depending on whether A_1 or A_2 sites are occupied (but patterns are identical for the two enantiomorphic A_2 or A_3 situations).

The comparison of the calculated and experimental X-ray data, in the (04, 22) domain,

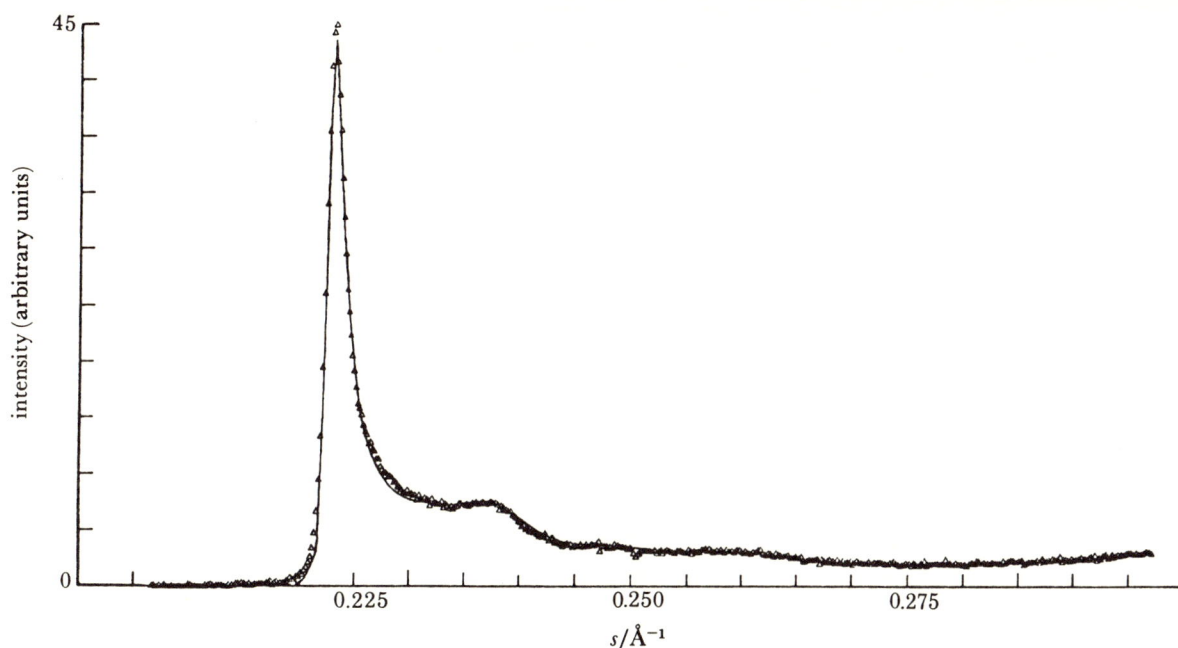

FIGURE 9. Two-water-layer Na-beidellite: agreement between the calculated and experimental patterns in the (02, 11) domain.

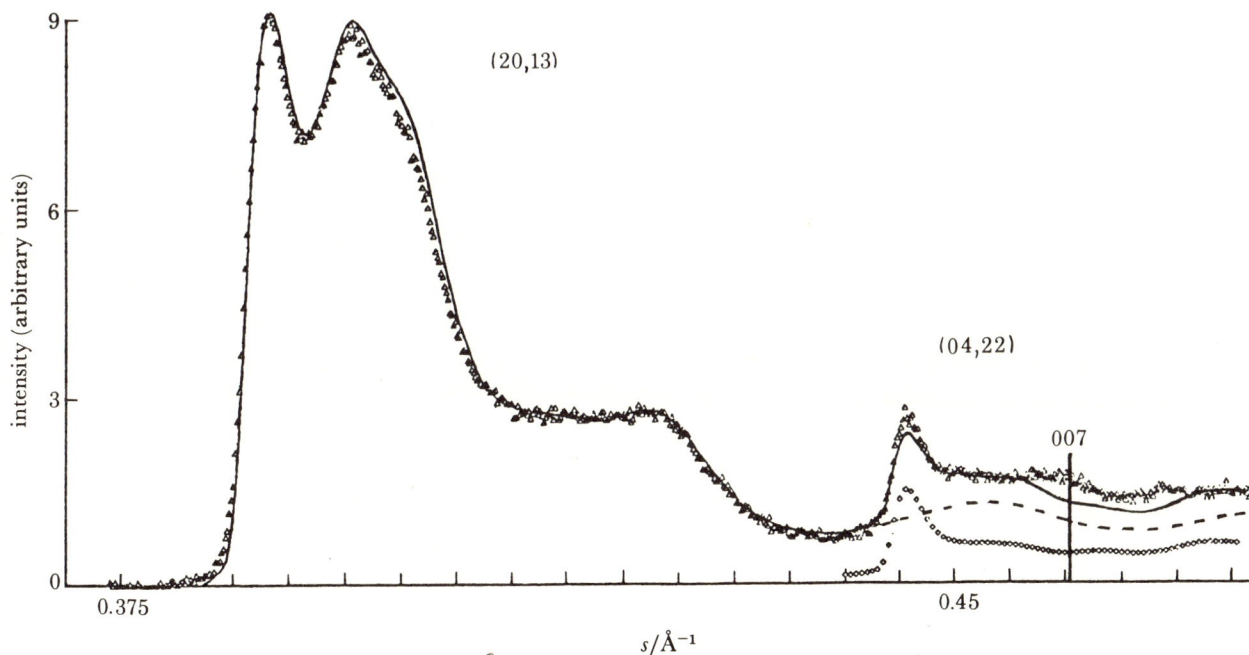

FIGURE 10. Two-water-layer Na-beidellite: agreement between the calculated and experimental patterns in the (20, 13) and (04, 22) domains.

[47]

leads to the conclusion that A_1 sites are vacant and A_2 (or A_3) sites are occupied; but simultaneous A_2 and A_3 occupations in the same water sheet is excluded.

The number of water molecules per unit cell and their z-coordinates are defined from the comparison of ratios I_{001}/I_{003} and I_{001}/I_{005} (Ben Brahim *et al.* 1983). And finally proportions of water molecules distributed between the B, C and A_2 (or A_3) sites are determined by fitting the profiles and intensities of the calculated pattern with the experimental one, simultaneously in the (02, 11), (20, 13) and (04, 22) diffraction domains.

Final agreement is given in figures 9 and 10 where calculated curves are full lines and experimental intensities, point by point, are represented by triangles. Figure 9 corresponds to the (02, 11) band and figure 10 to the (20, 13) and (04, 22) bands; the curve represented by squares is the calculated (04, 22) band before superposition of the (20, 13) band tail (dashed line). The agreement is obtained with the following structural characteristics:

(i) Probability of arbitrary stacking faults: $p_A = 0.74$;

(ii) z-coordinates of water molecules: $z = \pm 6.3$ Å (origin at the centre of the octahedral sheet);

(iii) total number of water molecules per unit cell: 10.8;

(iv) number of water molecules in A_2 (or A_3) sites: 7.5; in B sites: 2.6; in C sites: 0.7.

4. CONCLUSIONS

The modelling method is the only way that permits the interpretation of an important part of the problem connected with determination of defects in clays by X-ray powder pattern studies. It is effective even if the experimental pattern is reduced to successive poorly modulated (*hk*) bands. However, to find the concentration and nature of principal defects in each clay, it is necessary, in all cases (i) to obtain 'good' experimental patterns (that is, to control experimental conditions exactly); and (ii) to fit the calculated diagrams with the experimental ones quantitatively, not only on the basis of profile agreement but also by comparing the intensity values. Obviously, the better the ordering of each stacking, the more precise the solution will be.

It is also necessary to emphasize that imagining a model that contains structural defects does not mean making an arbitrary choice between an infinity of possibilities. In practice, this choice must always be controlled by crystallochemical considerations. It also depends on the set of data described in the literature obtained by means of different, not only diffractometric, techniques. In such conditions, the choice is considerably reduced and only a few models must be tested.

Finally it can be noted that it is always better not to limit a determination of defects to the X-ray pattern analysis only. It is preferable to associate other complementary techniques such as electron microscopy, electron and neutron diffraction, and different spectroscopic methods.

REFERENCES

Ben Brahim, J., Armagan, N., Besson, G. & Tchoubar, C. 1983 X-ray diffraction studies on the arrangement of water molecules in a smectite. I. Homogeneous two-water-layer Na-beidellite. *J. appl. Crystallogr.* **16**, 264–269.
Ben Brahim, J., Besson, G. & Tchoubar, C. 1984 Layer succession and water molecules arrangement in a homogeneous two-water-layer Na-smectite. *Proc. Vth Meet. Europ. Clay Groups.* (In the press.)
Besson, G. 1980 Structure des smectites dioctaédriques. Paramètres conditionnantl cs fautes d'empilement des feuillets. Thèse de Doctorat, Université d'Orléans, France.

Besson, G., Glaeser, R. & Tchoubar, C. 1983 Le césium, révélateur de structure des smectites. *Clay Miner.* **18**, 11–19.

Brindley, G. W. & Brown, G. 1980 '*Crystal structures of clay minerals and their X-ray identification* (ed. G. W. Brindley & G. Brown). London: Mineralogical Society, monogr. no. 5.

Brindley, G. W. & Mering, J. 1951 Diffraction des rayond X par les structures en couches désordonnées. *Acta crystallogr.* **4**, 441–447.

Brindley, G. W. & Robinson, K. 1946 'Randomness in the structures of kaolinitic clay minerals'. *Trans. Faraday Soc.* **42**, 198–205.

De Courville, J., Tchoubar, D. & Tchoubar, C. 1979 Détermination expérimentale de la fonction d'orientation; son application dans le calcul des bandes. *J. appl. Crystallogr.* **12**, 332–338.

Drits, V. A., Plançon, A., Sakharov, B. A., Besson, G., Tsipursky, S. I. & Tchoubar, C. 1983 Diffraction effects calculated for structural models of K-saturated montmorillonite containing different types of defects. *Clay Miner.* (Submitted.)

Drits, V. A. & Sakharov, B. A. 1976 *X-ray structural analysis of mixed-layer minerals. Moscow: Akad. Nauk S.S.S.R.*

Hendricks, S. B. & Teller, E. 1942 X-ray interference in partially ordered layer lattices, *J. chem. Phys.* **10**, 147–167.

Kakinoki, J. & Komura, Y. 1952 Intensity of X-ray diffraction by one dimensionally disordered crystal. I. General derivation in cases of the 'Reichweite' $S = 0$ and 1. *J. phys. Soc. Japan* **7**, 30–35.

MacEwan, D. M. C., Ruiz Amil, A. & Brown, G. 1961 Interstratified clay minerals. In *The X-ray identicfiation and crystal structures of clay minerals* (ed. G. Brown). London: Mineralogical Society.

Mamy, J. & Gaultier, J. P. 1976 Les phénomènes de diffraction des rayonnements X et électroniques par les réseaux atomiques. Application à l'étude de l'ordre cristallin dans les minéraux argileux. II. Evolution structurale de la montmorillonite associée au phénomène de fixation irréversible du potassium. *Ann. Agron.* **27-I**, 1–16.

Mering, J. 1949 L'interférence des rayons X dans les systèmes à stratification désordonnée. *Acta crystallogr.* **2**, 371–377.

Plançon, A. 1980 The calculation of intensities diffracted by a partially oriented powder with a layer structure. *J. appl. Crystallogr.* **13**, 524–528.

Plançon, A. 1981 Diffraction by layer structures containing different kinds of layers and stacking faults. *J. appl. Crystallogr.* **14**, 300–304.

Plançon, A. & Tchoubar, C, 1976 Etude des fautes d'empilement dans les kaolinites partiellement désordonnées. II. Modèle d'empilement comportant des fautes par rotation. *J. appl. Crystallogr.* **9**, 279–285.

Plançon, A. & Tchoubar, C. 1977 Determination of structural defects in phyllosilicates by X-ray diffraction. I. Principle of calculation of the diffraction phenomena. *Clays Clay Miner.* **25**, 430–435.

Reynolds, R. C. 1980 Interstratified clay minerals. In *Crystal structures of clay minerals and their X-ray identification* (ed. G. W. Brindley and G. Brown). London: Mineralogical Society, monograph no. 5.

Sakharov, B. A., Naumov, A. S. & Drits, V. A. 1982*a* X-ray diffraction by mixed layer structures with random distribution of stacking faults. *Dokl. Nauk Akad. SSSR* **265**, 339–343.

Sakharov, B. A., Naumov, A. S. & Drits, V. A. 1982*b* X-ray intensities scattered by layer structure with short range ordering parameters $S > 1$ and $G > 1$. *Dokl. Nauk Akad. SSSR* **265**, 871–874.

Taylor, R. M. & Norrish, K. 1966 The measurement of orientation distribution and its application to quantitative X-ray diffraction analysis. *Clay Miner.* **6**, 127–141.

Phil. Trans. R. Soc. Lond. A **311**, 271–285 (1984)   [271]

Printed in Great Britain

New ways of characterizing layered silicates and their intercalates

By J. M. Thomas, F.R.S.

Department of Physical Chemistry, University of Cambridge,
Lensfield Road, Cambridge CB2 1EP, U.K.

[Plates 1–3]

Four new methods of probing the atomic and microstructural characteristics of the clay minerals are described: solid-state, magic-angle-spinning ^{27}Al and ^{29}Si n.m.r. (along with ^{13}C n.m.r. of mobile, intercalated organic species); X-ray induced photoelectron studies encompassing photoelectron diffraction as a complement to conventional photoelectron spectroscopy; high-resolution electron microscopy either alone or in association with electron-stimulated X-ray emission microanalysis; and neutron scattering techniques. In reviewing the principles, scope and application of these methods specific case histories are selected from representative minerals belonging to the serpentines, kandites, smectites, micas, vermiculites, chloritoids, zeolites and intergrowths of these with one another or with other silicate minerals. Emphasis is placed on problems not readily amenable to solution by traditional, X-ray based procedures.

1. Introduction

A wide variety of techniques has, over the years, been used to elucidate the structure of clay minerals and the various complexes, usually intercalates, that they form with organic guests. Many of these techniques are classical and rather indirect; some of the more recent, spectroscopic ones are less model-dependent – as far as interpretation is concernerned – than their traditional predecessors, but only a few yield information of a more or less direct kind. If good crystalline specimens are available, X-ray crystallography reigns supreme as the single, most powerful direct technique inasmuch as it generally affords unambiguous information pertaining to bond lengths, bond angles, site occupancies and coordination numbers. By their very nature, the majority of clay minerals, and almost invariably the chemically more interesting ones, are not well ordered, they are not stoichiometrically well behaved, and they are not crystallographically or compositionally pure. Clay minerals, more so perhaps than most other minerals, occur as coexistent species, there being (as we shall see later) often quite marked tendencies for intergrowths of different structure and composition to occur within one particular microscopic region of a specimen. X-ray crystallographic techniques are not, therefore, universally suited for the study of many clay minerals and their compounds.

In this paper we focus on four relatively new methods of probing the structure of clay minerals: high-resolution (magic-angle-spinning) nuclear magnetic resonance (n.m.r.); X-ray photoelectron studies; high-resolution electron microscopy; and neutron scattering studies. Each of these techniques has brought to light fresh insights into the nature of silicates in general (see Thomas *et al.* 1978–9; Thomas 1977, 1982*a, b*, 1983; Fyfe *et al.* 1983; Thomas 1979; Rae-Smith *et al.* 1979; Cheetham *et al.* 1982). Some of them have also shed new light on the structure, properties and behaviour of the very many intercalates formed, especially by smectites (montmorillonite and hectorite in particular), with organic guests (see Thomas 1982*b*; Pinnavaia 1983; Ballantine *et al.* 1983, 1984; Tennakoon *et al.* 1983). These organic intercalates of clay minerals give rise to a rich diversity of chemical reactions (see Thomas 1982 b; Pinnavaia 1983; Ballantine *et al.* 1983). An illustration of the importance of interlamellar protonation and interlamellar isomerization of carbocations formed by, for example, the addition of protons to alk-1-enes is given in schemes 1 and 2.

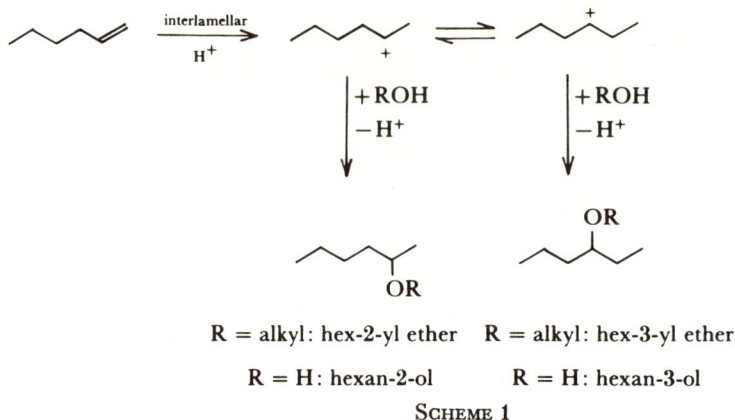

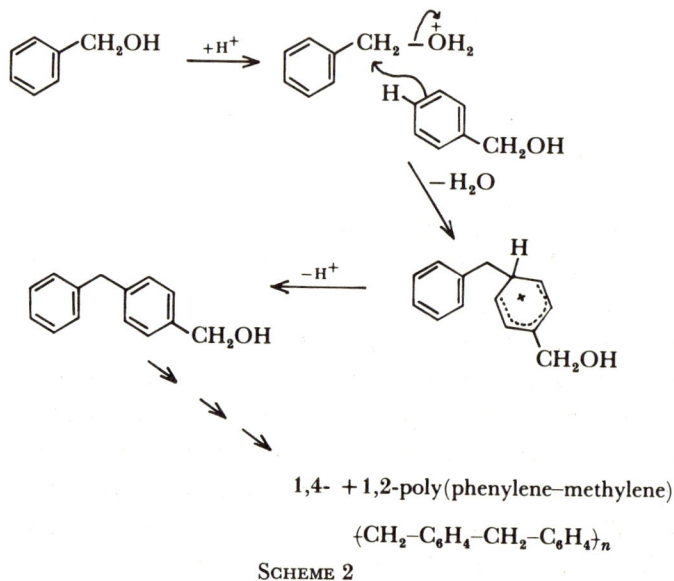

R = alkyl: hex-2-yl ether   R = alkyl: hex-3-yl ether

R = H: hexan-2-ol   R = H: hexan-3-ol

SCHEME 1

1,4- + 1,2-poly(phenylene–methylene)

$+CH_2-C_6H_4-CH_2-C_6H_4+_n$

SCHEME 2

2. HIGH-RESOLUTION SOLID-STATE NUCLEAR MAGNETIC RESONANCE

When molecules tumble freely, as they do in the isotropic liquid state, the associated n.m.r. absorption lines are very sharp because the broadening influences arising from dipolar and other interactions are averaged out by the motion of the dispersed species. Even with solid clay mineral catalysts, quite sharp ^{13}C and ^{1}H n.m.r. spectra can be obtained from intercalated entities provided the latter are in a more or less free state of rotational and translational motion in the interlamellar regions. Tennakoon *et al.* (1984) have shown how, using ^{13}C n.m.r., the course of a catalytic reaction taking place in the interlamellar environment may be traced. The proton-catalysed addition of water to 2-methyl-propene shown in scheme 3: yielding *t*-butanol is readily identified from the ^{13}C n.m.r. spectrum (figure 1).

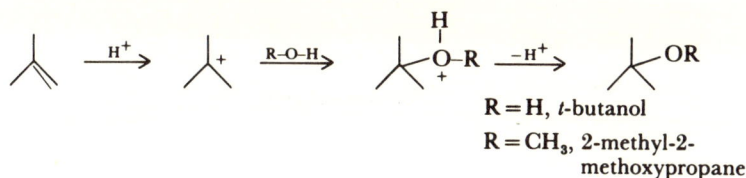

R = H, *t*-butanol

R = CH$_3$, 2-methyl-2-
methoxypropane

SCHEME 3

Al^{3+} ion exchanged synthetic hectorite exposed to isobutene

Al^{3+} ion exchanged synthetic hectorite exposed to *t*-butanol

$10^6 \times$ chemical shift from TMS

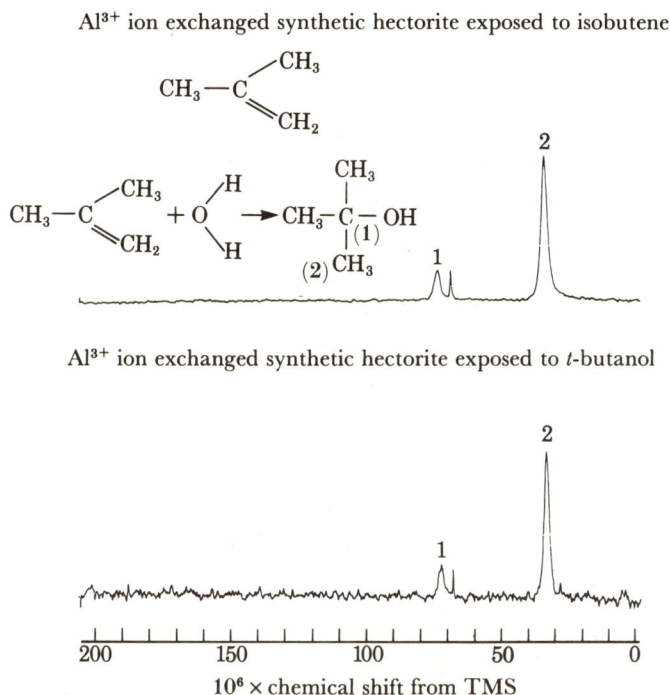

FIGURE 1. ^{13}C n.m.r. spectra showing that when 2-methyl propene (isobutene) is intercalated in a synthetic hectorite *t*-butanol is formed when the guest species reacts with the interlamellar water. The peaks labelled 1 and 2 refer to the two distinct types of carbon atom in the *t*-butanol.

So far as recording ^{27}Al and ^{29}Si n.m.r. spectra of clay minerals are concerned, however, it is necessary, in order to diminish by the necessary degree the considerable broadening influences brought about by the static, anisotropic dipolar and quadrupolar interactions, to resort to more sophisticated instrumental techniques. These sophisticated procedures have been reviewed by Andrew (1981), who, in 1958 was first responsible for introducing the principle of rapidly spinning the sample at the so-called magic-angle (54° 44′) to the magnetic field in order to produce high-resolution spectra. Reviews to the principles and application of this technique to aluminosilicate minerals have been given elsewhere (see Fyfe 1983; Fyfe *et al.* 1983).

The salient features of the magic-angle spinning n.m.r. (m.a.s.n.m.r.) technique in the context of clay mineralogy is that: (i) sixfold and fourfold coordination of Al and Si can be readily distinguished (see figure 2) by ^{27}Al and ^{29}Si m.a.s.n.m.r., respectively; (ii) non-crystalline (amorphous) or partially crystalline specimens are readily studied; (iii) time-dependent, solid-state phenomena are readily monitored (such as the distribution of Si and Al ions amongst eight distinct tetrahedral environments in annealed synthetic cordierite (see Fyfe *et al.* 1983),

[53]

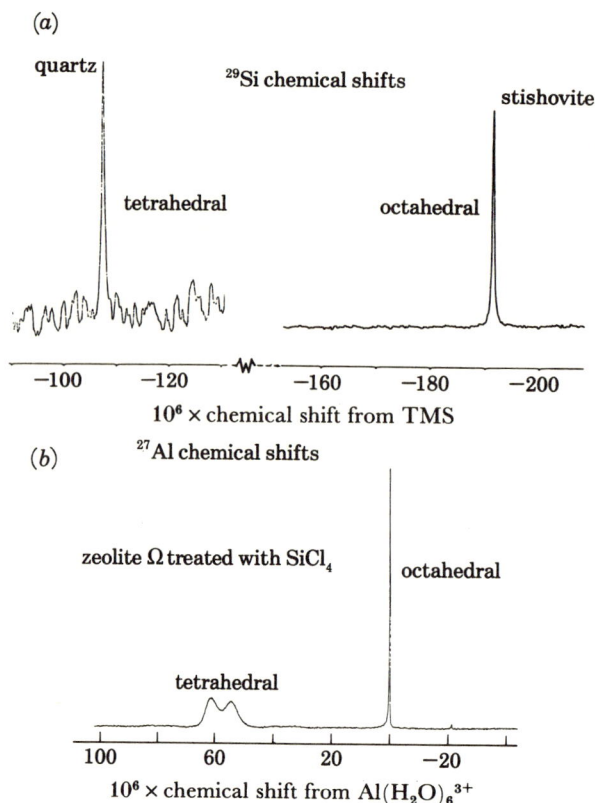

FIGURE 2a. ^{29}Si m.a.s.n.m.r. spectra showing the very different chemical shift values for silicon in SiO_2 in which there is fourfold, tetrahedral coordination, as in quartz (left), and silicon in SiO_2 in which there is sixfold, octahedral coordination, as in stishovite (right). (After Thomas *et al.* 1983.) (b) ^{27}Al m.a.s.n.m.r. spectrum of a dealuminated synthetic mazzite showing one sharp peak for an octahedral coordinated Al, and two broader peaks corresponding to two distinct tetrahedrally coordinated Al. (After Klinowski *et al.* 1983.)

TABLE 1. A SUMMARY OF SOME OF THE APPLICATIONS OF MAGIC-ANGLE-SPINNING N.M.R.
IN THE STUDY OF ZEOLITIC MATERIALS

^{29}Si

1. Discriminates between Si(OAl)$_4$
   Si(OAl)$_3$(OSi)
   Si(OAl)$_2$(OSi)$_2$
   Si(OAl)(OSi)$_3$
   Si(OSi)$_4$
2. Distinct peaks are obtained for non-equivalent Si(OSi)$_4$ groupings.
3. Enables framework Si/Al ratios to be obtained, for example, in zeolites,

$$\frac{Si}{Al} = \sum_{n=0}^{n=4} I_{Si(nAl)} \Bigg/ \sum_{n=0}^{n=4} \frac{n}{4} I_{Si(nAl)}$$

where n is the number of Al atoms joined, via oxygen, to the central silicon. A similar equation holds for sheet silicates.
4. In dealuminated zeolites, the number of distinct ^{29}Si peaks reflects the minimum number of non-equivalent Si sites in the unit cell.
5. A correlation exists between the chemical shift (δ, measured in parts per million from TMS) and the average TOT angle (θ measured in degrees): $\delta = -25.44 - 0.5793\,\theta$.

^{27}Al

1. Readily distinguishes Al in fourfold from Al in sixfold coordination.
2. Combined with intensities of ^{29}Si peaks, intensities of ^{27}Al peaks enable framework Si/Al ratios to be determined when these are very large (*ca.* 10000).
3. In aluminosilicate catalysts, under favourable circumstances, enables a direct determination of active sites (that is, when synonymous with Al sites in framework) to be determined.

[54]

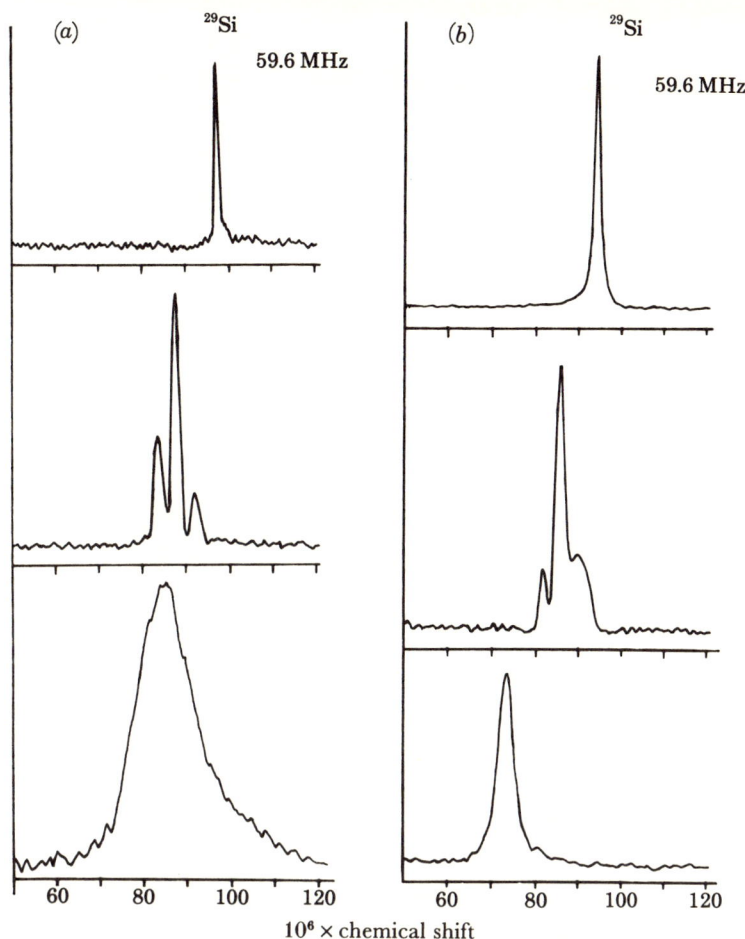

FIGURE 3. ^{29}Si m.a.s.n.m.r. (at 59.6 MHz, that is, a ^{1}H resonance of 300 MHz) of (a) talc, and two phlogopite samples one (P18) containing more magnetic impurities than the other (P8); and of (b) pyrophillite, muscovite and margarite. (After Sanz *et al.* 1984.) See text.

and (iv) in general, samples rich in paramagnetic or ferromagnetic constituents are not amenable to study because of unacceptably large line-broadening effects.

The power of m.a.s.n.m.r. in aluminosilicate chemistry, more particularly in characterizing zeolites, to which the ^{27}Al and ^{29}Si variants of this technique have so far been largely applied, is summarized in table 1. And the results of Sanz *et al.* (1984) and from our own studies (figures 3 and 4) show how promising ^{27}Al and ^{29}Si m.a.s.n.m.r. spectroscopy is likely to be so far as probing non-ferruginous samples of clay minerals is concerned. Serratosa's work (1984) has already shown how, using the procedures and equations employed in zeolite research (see Thomas *et al.* 1982; Klinowski *et al.* 1982) to evaluate Si/Al ratios and to test various models for Si, Al ordering, useful insights into the distribution of Al in the tetrahedral manifold of micas can be gained. What remains a little puzzling at present is the magnitude of the ratio of Al in tetrahedral and octahedral sites in, say, phlogopite. More exploratory work on, for example, differences in relaxation times and their influence upon peak intensities, is needed before we can gauge the trustworthiness of the m.a.s.n.m.r. technique as a general tool in clay

[55]

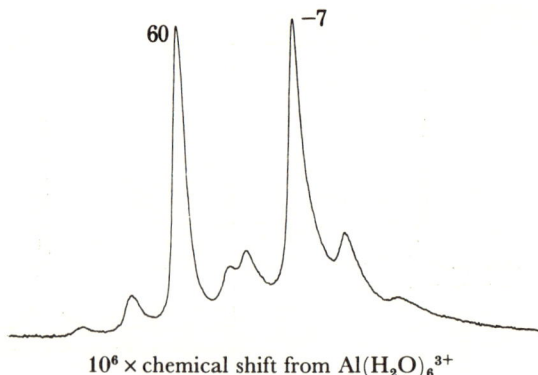

$10^6 \times$ chemical shift from $Al(H_2O)_6^{3+}$

FIGURE 4. ^{27}Al m.a.s.n.m.r. spectrum at 104.3 MHz showing, at $60/10^6$, a peak from tetrahedrally coordinated, and, at $-7/10^6$, a peak from octahedrally coordinated Al in muscovite. Other, subsidiary peaks are spinning side-bands of the major peaks.

mineralogy. What is already apparent, however, is its great utility in probing local environments in precursor gels (see Thomas *et al.* 1983; Engelhardt *et al.* 1983).

3. X-RAY INDUCED PHOTOELECTRON STUDIES

There are two aspects to the study of clay minerals by photoelectrons liberated by means of irradiation with soft, but monochromatic X-rays. It is well known that MgK_α ($h\upsilon = 1253$ eV) or AlK_α ($h\upsilon = 1486.6$ eV) X-rays will, according to the Einstein photoelectric equation, yield electrons possessing well-defined kinetic energies (Adams *et al.* 1972). Knowing the photoelectric cross sections of the various core levels (Evans *et al.* 1978) and escape depths of the various photoemitted electrons (Adams *et al.* 1977), it is possible to relate the intensity of peaks such as those shown in figure 5 to quantitative measures of the elements present in a sample. The first aspect, therefore, is simply to record the X-ray induced photoelectron spectrum.

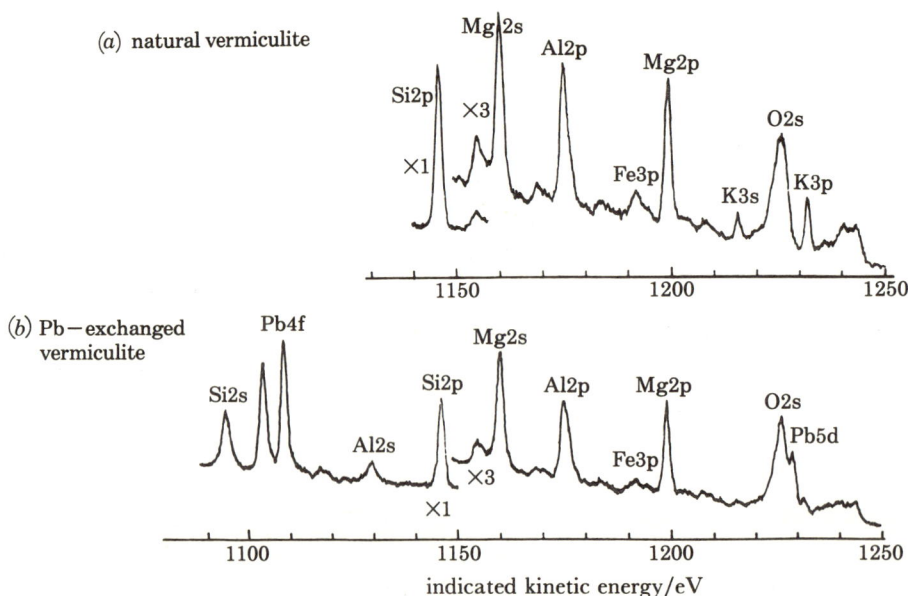

indicated kinetic energy/eV

FIGURE 5. X.p.s. spectra of (a) natural and (b) lead-exchanged vermiculites (see text).

In figure 5b, for example, we see how well this X-ray induced photoelectron spectroscopic procedure (often called X.p.s. or electron spectroscopy for chemical analysis (e.s.c.a.)) detects the cation exchange that takes place when naturally occurring vermiculite is exposed to a warm aqueous solution of Pb^{II}-containing ions. The disappearance of the K_{3s} and K_{3p} signals and the appearance Pb_{4f} peaks are noteworthy features which readily stand out in figure 5.

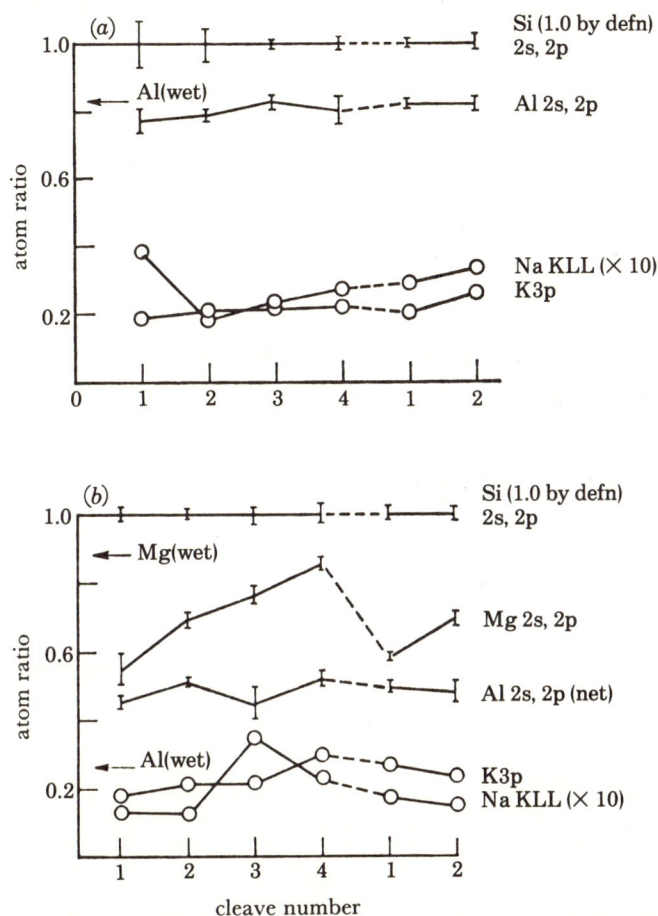

FIGURE 6. Variation in chemical composition over macroscopic distances as revealed by X.p.s. peak intensity ratios measured at a fixed take-off angle (see figure 7 and text). The numerals on the abscissa relate qualitatively to depth: the first four cleaves in each case were taken over about 1 mm of the crystal. All cleavages were performed *in vacuo*. Bulk analyses for Mg and Al are indicated by the word 'wet', with arrows, in each plot: (a) muscovite, (b) phlogopite. (After Evans *et al.* 1979.)

When rather large single-crystal specimens of sheet silicates are available (e.g. muscovite and phlogopite) it is relatively straightforward to carry out successive cleavage, *in situ*, inside the sample chamber of the X-ray photoelectron spectrometer (see Evans *et al.* 1979) so that detailed surface compositional analyses (to a depth of approximately 4–10 nm below the exposed cleavage-face) can be carried out by X.p.s. measurements following successive acts of cleavage (see figure 6). What is noticeable here is that, whereas the composition of successive exposed surfaces of muscovite agree well with one another, and with the average bulk composition, for surfaces of phlogopite there are significant variations one with respect to the other and with respect also to the mean composition of the bulk specimen. The magnesium content

[57]

is particularly different at each cleavage surface from that of the bulk and so, but to a rather lesser degree, is the aluminium content of phlogopite.

(a) X-ray photoelectron diffraction (X.p.d.)

In this technique, which constitutes the second aspect of the study of clay minerals by photoelectrons, one monitors the 'diffraction' of the X-ray induced photoelectrons as a function of the take-off angle of the emitted electron. Although the existence of such diffraction has been known for some time (see Siegbahn *et al.* 1970), its use as a tool for structural elucidation of the near-surface regions of solids is quite recent (Adams *et al.* 1978; Evans *et al.* 1979; Thomas 1979; Evans *et al.* 1979). To illustrate the principles of the technique, consider the

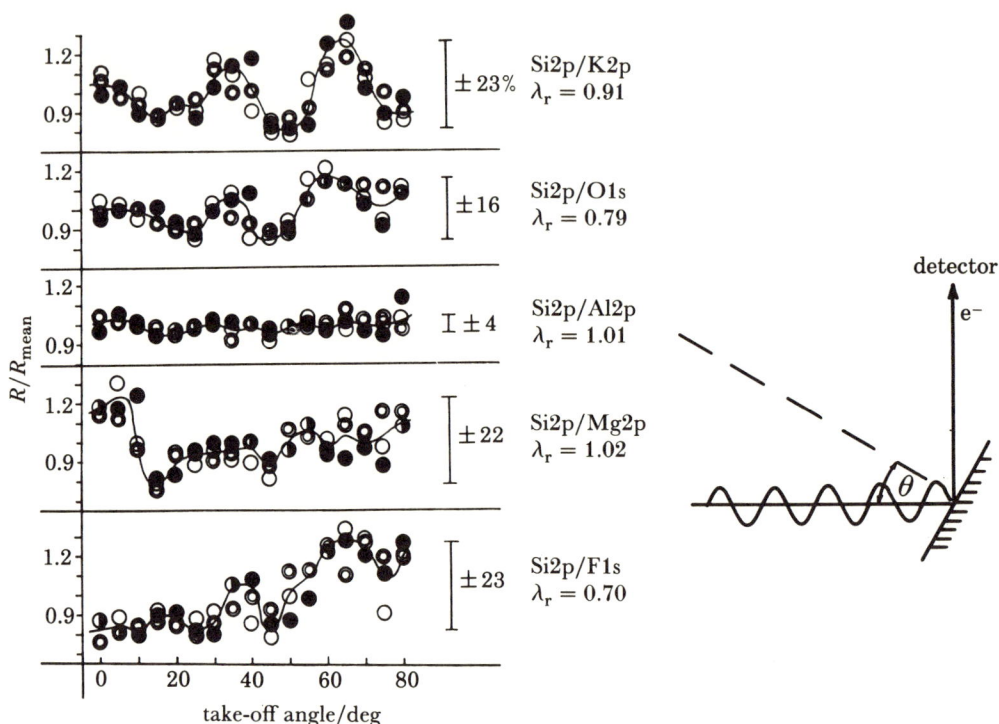

FIGURE 7. Schematic illustration of orientation of crystal and the set-up required to record X.p.d. spectra (see text). The X.p.d. patterns for phlogopite (where three cleaves in the same orientation are distinguished by single circles, double circles and solid circles) show that Si and Al atoms in this silicate are in essentially identical crystallographic environments, since there is little variation in intensity ratio with take-off angle.

variation of the intensity of the Si_{2p} photoemitted electrons of phlogopite mica as a function of take-off angle. The intensity will fluctuate because of the multiple diffraction that the electrons experience between the region where they are generated (inside the crystal) and the region where they escape (at the exterior surface) into the vacuum and into the electron spectrometer (the setting of which is fixed so as to register, in this case, Si_{2p} photoelectrons only). If, as is generally supposed, Al^{3+} ions occupy some of the tetrahedral sites normally taken up by Si^{4+}, then the angular variation of Al_{2p} and Si_{2p} photoelectrons should be identical, a situation which would not hold good for a mineral in which Al^{3+} is preferentially resident in octahedral sites. (There are a few other subtleties that have to be considered in a rigorous treatment of the factors governing angular variation in intensity of photoemission; see Evans *et al.* (1979), for a full discussion.)

[58]

Figure 7, which has, as an inset, a diagrammatic representation of the experimental set-up required to record X-ray photoelectron diffractograms, shows that, in phlogopite, there is indeed accommodation of the Al^{3+} in the tetrahedral manifold. (Ratios of intensity such as Si_{2p}/Al_{2p} constitute a more reliable and useful means of plotting these diffractograms: instrumental and other difficulties are largely compensated in this fashion.) Note that, for two elements (such as Si and K) which are separately confined to unique and different crystallographic sites, the ratio of their photo-peak intensities exhibit maximal variation with take-off angle (see the Si_{2p}/K_{2p} variation in figure 6).

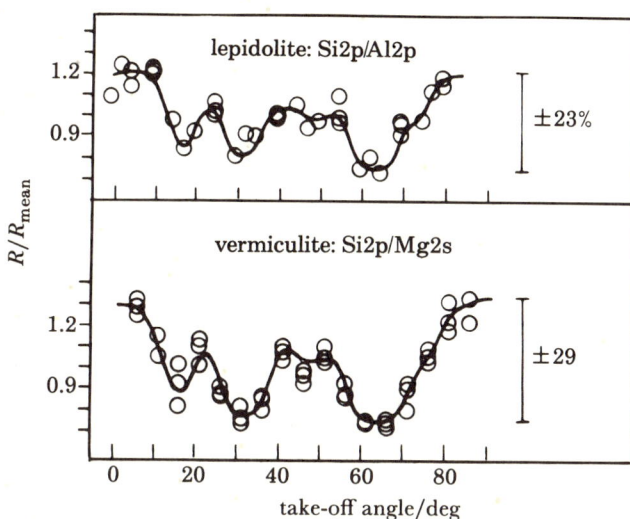

FIGURE 8. X.p.d. patterns for the major tetrahedrally (that is, Si and Al) and octahedrally (Al, Mg) coordinated elements in (a) lepidolite and (b) vermiculite.

By using procedures of this kind it has been shown that Al^{3+} ions in lepidolite and Mg^{2+} ions in vermiculite are in identical crystallographic environments, both being situated at octahedral sites (figure 8) other useful information to emerge from X.p.d. measurements on clay minerals are that Na^+ ions present in muscovite in quantities too small to be detected by conventional X-ray methods (see Brown, this symposium), occupy sites that are similar, but not quite equivalent, to those occupied by the interlamellar ions for which they substitute; and that the degree of hydration of exchangeable cations in clay minerals can be ascertained. Thus, whereas interlamellar Ca^{2+} and Pb^{2+} ions in weathered or solution-treated vermiculite are hydrated, interlamellar K^+ ions are not. X.p.d. measurements have so far been restricted to those clay minerals that occur as rather large crystals. There is no reason why this technique can also be applied to much smaller specimens provided the necessary instrumental modifications and improvements in detection of photoelectrons (both already feasible) are made.

4. High-resolution electron microscopy

The principles and practical considerations relating to the use of this powerful technique for the study of silicate minerals in general and for clay minerals in particular have been previously described (see Thomas et al. 1978–9, 1982; Thomas 1980, 1982, 1983; Millward & Thomas 1982; Beer et al. 1981; Jefferson et al. 1981). Briefly, very thin specimens are examined and the image is recorded at several, so-called defocus settings and also using several thicknesses in the range 5–50 nm. Methods are available for the calculation of the expected image contrast for a given supposed structure. The trustworthiness of a supposed model structure is assessed according to its degree of matching with the observed image using several thicknesses and several defocus settings. If possible, also, the matching operation between observed and calculated image is carried out employing as many different crystallographic projection directions as possible. For the anisotropic clay minerals it is generally much better to record images in directions that lie along the layer planes. Although this is, in practice, not easy; since appropriate fractured, or ion-beam thinned specimens can be produced only with some difficulty, more discriminating, interpretable and less ambiguous, images are thus obtained. Occasionally, however, as with the serpentines (see below), valuable information can be obtained by imaging in directions perpendicular to the layer planes. (For a summarizing account of the especial merit of high-resolution electron microscopy (h.r.e.m.), which produces real-space images, in the structural elucidation of minerals containing light elements, see Pring *et al.* (1983), who describe the matching of calculated and observed images along all three principal, high-symmetry directions in the mineral rhodizite.)

One of the supreme advantages of h.r.e.m. in the study of silicate minerals is that it records, in real-space, the presence and extent of structural intergrowths. Much valuable information pertaining to zeolites, both faujasitic (Millward & Thomas 1982) and the synthetic ones known as pentasils, that is, ZSM-5 and ZSM-11 (see Millward & Thomas 1982*b*; Thomas *et al.* 1983), can be obtained in this way. New types of boundaries including those known as coincidence boundaries (Terasaki *et al.* 1984) have been found in other zeolitic structures where there are one-dimensional tunnels, as in zeolite L.

(a) *The study of serpentine minerals: their microstructure and polymorphism*

There are three forms of serpentine (ideal formula $Mg_3Si_2O_5(OH)_4$): lizardite, which has a platy morphology; chrysotile, which is tubular; and antigorite, which is also platy but corrugated (see figure 9). All three polymorphs tend to coexist with one another. All are made up of layers separated by *ca.* 0.7 nm, the components of which are a two-dimensional sheet of corner-linked SiO_4 tetrahedra, and, coherently attached to this sheet; a two-dimensional network of $Mg(O, OH)_6$ edge-sharing octahedra. The upper vertices of the SiO_4 tetrahedra are linked as shown in figure 9, to the $Mg(O, OH)_6$ octahedra: some of the oxygens are common to both networks.

Chrysotile is tubular, so it is argued, because the ideal (Mg-containing) octahedral sheet is too large to enmesh exactly with the adjoining SiO_4 sheet. Hence, the tubular morphology is adopted because, in this way, the structure can neatly accommodate the larger radius of curvature of the outer octahedral layer as well as the smaller curvature of the inner tetrahedral layer.

It has been possible to test the hypothesis that, by substituting Al for Mg in octahedral sites and, or, Al for Si in tetrahedral sites, the dimensions of the octahedral sheets can then be equalized with those of the tetrahedral ones. Combined h.r.e.m. and energy-dispersive X-ray emission microanalysis confirms that, in chrysotile there is no aluminium (figure 10, plate 1). The situation is essentially the same in antigorite where a corrugated structure (figure 9) is adopted as an alternative to the tubular one so as to accommodate the discrepant dimensions of the contiguous octahedral and tetrahedral sheets. The flat, platy lizardite does indeed contain aluminium, and it has been shown by Crawford *et al.* (1978), that 16% of the Mg in octahedral sites and 12% of the Si in tetrahedral sites are replaced by Al. Note that, in the high-resolution image of antigorite, the postulated 'corrugated' structure of, with its long wavelength periodicity of *ca.* 43 nm is directly visible, as is the 0.46 nm unit-cell repeat distance.

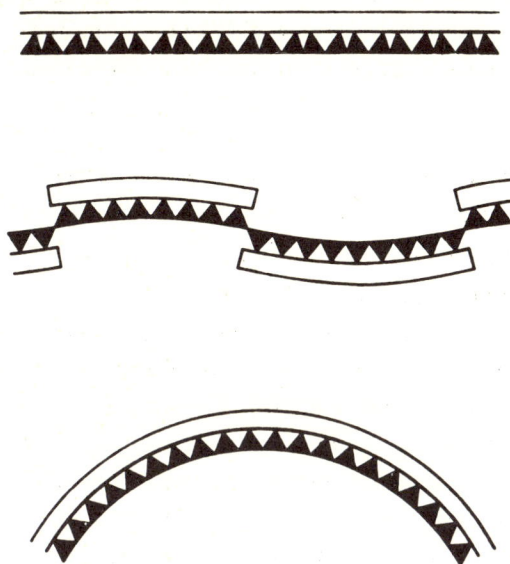

FIGURE 9. Schematic illustration of the morphological features of the serpentine minerals: chrysotile (bottom), antigorite (middle) and lizardite (top). The tetrahedral (SiO_4) layer is shaded and the octahedral sheet (Mg(O, OH)) or Mg, Al(O, OH) is left unshaded, for clarity.

In an elegant h.r.e.m. study of the serpentine minerals (from the Cascade Mountains, Washington, U.S.A., and from Ocna de Fier, Romania) Veblen & Buseck (1979) found, intimately, intergrown within each other the serpentine minerals chrysotile, lizardite and antigorite as well as talc, chlorite and amphibole. They also found evidence for mixed-layer silicate consisting of serpentine and talc layers.

(b) The study of chloritoid: an unusual layer silicate

This silicate (idealized formula $(Mg_1Fe)_2Al(Al_3Si_2O_{10})(OH)_4$) consists of two kinds of quite complicated sheets; and the several polytypic forms of the mineral arise from the various possible stacking sequences of these component sheets. It so happens that specimens of chloritoid are rather readily prepared for h.r.e.m. studies because it possesses a cleavage perpendicular to the basal plane (see Jefferson & Thomas 1978) which make it readily possible for the stacking of the sheets to be directly recorded.

Three main points emerge from the h.r.e.m. study of chloritoid. First, two new polytypes were discovered, one of monoclinic the other of triclinic symmetry. Figure 11, plate 2, shows a typical example of the way in which the various polytypes intergrow coherently with one another: four of the known five polytypes of chloritoid can be seen to coexist in a small region of one sample. All these polytypes (the $2M_1$, $2M_2$ monoclinic variants and the 1Tc 3Tc triclinic variants) were all 'encountered' in the h.r.e.m. study of Jefferson & Thomas (1979a) in traversing a distance of *ca.* 50 nm.

Secondly, just as with other silicate minerals, notably the chain silicates typified by the pyroxenoids such as wollastonite, rhodonite and pyroxmangite, it has been found that poly-typic strips can terminate within others, thereby creating local structural discontinuities which are likely to function both as centres of preferential chemical attack and as points of mechanical weakness. Thirdly, energy dispersive X-ray emission microanalysis, down to volumes of ca. 10^7–10^8 nm^3, show no correlation between elemental composition and the particular polytype peculiar to the region in question. In the course of a microscopic study of topotactic dehydration of chloritoid (Jefferson & Thomas 1979 b) it emerged that there could be a correlation between a given polytype and the oxidation state of the Fe ions present in the clay mineral.

(c) Exsolution and the crystal chemistry of the recently discovered mica mineral wonesite

In a very recent study, Veblen (1983 a) has demonstrated well the merit of h.r.e.m. as a tool in clay mineral research. He has focused attention on the sodium mica wonesite from the Post Pond Volcanics, Vermont, and has discovered the first known example of exsolution in a mica. It transpires that the wonesite (see Spear et al. (1981) for a description of this new rock-forming silicate) exsolves partially into a lamellar intergrowth of talc and a sodium mica having fewer interlayer site vacancies than the initial wonesite. The mica is found to be enriched in Na, Al, K, Ti, Cr and Fe relative to the talc. The lamellae, the wavelength of which varies from tens of nanometres to about 0.5 μm, are inclined to the layers of the mica and talc structures at a variable angle that averages about 37° (that is, about a third of the tetrahedral angle). This microstructure explains why wonesite is not expandable in water, in contrast to the behaviour of many synthetic and naturally occurring intergrowths of smectites and micas. A schematic illustration of the exsolved structure, along with the low-resolution electron micrograph on which it is based is shown in figure 12.

From Veblen's h.r.e.m. studies (Veblen 1983 b) of the microstructures and mixed layering in intergrown wonesite, chlorite, talc, biotite and kaolinite, several noteworthy features emerge. The chlorite occurs in intergrown one-layer and disordered polytypes; the wonesite occurs in one-layer, two-layer, three-layer and disordered forms, and the biotite occurs as one-layer and two-layer polytypes with minor stacking disorder. Some of the wonesite also exhibits turbo-stratic stacking, of the kind that is common in incompletely crystallized specimens of graphite. Parts of the chlorite exhibit perfect alternation of the talc-like and brucite-like layers. Likewise, occasional brucite-like layers are found in most of the wonesite (see figure 13, plate 3). In both chlorite and wonesite, these extra layers are in some cases observed to terminate (compare chloritoid described in (b) above). In some places, the chlorite and mica structures intergrow freely, either in almost total disorder or as slabs of chlorite and mica up to tens of nanometres thick. Thin slabs of kaolinite, talc and potassium biotite also intergrow with the chlorite and wonesite. The mixed layering phenomena in this specimen demonstrates that chlorite which appears normal in thin section can occasionally deviate substantially from ideal chlorite stoichiometry.

Description of plate 1

Figure 10. High-resolution electron microscopy micrographs and X-ray emission spectra of serpentines: (a) lizardite, (b) antigonite, showing the corrugation periodicity of ca. 4.3 nm, and (c) chrysotile, showing the interlayer spacing (0.74 nm).

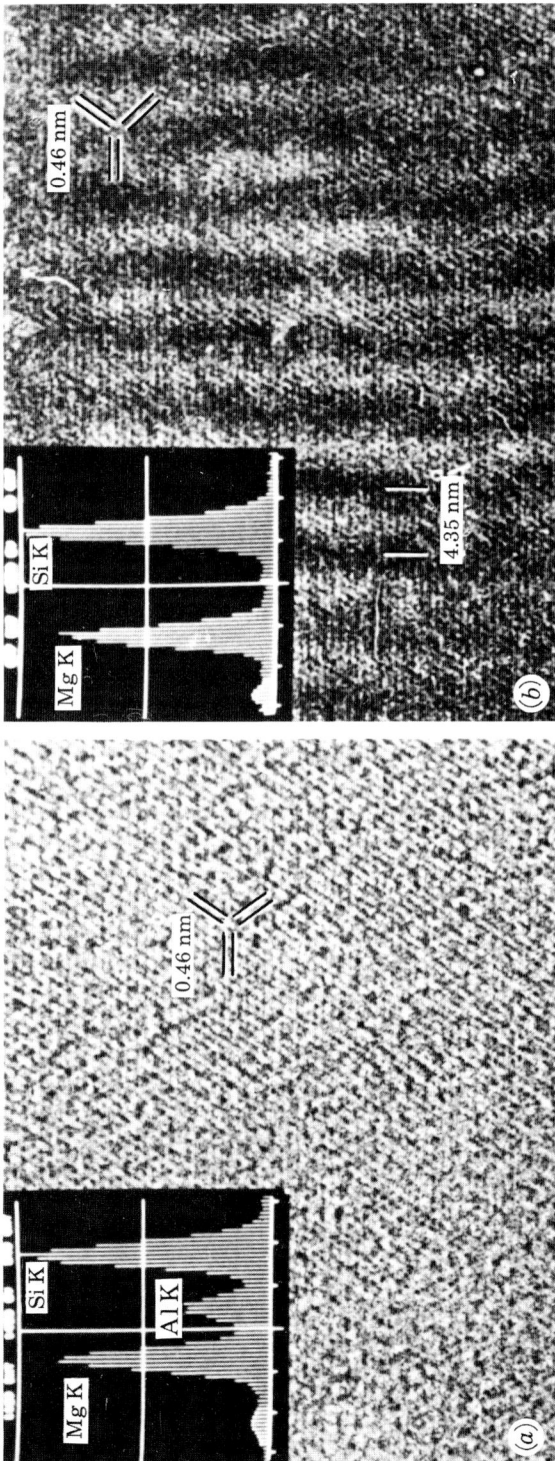

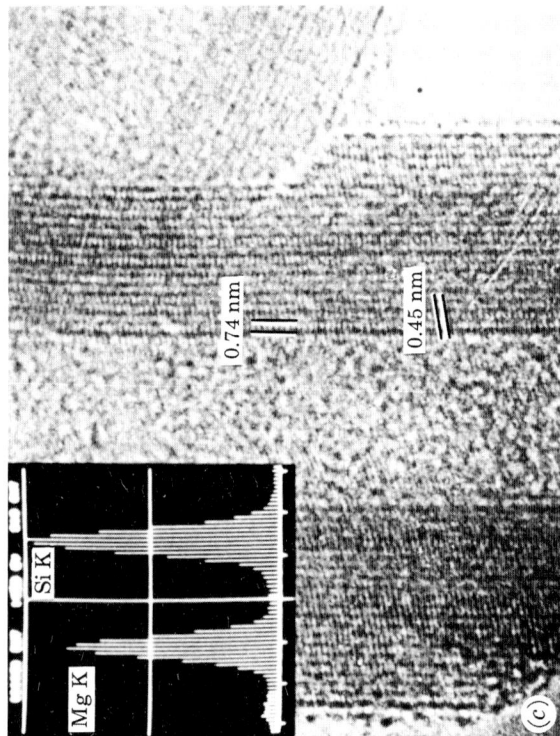

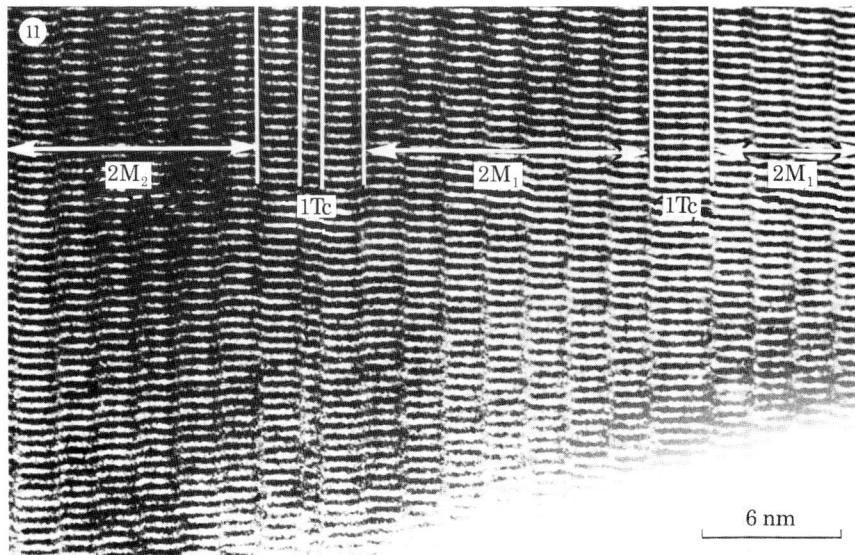

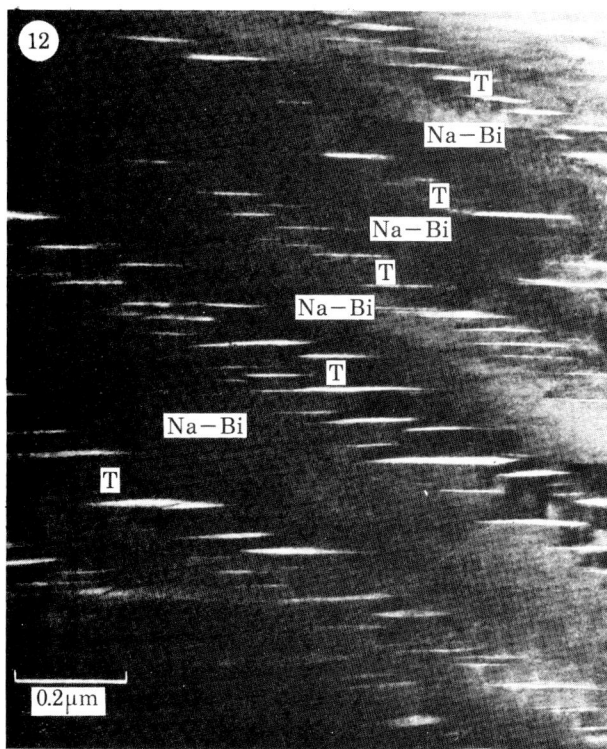

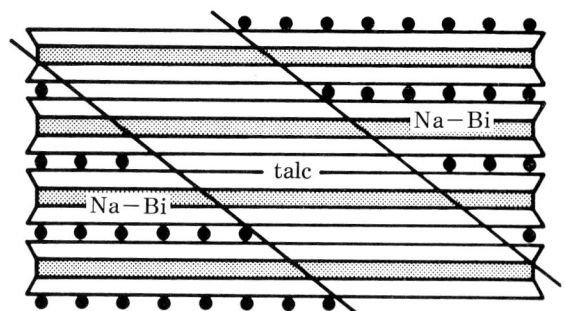

FIGURES 11 AND 12. For description see opposite.

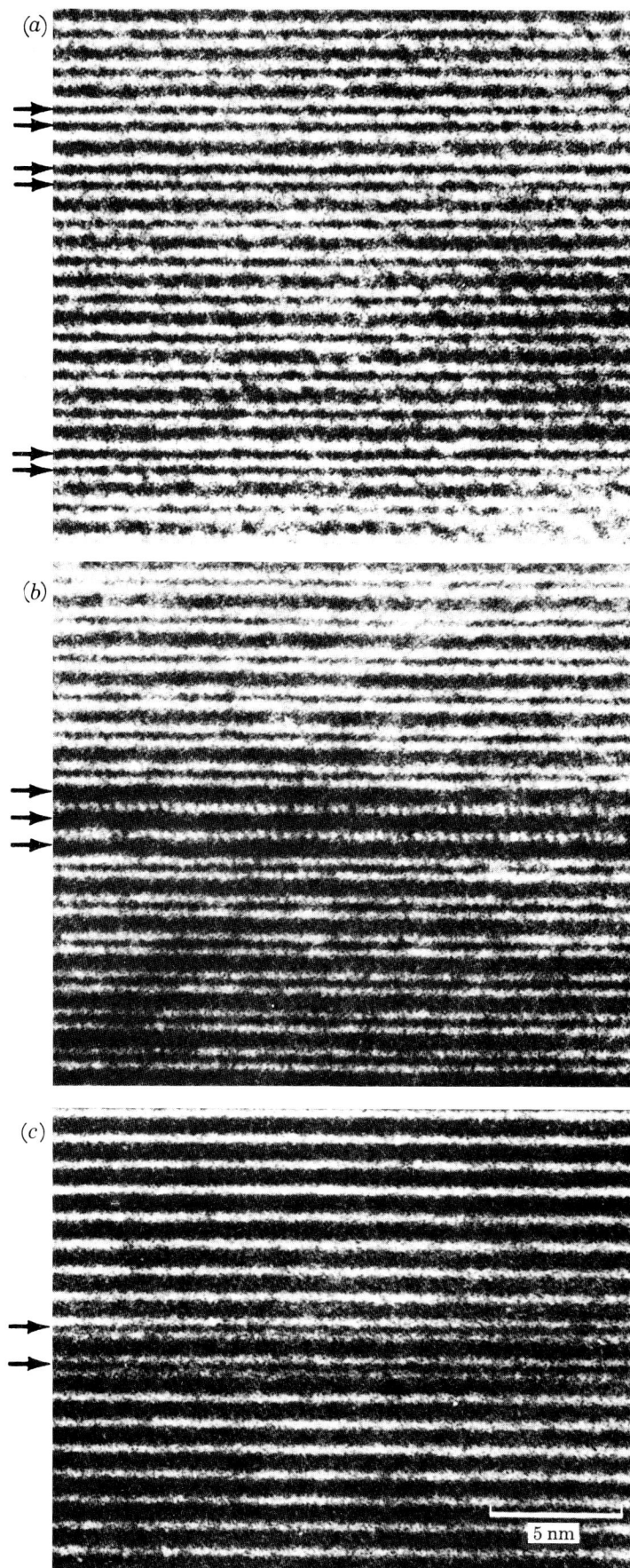

FIGURE 13. For description see opposite.

The unusual material designated 'poorly crystallized material' abundant in Mighei CM_2 carbonaceous chondrite (see Fuchs *et al.* 1973) has recently been shown by Tomeoka & Buseck (1983), using h.r.e.m., to be largely composed of a Fe–Ni–S–O layered mineral which occurs in coherent intergrowths with smectites in various ordered and disordered sequences.

(d) Other possibilities

Already it has proved possible to image some of the pillared clays (otherwise known as cross-linked smectites) which are proving to be so promising as catalytic agents intermediate in their performance between zeolites on the one hand and clay minerals on the other (see Vaughan & Lussier 1980; Vaughan 1984; Thomas 1982*b*). Imaging of this kind by h.r.e.m. should shed much light on the structure and properties of this new class of novel, pillared catalysts, in which the pillars themselves can be rich in aluminium oxide, or gallium oxide or zirconium (as well as other metal) oxides. So far as the micas themselves are concerned, Iijima & Buseck (1978) and Iijima & Zhee (1982) have shown the value of h.r.e.m. as a tool for structural characterization.

5. Neutron scattering

Most clay minerals exhibit a high degree of preferred orientation parallel to the layer planes. This being so it is readily possible, using appropriately mounted thin films of the minerals, to obtain (00*l*) reflections from X-rays or neutrons up to very high order (for example, $l = 16$). This, in turn, yields one-dimensional Fourier plots of nuclear scattering density of the constituent atoms in the framework as well as of the ions in the interlamellar regions of the clay mineral in question (see Thomas *et al.* 1976; Adams *et al.* 1976).

For the study of intercalates, including time-dependent monitoring of transformations from 3-layer to 2-layer to 1-layer intercalates of organic guests (Adams *et al.* 1977), this approach has been of considerable value. The subtle changes in orientation of, for example, tetrahydrofuran in proceeding from Ni^{2+}-exchanged to Co^{2+}-exchanged montmorillonite can be readily followed in this way (see Thomas *et al.* 1976). So also can the positions of hydrogen atoms in the kaolinite: formamide intercalate (Adams *et al.* 1976).

A quite different kind of neutron-based study of aluminosilicates is also possible. This entails the so-called Rietveld powder profile method in which, using routines described fully elsewhere (see Stewart 1978; Rae-Smith *et al.* 1979), the atomic coordinates and occupancy factors for a powdered (but ordered) structure is evaluated by minimizing the differences between observed and computed (on the basis of the refined structure) profiles for the neutron-diffractogram.

This approach has been particularly successful for various cation-exchanged zeolites (Cheetham *et al.* 1982; Cheetham *et al.* 1983) – see figure 14. Si–O and Al–O bond distances within the zeolite framework can be reliably extracted using this technique. Recently evidence has been obtained, using the Rietveld method, to suggest that in dealuminated faujasitic zeolites

---

Description of plate 3

Figure 13. Mixed layering defects. (*a*) Extra brucite-like layers (or missing talc-like layers) in chlorite. The arrows indicate pairs of adjacent brucite-like layers. (*b*) Extra talc-like layers in chlorite (or missing brucite-like layers). The arrows indicate three adjacent talc-like (TOT) layers. (*c*) Brucite-like layers (arrowed) intercalated in wonesite.

[63]

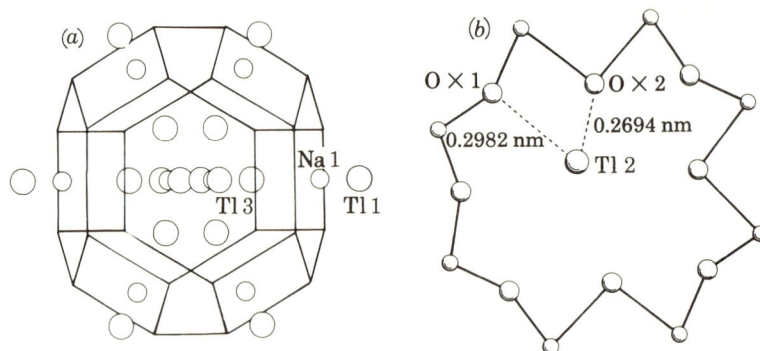

FIGURE 14. The location of the exchangeable cations Na$^+$ and Tl$^+$ in zeolite A. (a) The Tl(1), Tl(3) and Na(1) sites and their closeness to the six-membered face of the β-cage are emphasized. (b) Location of Tl(2) in the so-called eight-ring aperture. (After Cheetham et al. 1982.)

AlO$_4^{5-}$ units are arranged, in an ordered fashion within the so-called β-cages (sodalite cages). (G. D. Stucky, personal communication.)

In those clay minerals where crystallographic order is well-developed (that is, in those that have already been amenable to X-ray crystallographic techniques) there is every prospect that the Rietveld neutron profile method (or indeed, its more recently developed X-ray Rietveld method) will prove to be an important additional structural tool especially in that the method dispenses with the need to obtain single-crystal specimens.

I acknowledge with gratitude support from the S.E.R.C. and the University of Cambridge.

REFERENCES

Adams, J. M., Evans, S., Reid, P. I., Thomas, J. M. & Walters, M. J. 1977a Analyt. Chem. 49, 2001.
Adams, J. M., Evans, S. & Thomas, J. M. 1978 J. Am. Chem. Soc. 100, 3260.
Adams, J. M., Lukawski, K. S., Reid, P. I., Thomas, J. M. & Walters, M. J. 1977b J. chem. Res. (M) 0301.
Adams, J. M., Reid, P. I., Thomas, J. M. & Walters, M. J. 1976 Clays Clay Miner. 24, 267.
Adams, I., Thomas, J. M. & Bancroft, G. M. 1972 Earth Planet Sci. Lett. 16, 429.
Andrew, E. R. 1981 Int. Rev. phys. Chem. 1, 195.
Bailey, S. W. 1983 Clay minerals (ed. G. Brown and G. W. Brindley). Mineralogical Society.
Ballantine, J. A., Purnell, J. H. & Thomas, J. M. 1983 Clay miner. (In the press.)
Ballantine, J. A., Purnell, J. H. & Thomas, J. M. 1984 J. Molec. Catalysis. (In the press.)
Beer, M., Carpenter, R. W., Lyman, C. E., Eyring, L. E. & Thomas, J. M. 1981 Chem. Eng. News 59, 40.
Cheetham, A. K., Eddy, M. R., Jefferson, D. A. & Thomas, J. M. 1982a Nature, Lond. 299, 24.
Cheetham, A. K., Eddy, M. R., Jefferson, D. A. & Thomas, J. M. 1982b Nature, Lond. 299, 924.
Cheetham, A. K., Eddy, M. R., Klinowski, J. & Thomas, J. M. 1983 J. chem. Soc. chem. Commun. 23.
Crawford, E. S., Jefferson, D. A., Thomas, J. M. & Bishop, A. C. 1978 J. chem. Soc. chem. Commun. 986.
Engelhardt, G., Fahlke, B., Magi, M. & Lippmaa, E. 1983 Zeolites 3, 292.
Evans, S., Adams, J. M. & Thomas, J. M. 1979 Phil. Trans. R. Soc. Lond. A 292, 563.
Evans, S., Raftery, E. & Thomas, J. M. 1978 Surface Sci. 89, 64.
Fyfe, C. A. 1983 In Inorganic chemistry: towards the 21st century (ed. M. H. Chisholm), p. 405. A.C.S. Symposium series, no. 211. Washington D.C.: American Chemical Society.
Fyfe, C. A., Gobbi, G., Klinowski, J., Putnis, A. & Thomas, J. M. 1983 J. chem. Soc. chem. Commun. 23, 556.
Fyfe, C. A., Thomas, J. M. & Lyerla, J. R. 1981 Angew. Chem. 20, 96.
Hewat, A. W. 1978 Powder Rietveldt and refinement system (I.L.L. Grenoble Report). Grenoble: I.L.L.
Iijima, S. S. & Buseck, P. R. 1978 Acta Cryst. A 34, 709.
Iijima, S. S. & Zhu, J. 1982 Am. Miner. 67, 1195.
Jefferson, D. A. & Thomas, J. M. 1978 Proc. R. Soc. Lond. A 361, 399.
Jefferson, D. A. & Thomas, J. M. 1979 Acta. Cryst. A 35, 416.
Jefferson, D. A., Thomas, J. M. & Egerton, R. 1981 Chemy Brit. 17, 514.
Jenkins, H. D. B. & Hartman, P. 1982 Phil. Trans. R. Soc. Lond. A 304, 397.

Klinowski, J., Anderson, M. W. & Thomas, J. M. 1983 *J. chem. Soc. chem. Commun.* 525.

Klinowski, J., Ramdas, S., Thomas, J. M., Fyfe, C. A. & Hartman, J. S. 1982*b* *J. Chem. Soc., Faraday Trans. II.* **78**, 1025.

Klinowski, J., Thomas, J. M., Fyfe, C. A. & Gobbi, G. 1982*a* *Nature, Lond.* **296**, 533.

Millward, G. R. & Thomas, J. M. 1983 In *Surface properties and catalysis by non metals and oxides* (ed. J. P. Bonelle), p. 197. Proc. NATO Advanced Study Institute. Dordrecht: D. Reidel.

Pinnavaia, T. J. 1983 *Science, Wash.* **220**, 365.

Pring, A., Jefferson, D. A. & Thomas, J. M. 1983 *J. chem. Soc. chem. Commun.* 734–736.

Rae-Smith, A. R., Cheetham, A. K. & Skarnulis, A. J. 1979 *J. appl. Cryst.* **12**, 485.

Sanz, J., Förster, H. G. & Serratosa, J. M. 1984 *Nature, Lond.* (In the press.)

Siegbahn, K., Gelius, U., Siegban, P. & Olson, D. 1977 *Phys. Lett.* A **32**, 221.

Spear, F. S., Hazen, R. M. & Rumble, E. 1981 *Am. Miner.* **66**, 100.

Tennakoon, D. T. B., Jones, W., Rayment, T., Schlögl, R., Klinowski, J. & Thomas, J. M. 1983*b* *Clay Miner.* (In the press.)

Tennakoon, D. T. B., Jones, W., Thomas, J. M., Ballantine, J. A. & Purnell, J. H. 1984 (In preparation.)

Tennakoon, D. T. B., Jones, W., Thomas, J. M., Williamson, L. J., Ballantine, J. A. & Purnell, J. H. 1983*a* *Proc. Indian Acad. chem. Sci.* **92**, 27.

Terasaki, O., Thomas, J. M. & Ramdas, S. 1984 *J. chem. Soc. chem. Commun.* 216.

Thomas, J. M. 1983*a* In *Inorganic chemistry: towards the 21st century* (ed. M. H. Chisholm), p. 445. A.C.S. Symposium Series no. 211.

Thomas, J. M. 1977 *Chem. Brit.* **13**, 175.

Thomas, J. M. 1979 *Nature, Lond.* **2, 79**, 755.

Thomas, J. M. 1980 *New Scient.* 580.

Thomas, J. M. 1982*a* *Ultramicroscopy* **8**, 13.

Thomas, J. M. 1982*b* In *Intercalation chemistry* (ed. A. J. Jacobson & M. S. Whittingham), p. 56. New York: Academic Press.

Thomas, J. M., Adams, J. M., Tennakoon, D. T. B. & Graham, S. H. 1976 *A.C.S. Symposium* **163**, 298.

Thomas, J. M., Gonzalez-Calbet, J. J., Fyfe, C. A., Gobbi, G. C. & Nicol, M. 1983*c* *Geophys. Res. Lett.* **10**, 91.

Thomas, J. M., Jefferson, D. A., Mallinson, L. G., Smith, D. J. & Crawford, E. S. 1978–9*e* *Chemica Scripta* **14**, 167.

Thomas, J. M., Jefferson, D. A. & Millward, G. R. 1982*d* *J. Microscop. Spectroscop. Electron.* **7**, 315.

Thomas, J. M., Klinowski, J., Wright, P. A. & Roy, P. A. 1983*b* *Angewandte Chemie* **22**, 614.

Thomas, J. M. & Millward, G. R. 1982*e* *J. chem. Soc. chem. Commun.* **22**, 1380.

Thomas, J. M., Millward, G. R., Ramdas, S. & Audier, M. 1983*d* *A.C.S. Symposium Series* 218 (ed. G. D. Stucky & F. G. Dwyer), p. 181.

Thomas, J. M., Ramdas, S., Klinowski, J., Fyfe, C. A. & Gobbi, G. 1982*c* *J. phys. Chem.* **86**, 1247.

Tomeoka, K. & Buseck, P. R. 1983 *Nature, Lond.* **306**, 354.

Vaughan, D. E. W. 1984 *J. molec. Catal.* (In the press.)

Vaughan, D. E. W. & Lussier, R. 1980 *5th Zeolite International Conference, Naples* (ed. L. V. Rees), p. 94. London: Heydon.

Veblen, D. R. 1983*a* *Am. Miner.* **68**, 594.

Veblen, D. R. 1983*b* *Am. Miner.* **68**, 566.

Veblen, D. R. & Buseck, P. R. 1979 *Science, Wash.* **206**, 1398.

Phil. Trans. R. Soc. Lond. A **311**, 287–299 (1984)   [287]
Printed in Great Britain

Interaction of water with clay surfaces

By J. J. Fripiat, M. Letellier and P. Levitz
Centre de Recherche sur les Solides à Organisation Cristalline Imparfaite, 1B, rue de la Férollerie 45045 Orléans, France

[Plate 1]

In clay suspensions and gels water molecules are distributed between two populations in fast exchange. The physical meaning of this distribution is discussed here. From the molecular dynamics point of view it is shown that the surface force fields perturb the rotational motion on a distance z^* extending to approximately 10^{-9} m. The product z^* times the surface area available to water defines the volume of a surface phase b. In the bulk phase a the rotational diffusional correlation time is not affected and the pulsed field gradient nuclear magnetic technique shows that the self-diffusion coefficient of water in that phase is the same as in pure liquid.

The effects of salt concentration, pressure and temperature on phase b are studied, the main finding being that in the surface phase the well-developed random hydrogen bond network characteristic of liquid water is, at least, partly destroyed. Freezing the the gel suppresses the rapid exchange between the two phases.

From the thermodynamic point of view it is not possible with the present data to decide whether phase b has thermodynamic significance or if it constitutes a heterogeneity within the aqueous phase.

1. Introduction

It was recently shown (Fripiat *et al.* 1982) that water molecules in clay gels and suspensions are distributed between two populations: phase a contains the so-called free water in the pores between the clay tactoids, whereas phase b is made from about 3 layers of molecules directly affected by surface force fields. Thus these fields extend only to approximately 10^{-9} m. This conclusion was reached by measuring the variation of the heat of immersion with respect to the average degree of coverage with water or the variation of the ^{1}H or ^{2}H spin–lattice n.m.r. relaxation time (T_1) with respect to the concentration Cf of the gel ($C = m_s/m_w$ with m_s: mass of air-dried solid and m_w mass of water respectively; f is a correction factor accounting for the hydration of the air-dried sample).

The heat of immersion decreases rapidly as the degree of coverage is increased until it becomes equal to the internal energy of bulk water in equilibrium with its vapour, namely 119.5 mJ m^{-2}. This occurs when the solid is covered with about 3–4 layers of water.

The variation of the relaxation rate T_1^{-1} with Cf is linear in a large range of concentration in solid. The intercept is proportional to the rotational–diffusional correlation time in phase a and therefore inversely proportional to the self-diffusion coefficient of water in that phase. The diffusion coefficient computed in that way is very close to that of pure liquid water, irrespective of C. This is rather surprising considering that clay gels with very high shear strength are eventually obtained. The slope of the linear variation T_1^{-1} against Cf is proportional to the volume of phase b. Knowing the surface available to water, the average number of water layers in phase b is determined. It was practically equivalent to that obtained from heat of immersion measure-

[67]

ments. This also was not necessarily expected since these two techniques operate in domains of energy that are about 7 orders of magnitude apart.

The linear behaviour of T_1^{-1} against Cf results from the fast molecular exchange between populations a and b

$$T_1^{-1} = x_a\,T_{1a}^{-1} + x_b\,T_{1b}^{-1} \approx T_{1a}^{-1} + x_b\,T_{1b}^{-1}, \tag{1}$$

with $x_a + x_b = 1$ and $x_b \ll x_a$; x_a and x_b are the molar fractions of water in the two phases and T_{1a}^{-1} and T_{1b}^{-1} are specific relaxation rates in pure water and in the first layers of water in contact with the surface respectively. The latter can be estimated from relaxation rate measurements on clay surfaces that have adsorbed one or a few layers of water.

In operating in this way it was assumed that the surface effects are ruled by a step function. Phase b would be defined in a thermodynamic sense similar to that used according to the concept of the Gibbs dividing plane.

The aim of this contribution is to check to what extent such a concept is valid. From the experimental point of view it should be first established that the direct measurement of the self-diffusion coefficient by the n.m.r. pulsed field gradient gives for phase a values in good agreement with those of pure liquid water. Then using a non-microporous solid with a well known surface area fully available to water, the slope of the T_1^{-1} against Cf should permit one to obtain the volume of phase b and therefore its surface area by assuming an average thickness of 10^{-9} m for the surface phase.

If reasonable agreement were obtained for these two sets of experiments, the principle on which (1) is founded would be considered as valid and various consequences for water–clay systems could be developed.

2. EXPERIMENTAL

(a) Samples

Kaolinite from Georgia (U.S.A.), a synthetic magnesium phyllosilicate (trade name laponite (Laporte Industrie G.B.)), and hectorite from Hector (California, U.S.A.) were described earlier (Fripiat et al. 1982). Their properties pertinent to the present subject can be found also in the same reference.

Hectorite and laponite are swelling clays, kaolinite is not. Kaolinite could have been an interesting sample for measuring the surface area available to water. However, xonotlite, a synthetic non microporous calcium silicate, was preferred for that purpose because of two essential chacteristics, namely, a low iron content and a high surface area.

Its classical composition was the following (percentage mass); SiO_2: 49.0; Al_2O_3: 0.25; CaO: 42.9; MgO: 0.35; NaO_2: 0.05; K_2O: 0.1; CO_2: 1.55. Constitutional and hydration water content of the air-dried material was ca. 7 %. The iron content in the particularly pure sample used here was about $200\,\mu g/g$. It was most probably homogeneously spread into the bulk of the solid and not mainly concentrated on the surface, as is most often the case for kaolinite. The b.e.t. surface area was $62 \pm 2\,\mathrm{m^2}$: it was equal to that obtained by the Harkins & Jura technique (Harkins & Jura 1944), using the heat of immersion in water.

From the extent of the surface area, the iron content and a specific mass of $2.6\,\mathrm{g\,cm^{-3}}$, the number of paramagnetic centres per unit surface is $k = 0.88\ 10^{12}$ Fe atoms per square centimetre. This parameter is important for the interpretation of the ^{1}H n.m.r. relaxation rate.

The xonotlite material is made from fibres aggregated in hollow spheres with average diameter of about $80\,\mu m$, as shown in figure 1, plate 1.

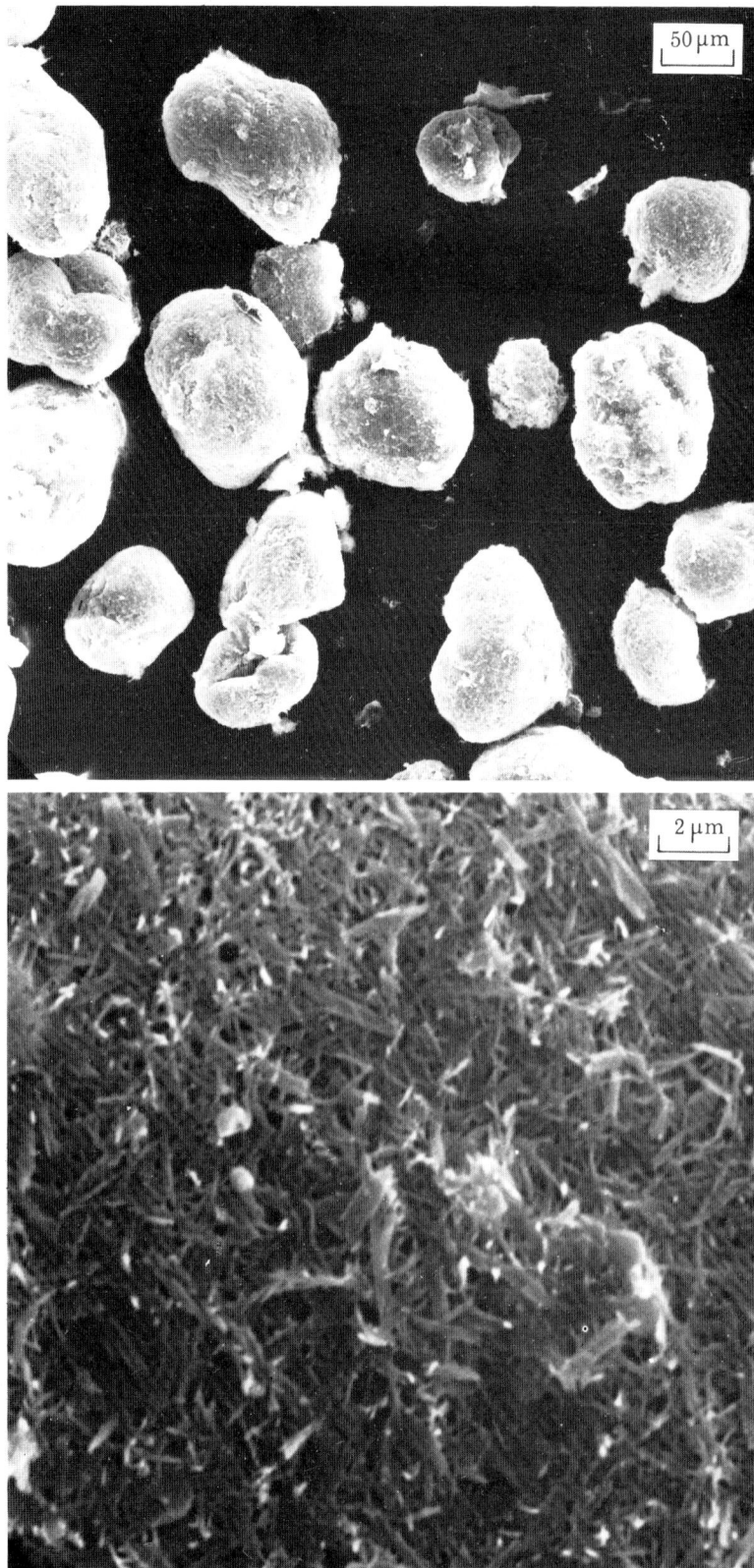

FIGURE 1. Scanning electron micrographs of xonotlite particles: two magnifications
(thanks due to the courtesy of Dr O. Anton, Redco, Belgium).

For self-diffusion coefficient measurements, a defibrated sepiolite (trade name Pangel, Tolsa, Spain), forming good thixotropic gel for $C > 0.05$ was also used. Sepiolite is a fibrous non-swelling mineral which contains zeolitic channels parallel to the fibre axis. The Spanish sample is very pure from the mineralogical point of view. After outgassing at $100°$ during 20h the N_2 b.e.t. surface area is $330 \pm 10\,m^2\,g^{-1}$. Because of the microporous nature of this material the significance of this value is surely questionable as will be discussed later on.

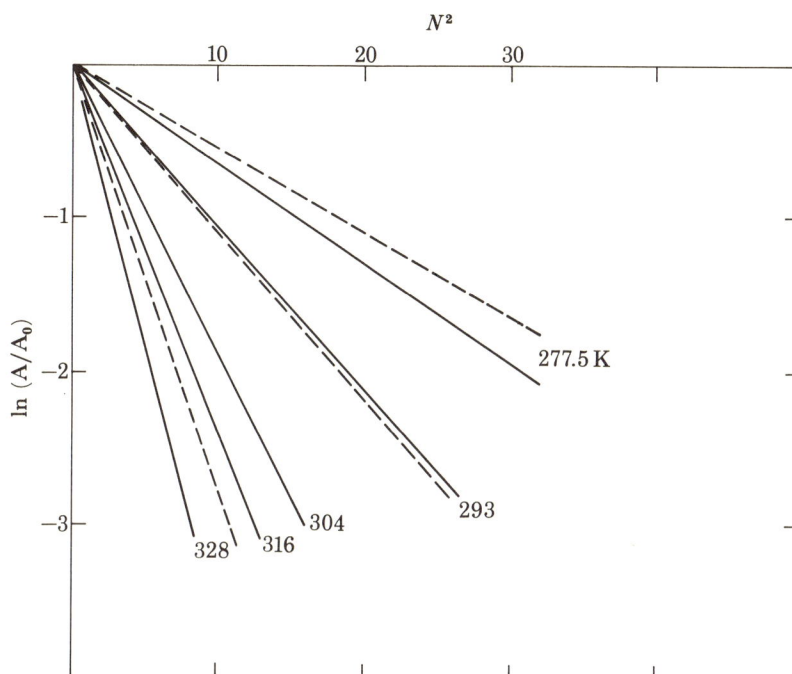

FIGURE 2. Example of application of the pulsed field gradient technique to the measurement of the self-diffusion coefficient of water in sepiolite suspension and in pure liquid at different temperatures (see equation (2)). N is proportional to the field gradient G_t.

(b) Measurements

^{1}H or ^{2}H relaxation rates T_1^{-1} were measured by the classical $\pi, \tau, \frac{1}{2}\pi$ pulses sequence at 90 MHz or 13.85 MHz, respectively. The pulsed field gradient (p.f.g) measurements were done with the $\frac{1}{2}\pi, t_1$, p.f.g., π, p.f.g. sequence (Stejskal et al. 1964). The echo amplitude in presence (A) or absence (A_0) of field gradient G_t is given by the relation

$$\ln (A/A_0) = -\alpha D G_t^2,\tag{2}$$

where α contains time parameters adjusting the position of the p.f.g. with respect to the $\frac{1}{2}\pi$ and π pulses. An example of the practical determination of D is shown in figure 2.

The self-diffusion coefficient of water at 293 K, i.e. $2.2 \times 10^{-5}\,cm^2\,s^{-1}$ was taken as reference to obtain α. Then with α constant, self-diffusion cofficients in the pure liquid at other temperature or in the suspensions were obtained.

3. RESULTS

(a) Self-diffusion coefficient measurements

Diffusion coefficients obtained for water are compared in figure 3 with those obtained by other experimentors, whereas table 1 gives the ratio D_s/D_0 of the diffusion coefficient in the suspensions (or gels) to that of water at the same temperature, and the average values M of these ratios. Knowing that the p.f.g. technique has an accuracy of about $\pm 10\%$ the clear conclusion is that D_s is effectively equal to D_0 within the experimental errors. The main source of experimental uncertainty here is the temperature stability of the n.m.r. probe which is surely not better than ± 1 K. Also the non-Arrhenius behaviour of D compared with $1/T$ looks very similar for water and for the suspensions or gels.

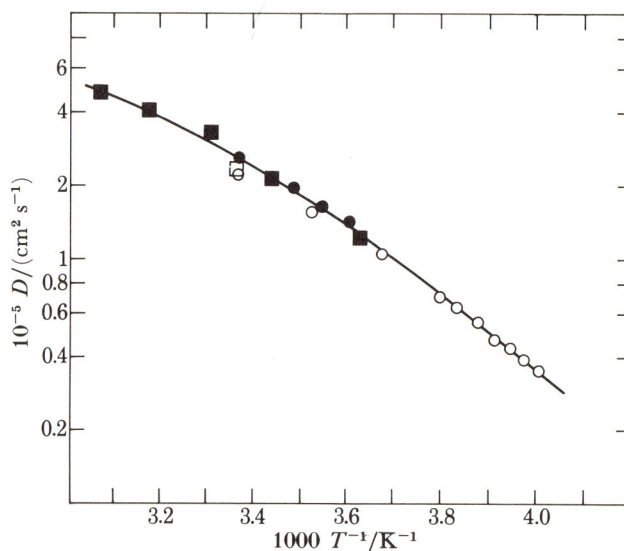

FIGURE 3. Self-diffusion coefficient of water with respect to $1000/T$. Experimental data of: ○ Gillen (1972); ● Wang (1965); □ Goldammer (1970); ■ this work.

The agreement between D_s and D_0 is striking for gels with high viscosity such as those obtained for laponite or sepiolite. When the diffusion coefficient was obtained indirectly from T_1^{-1} by using the rotational diffusional correlation time τ_{ra} in phase a and Einstein's equation $D = \langle l^2 \rangle / 6\tau_{ra}$, $\langle l^2 \rangle$ being the mean quadratic jump distance approximated as the square of the molecular diameter, the agreement was not surprising because at the molecular scale, the thermal fluctuations that allow diffusion could be considered easily as not influenced by the intricate network of solid particles. With the p.f.g. technique the agreement is more surprising since during the duration of the field gradient pulses, the water molecules diffuse through distances much larger than the molecular diameter. Thus from the molecular dynamics point of view, even in thick gels, the behaviour in phase a is identical with that in the pure liquid.

(b) Surface area available to water

Figures 4 and 5 show the variation of $^2H\,T_1^{-1}$ and $^1H\,T_1^{-1}$ measured for the system xonotlite–water at increasing solid content (Cf). The linear relation predicted by (1) is well obeyed. This agreement confirms the previous findings recalled in §1.

[70]

TABLE 1. RATIO OF THE DIFFUSION COEFFICIENT OF WATER IN THE SUSPENSIONS (D_s) TO THE DIFFUSION COEFFICIENT OF THE PURE LIQUID (D_0) BETWEEN 328 AND 277.5 (± 1) K

suspension	temperature	D_s/D_0
kaolinite, $C = 0.65$	328	0.74
	316	0.90
	304	0.96
	293	1.04
	283	1.10
	277.5	1.12
		$M = 0.98$
hectorite, $C = 0.12$	328	1.09
	316	1.04
	304	1.15
	293	1.13
	283	1.33
	277.5	0.92
		$M = 1.11$
$C = 0.07$	316	1.13
	304	1.25
	293	1.36
	283	0.98
	277.5	0.87
		$M = 1.12$
laponite, $C = 0.1$	316	0.85
	293	1.05
	283	1.13
	277.5	1.31
		$M = 1.08$
$C = 0.05$	316	0.94
	304	1.04
	293	1.0
	283	1.08
		$M = 1.02$
sepiolite, $C = 0.1$	328	1.27
	316	1.20
	304	1.26
	293	1.0
	277.5	1.04
		$M = 1.15$
$C = 0.06$	328	0.66
	316	0.64
	303	0.82
	293	1.0
	283	0.98
	277.5	0.86
		$M = 0.828$

The straightforward treatment for obtaining the surface area available to water is to use the the ^{2}H relaxation rate because of the intramolecular mode of relaxation. Indeed for rapid exchange between the two phases, the observed relaxation rate is

$$(^2\text{H})\, T_1^{-1} = F_a x_a \tau_{ra} + F_b x_b \tau_{rb} \approx F_a \tau_{ra} + F_b x_b \tau_{rb}, \tag{3}$$

where $F = \frac{3}{8}(2\pi \text{ q.c.c.})^2$, q.c.c. being the quadrupole coupling constant. The subscripts a or b denote phases a or b respectively; τ_r is the correlation time; the q.c.c. of pure liquid D_2O is

2.26×10^5 Hz at 298 K. In the one layer hydrate of halloysite (Cruz *et al.* 1978), q.c.c. = 1.65 $\times 10^5$ Hz. By using this value, F_b can be computed.

The correlation time in phase a is $\tau_{ra} = 3 \pm 0.2 \times 10^{-12}$ s^{-1} at 298 K (Eisenberg & Kauzman 1969) whereas the rotational correlation time obtained for the one- or two-layer hydrates of Na and Ca montmorillonites, the two-layer hydrate of Na vermiculite and the one-layer hydrate of Li hectorite is about 10^{-10} s^{-1} (Fripiat 1980) at this temperature; τ_{rb} can be approximated by that value.

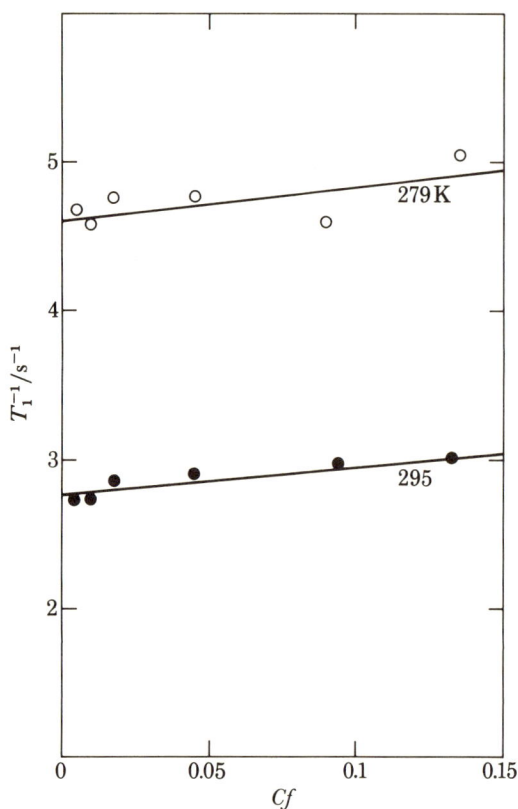

FIGURE 4. Deuteron relaxation rate against Cf (the concentration of solid in water, grams per gram) for xonotlite suspensions at two temperatures.

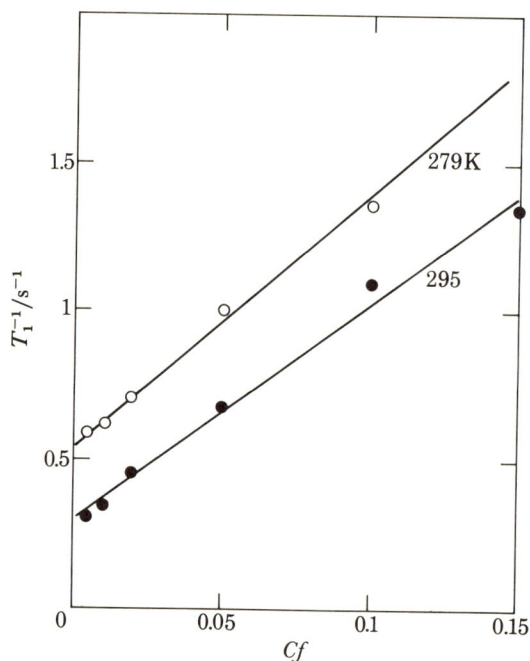

FIGURE 5. Proton relaxation rate against Cf (the concentration of solid in water, grams per gram) for xonotlite suspensions at two temperatures.

The average experimental slope (figure 4) obtained from the plot ^{2}HT_1^{-1} against Cf is 2.2 ± 0.2 s^{-1}. The relation between x_b and Cf is

$$\rho_w S_b z^* Cf = V_b/V_a \approx x_b, \qquad (4)$$

obtained by neglecting V_b with respect to the total volume of water and assuming that the molar volume is the same in the two phases. In (4) S_b is the specific surface area available to water, z^* the average thickness of phase b and ρ_w the specific mass of water. It is assumed that $\rho_{wa} \equiv \rho_{wb}$. Since $F_b \tau_{rb} = 40, S_b z^* = 0.055$ cm^3 g^{-1} of xonotlite, and $S_b = 55$ m^{2}g$^{-1} \pm 10\%$, with $z^* = 10^{-9}$ m. This too perfect agreement with the b.e.t. surface area can be checked by using the proton relaxation data. The situation here is more complex because relaxation in the surface phase is mainly controlled by the paramagnetic impurities in the surface layers. It was shown in that case that the slope K' of the experimental relation

$$^1H T_1^{-1} = K + K'Cf \tag{5}$$

was

$$K' = 1.7 \times 10^{-10} S_b z^* k. \tag{6}$$

The average value of the experimental slopes in figure 5 at 295 and 279 K is $8.9\,\mathrm{s}^{-1} \pm 10\,\%$. Using $k = 0.88\ 10^{12}$ paramagnetic (Fe^{3+}) centres per square centimetre yields $S_b z^* = 0.058\,\mathrm{cm}^3\,\mathrm{g}^{-1}$ with $S_b = 59\,\mathrm{m}^2\,\mathrm{g}^{-1} \pm 10\,\%$. The main source of uncertainty here is the actual value of k, since it has been assumed that the paramagnetic centres are homogeneously distributed.

The very good agreement between the specific surface areas calculated from by the ^2H and ^1H relaxation rates and those observed by the N_2 b.e.t. method and the Harkins & Jura technique may be partly fortuitous. It shows at least that the thickness z^* assigned to the surface phase b of the right order of magnitude.

4. DISCUSSION

The two main conclusions from §3(a) and §3(b) reinforce those obtained previously and the rest of this paper will be devoted to the analysis of the significance of the distribution in clay suspensions and gels of water molecules between two phases and of the various implications of this situation.

(a) Thermodynamic aspect

We know that at 298 K the life-time in phase b is of the order of magnitude of, or longer than, $10^{-9}\,\mathrm{s}$ (Fripiat et al. 1982), and that the rotational correlation time in phase b is about $10^{-10}\,\mathrm{s}$ whereas it is $3 \times 10^{-12}\,\mathrm{s}$ in phase a. What we do not know is whether or not the so-called 'phase b' has a thermodynamic meaning!

Suppose that the life time in phase b is long (with respect to τ_{rb}), so that phase b can be considered as distinct from phase a. The free energy of the suspension is

$$F = F_a + F_b. \tag{7}$$

Then, from the cell theory of liquids (Hill 1962)

and
$$\left.\begin{array}{l} F_a = -N(1-x_b)\,kT\ln(\bar{V}eZ_b), \\ F_b = -Nx_b kT\ln(\bar{V}eZ_b), \end{array}\right\} \tag{8}$$

where $N = N_a + N_b$, $\bar{V}$ is the molecular volume of water or the volume of the cell (equal in phases a and b), Z_a and Z_b are the partition functions of the water molecule in phases a and b respectively and $\ln e = 1$. The chemical potential of water in the system is $\mu(x_b, T) = (\partial F/\partial N)_{T, x_b}$:

$$\mu(x_b, T) = -(1-x_b)\,kT\ln(\bar{V}eZ_a) - x_b kT\ln(\bar{V}eZ_b). \tag{9}$$

Moreover
$$\mu(x_b, T) = \mu_0(T) + kT\ln P,$$

whereas for $x_b = 0$
$$\mu(0, T) = \mu_0(T) + kT\ln P_0.$$

Thus
$$RT\ln(P/P_0) = x_b RT\ln(Z_a/Z_b) = x_b[(\bar{E}_b - \bar{E}_a) - T(\bar{S}_b - \bar{S}_a)] = x_b \overline{\Delta F}, \tag{10}$$

where $\bar{E}$ and $\bar{S}$ denote the molar energy and entropy, respectively. It follows that

$$RT\ln(P/P_0) = \overline{\Delta F} x_b, \tag{11}$$

where $\overline{\Delta F}$ is the transfer molar free energy of water from phase a to phase b.

The presence of a solute or of colloid particles decreases the free energy and thus the vapour pressure $(P < P_0)$. Equations (11) and (4) predict that $RT\ln(P/P_0)$ must be a linear function of x_b and thus of Cf for a specified clay suspension (constant $S_b z^*$).

Now suppose that the lifetime in phase b is too short (with respect to τ_{rb}) such that phase b can no longer be considered as distinct from phase a. The clay suspension is then, from the solvent point of view, a heterogeneous medium to which the mixture model approach of Ben-Naim (1974) can be applied.

Then

$$F = kTN\ln[e(Z_a V_a + Z_b V_b)/N], \tag{12}$$

where

$$V_a = (1 - x_b)\bar{V}N, \tag{13}$$

$$V_b = x_b \bar{V}N, \tag{14}$$

and

$$\mu(x_b, T) = -kT\ln[(Z_a\bar{V}(1 - x_b) + Z_b\bar{V}x_b)e]. \tag{15}$$

Using the same procedure as above yields

$$RT\ln(P/P_0) = -RT\ln[1 - x_b(1 - Z_b/Z_a)]. \tag{16}$$

Since $P < P_0$, $\overline{\Delta F}$ must be negative, and for small x_b (15) becomes

$$RT\ln(P/P_0) = RTS_b z^* \rho_w[1 - \exp(-\overline{\Delta F}/RT)]Cf. \tag{17}$$

For small values of $\overline{\Delta F}$ (17) is indistinguishable from (11). In both cases $RT\ln(P/P_0)$ should be proportional to Cf.

A gel made from sepiolite was used to check these models. Two types of measurements were carried out. The gels were made by stirring various amounts of sepiolite in D_2O or H_2O at about 10 000 r.p.m. for a few minutes.

Figure 6 shows the variation of the ^2H T_1^{-1} relaxation rates with respect to Cf at 297 and 279 K. The slopes are 15.7 and 18.7 s^{-1} respectively. Applying equation (3), one obtains $S_b z^* \approx 0.5$ cm³ g⁻¹ which means that the $S_b \approx 500$ m² g⁻¹. This is the highest value so far obtained for a surface phase in rapid exchange with bulk water. Because of the formation of tactoids, the corresponding specific surface areas are smaller for swelling clays (under 100 m² g⁻¹).

On the other hand $(RT/N_A\bar{V})\ln(P/P_0)$ (N_A is the Avogadro number) was measured for $C = 0.03$ and $C = 0.06$ sepiolite gels with a thermocouple psychrometer, as described by Brown (1970).

The plot against Cf was observed to be linear and passed through the origin. At 295 K, the slope was 7 bar.† From (11) or (17), it may be easily calculated that $\overline{\Delta F} \approx 5$ cal mol⁻¹‡ or ca. 10^{-2} kT per molecule.

This is a rather astonishing result since, from heats of immersion measurements, we know that $\bar{E}_b - \bar{E}_a$ is far from negligible. Therefore in order for $\overline{\Delta F}$ to be that small, the entropic term must cancel most of the energetic contribution. Thus on the basis of the present data it is impossible to draw any conclusion about which of the two thermodynamic models is applicable to the clay–water system.

† 1 bar = 10⁵ Pa.   ‡ 1 cal = 4.184 J.

(b) Effects of salt concentration, pressure and temperature on the surface phase volume

It could have been expected that compressing the electrical double layer by increasing the ionic strength of the solution may affect the volume of phase b.

Sepiolite was chosen for these experiments because of its high surface area and because an electrolyte concentration as high as 1 M KCl does not affect the viscosity of the gel to a noticeable extent. Therefore the fibrous network should not be considerably modified. The pH of the suspension in presence or absence of KCl was about 8.2. Note that sepiolite has a rather small cation exchange capacity (c.e.c. ≈ 0.2 meq g^{-1}).

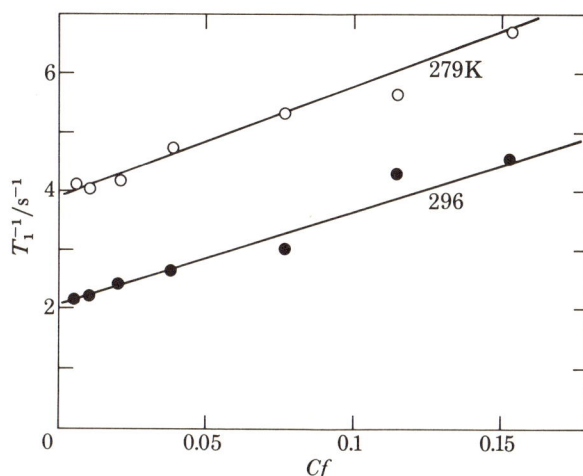

FIGURE 6. Deuteron relaxation rate against Cf (the concentration of solid in water, gram per gram) for sepiolite suspensions at two temperatures.

TABLE 2. COMPARISON OF THE SLOPE OF THE EMPIRICAL EQUATION (3) OBSERVED FOR 1 M KCl SEPIOLITE D_2O SUSPENSIONS WITH SEPIOLITE SUSPENSIONS IN PURE D_2O

temperature/K	$(KCl + D_2O)/s^{-1}$	$(D_2O)/s^{-1}$
299	16	15.7
279	19	18.7

In a 1 M KCl aqueous solution, according to Endom *et al.* (1967), T_1^{-1} is not appreciably affected. For instance at 299 K, ^{1}H T_1^{-1}(KCl 1 M) = 0.273 s^{-1} whereas it is 0.28 s^{-1} in water. In table 2, the slopes of the empirical equation (3) observed for sepiolite suspensions with solid contents up to about 0.15 are shown to be the same.

It may be concluded that the volume of phase b is not influenced by the ionic strength in spite of the fact that the osmotic pressure in a 1M KCl (H_2O) solution is 44.5 bar. That the effect of such a pressure is too small to be observed agrees with the conclusions of an experimental study by Jonas *et al.* (1982) on the effect of a high mechanical pressure on the deuterium relaxation rate in the kaolinite D_2O system. Indeed it was shown that T_{1b}^{-1} increases only by about 20 % when the pressure increases from 1 bar to 5 kbar.

The temperature effect was also determined for the same kaolinite–water system by measuring ^{2}H(T_1^{-1}) or ^{1}H(T_1^{-1}) with respect to temperature. The activation energy obtained in that way is that of τ_{ra} or of τ_{rb} as shown in table 3. The activation energy in phase b is noticeably different from that in phase a.

As the main intermolecular interactions are due to hydrogen bonding, it may be concluded that the energy required to activate these bonds is lower in phase b than in phase a. This would suggest that the kaolinite surface destroys the well developed random hydrogen bond network characteristic of liquid water. Jonas *et al.* (1976), Wilbur *et al.* (1976) and De Fries & Jones (1977) have shown that the compression of pure liquid water distorts the optimal tetrahedral order. An increase in temperature is an efficient way to produce these perturbations. The surface force field is thus at the same time more efficient than an increase in pressure and temperature in that respect.

TABLE 3. ACTIVATION ENERGY OF THE ROTATIONAL CORRELATION TIMES IN THE TEMPERATURE RANGE 10–60 °C FOR PHASE a AND PHASE b IN A KAOLINITE SUSPENSION ($C = 2.65$)

solvent	pressure/bar	phase a/(kcal mol^{-1})	phase b/(kcal mol^{-1})
D_2O	1	5.2	1.8
	5×10^3	4.6	1.8
H_2O	1	4.1	1.1
	5×10^3	4.0	0.7

This conclusion can be generalized to hydrophylic surfaces in general since the lengthening of the rotational correlation time in phase b by about two orders of magnitude with respect to that in phase a is observed for all the minerals studied so far.

On the other hand the insensitivity with respect to the ionic strength of phase b volume in a sepiolite gel cannot be generalized to swelling silicates. In fact, as shown by Fripiat *et al.* (1982), the volume of phase b for a suspension of Ca-hectorite is about $\frac{2}{3}$ of that observed for a suspension of Na-hectorite. Since z^* seems to be constant, this observation could be interpreted as an increase in the number of clay layers forming the tactoids with the simultaneous decrease of the specific surface area available to water (S_b).

(c) Phase transition upon freezing

Anderson & Tice (1971) have shown that below the freezing point (f.p.) a fraction of the water molecules in a clay gel remains unfrozen. This fraction contains the water molecules influenced by the surface force fields and also those that are in very narrow pores between clay tactoids. As expected, the larger the freezing point depression, the narrower are the pores in which freezing occurs. The interlayer water within the tactoids freezes at the lowest temperature. The water in phase b on the external surface of the tactoid should behave differently from that in pores with diameters over 2×10^{-9} m (i.e. over $2z^*$).

Suppose an ideal model where phase a is frozen whereas phase b is not. It may be expected that the rapid exchange averaging the n.m.r. relaxation rate T_1^{-1} (equation (1)) should be suppressed and that the decay of the longitudinal component of the magnetization could be the sum of two exponentials:

$$(M(\infty) - M(t))/2M_\infty = P'_a \exp(-t/T'_{1a}) + P'_b \exp(-t/T'_{1b}), \qquad (18)$$

where the apparent proton populations and relaxation rates are P'_a and P'_b, $1/T'_{1a}$ and $1/T'_{1b}$, respectively.

This was actually observed by Fripiat & Letellier (1984) for a Na-hectorite H_2O suspensions with $0.05 < C < 0.18$. T'_{1b} was near that observed for the monohydrate of Li-hectorite (Fripiat *et al.* 1980), whereas T'_{1a} was about 4 times larger than T'_{1b}. For the proton, the suggested relaxation mechanism in the frozen phase was spin diffusion toward 'freely' rotating water molecules either in phase b and, or within the tactoids. For D_2O suspensions with $0.05 \leqslant C \leqslant 0.18$, a single T_1^{-1} was observed and the deuteron resonance line intensity below f.p. was not more than 2 % that above f.p. The deuterium signal was hardly observable below f.p. for more diluted suspensions.

Below f.p. the ^2H resonance line was a doublet which splitting increased with temperature, in contrast with what was observed for pure D_2O polycrystalline ice (Jackson & Rabideau 1964 *a, b*). The T_1 behaviour with respect to temperature is represented in figure 7.

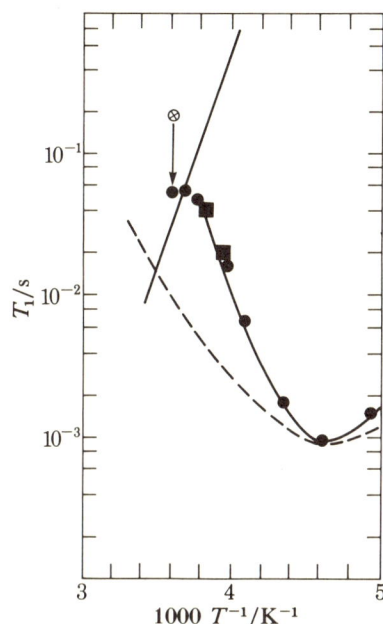

FIGURE 7. Deuteron relaxation rate against 1000/T. The solid straight line is computed from the ^1H spin lattice relaxation time and q.c.c. = 195 KHz, (Jackson *et al.* 1964 *b*). The dashed line represents the values obtained for the one-layer D_2O hydrate of halloysite by Cruz *et al.* (1978). Experimental values observed for a Na-hectorite gel with *Cf* = 0.05, ● and *Cf* = 0.18, ■, ⊗ indicates the drop of T_1 at freezing point.

These results were interpreted by assuming that below f.p. the resonance of unfrozen D_2O molecules (in phase b or within the clay tactoids) was the only one observed with the n.m.r. instrument used for this study. The frozen D_2O molecules in phase a would display a signal too large to be recorded. The relaxation time fitting the experimental T_1 shown in figure 7 is

$$\tau_r(s) = 1.9 \times 10^{-20} \exp(11\,700\,\text{cal}/RT), \tag{19}$$

the activation energy being about 2 kcal mol^{-1} lower than that reported for dielectric relaxation and for the proton T_1 in H_2O polycrystalline ice, (Barnall & Lowe 1968). The q.c.c. at the $^2H T_1^{-1}$ minimum is about 160 kHz, in agreement with the value given in § 3 (*b*). This value is about $\frac{3}{4}$ that of polycrystalline D_2O (Jackson & Rabideau 1964 *b*).

Using the activation energy for D_2O, shown in table 3, the correlation time τ_{rb} in phase *b* is 0.8×10^{-10} s at f.p. (277 K). At the same temperature, τ_r calculated from (19) is 0.4×10^{-10} s. This discrepancy is within the limits of the experimental errors but what is amazing is the large increase of the activation energy (equation (19)) with respect to that observed for phase b above f.p. It may be suggested that hydrogen bonding at the interface between unfrozen and frozen molecules is stronger than at above f.p.

[77]

(e) Conclusions

From the preceding discussions it may be safely concluded that from the microdynamic point of view the surface and bulk phases have a physical significance in clay suspensions and gels. However, it cannot be concluded that these phases have a real thermodynamic meaning since the transfer molar free energy is undoubtedly very small (see § 4 (a)). Since the lifetime in the surface phase is between ten to one hundred times longer than the rotational correlation time in phase b which itself is about one hundred times larger than the rotational correlation time in phase a, the use of a step function differentiating phase b from phase a is justified.

For minerals without internal surface (kaolinite and xonotlite), the surface phase has an area that is well approximated by the N_2 b.e.t. or by the Harkins & Jura techniques. For minerals with internal surfaces, the situation is more ambiguous.

In suspensions, swelling clays such as hectorite and laponite display surface areas available to water much less in extent than the sum of internal and external surface areas. This is because the internal surface is not available for rapid exchange. This observation supports the concept of tactoids, the interesting point being that these tactoids can be considered as textural units already present in dilute suspensions, $(C < 0.01)$.

In suspensions, the non-swelling sepiolite develops a very high surface area available to rapid exchange in spite of its zeolitic fabric. According to Cases & François (1983), personal communication) the surface area available to CO_2 and to N_2 (after outgassing at 100 °C for 20 h) is of the order of $350 \, m^2 g^{-1}$ whereas that computed from the product $S_b z^*$ (assuming that $z^* = 10 \times 10^{-10} \, m$) is between 450 and $500 \, m^2 g^{-1}$. The observation of a surface area available to water higher than that measured with N_2 of CO_2 may find its origin in the fact that the zeolitic channels ($\phi \approx 4 \times 10^{-10} \, m$) are open on larger pores ($10 \times 10^{-10} \, m < \phi < 30 \times 10^{-10} \, m$). Therefore a large fraction of the zeolitic water could participate in the exchange.

As far as the mechanism of gel formation is concerned, this contribution has again emphasized the fact that the influence of the surface force field on the molecular dynamics does not extend to more than about $10 \times 10^{-10} \, m$ and that the diffusion coefficient of water in phase a is, within the experimental error, the same as in pure liquid. Thus the formation of a clay gel cannot be attributed to surface forces ordering water molecules on long distances.

This research was supported by a D.G.R.S.T. contract no. 81.D.1109. The authors are very grateful to Dr J. M. Cases and Miss M. François for specific surface area measurements as well as for numerous discussions. Tolsa (Spain) and Redco (Belgium) gave well characterized samples of pure sepiolite and xonotlite respectively.

REFERENCES

Anderson, D. M. & Tice, A. R. 1971 *Proc. Am. Soil Sci. Soc.* **35**, 47–54.
Barnall, D. E. & Lowe, I. J. 1968 *J. chem. Phys.* **48**, 4614–4618.
Ben-Naim, A. 1974 *Water and aqueous Solutions*, p. 201. New York, London: Plenum.
Brown, R. W. 1970 *Measurement of water potential with thermocouple psychrometers: construction and application.* U.S.D.A. Forest Serv. Res. pap. INI-80 (27 p.)
Cruz, M. I., Letellier, M. & Fripiat, J. J. 1978 *J. chem. Phys.* **69**, 2018–2027.
De Fries, T. & Jonas, J. 1977 *J. chem. Phys.* **66**, 896–901.
Eisenberg, D. & Kauzman, W. 1969 *The structure and properties of water*, p. 215. Oxford University Press.
Endom, L., Hertz, H. G., Thul, B. & Zeidler, M. D. 1967 *Ber. BunsenGes. phys Chem.* **71**, 1008–1031.

Fripiat, J. J. 1980 *Bull. Soc. Française Mineral.* **103**, 440–443.

Fripiat, J. J., Kadi-Hanifi, M., Conard, J. & Stone, W. E. E. 1980 In *Magnetic resonance in colloid and interface science* (ed. J. Fraissart & H. A. Resing), pp. 529–535. Dordrecht: Reidel.

Fripiat, J. J., Cases, J., François, M. & Letellier, M. 1982 *J. Colloid Interface Sci.* **89**, 378–400.

Fripiat, J. J. & Letellier, M. 1984 *J. magn. Reson.* (In the press.)

Gillen, K. T., Douglass, D. C. & Hoch, M. J. R. 1972 *J. chem. Phys.* **57**, 5117–5119.

Goldammer, E. V. & Hertz, H. G. 1970 *J. phys. Chem.* **74**, 3734–3755.

Harkins, W. D. & Jura, G. 1944 *J. Am. chem. Soc.* **5**, 1366–1373.

Hill, T. L. 1962 *An introduction to statistical thermodynamics*, 2nd edn, p. 286. Reading, Massachusetts: Addison Wesley.

Jackson, J. A. & Rabideau, S. W. 1964a *Bull. Am. phys. Soc.* **9**, 142–147.

Jackson, J. A. & Rabideau, S. W. 1964b *J. chem. Phys.* **41**, 4008.

Jonas, J., De Fries, T. & Wilbur, D. J. 1976 *J. chem. Phys.* **65**, 582–588.

Jonas, J., Brown, D. & Fripiat, J. J. 1982 *J. Colloid. Interface Sci.* **89**, 374–377.

Stejskal, E. O. & Tanner, J. E. 1964 *J. chem. Phys.* **42**, 288–292.

Wang, J. H. 1965 *J. phys. Chem.* **65**, 4412.

Wilbur, D. J., De Fries, T. & Jonas, J. 1976 *J. chem. Phys.* **65**, 1783–1786.

Discussion

J. M. THOMAS, F.R.S. (*University of Cambridge, Department of Physical Chemistry, Cambridge, U.K.*). Could Professor Fripiat tell us how the rotational correlation times that he has measured for the rather strongly bound (so-called 'phase b') H_2O and D_2O in sepiolite compare with corresponding values for water in:

> (i) one-layer, two-layer and three-layer montmorillonites; and
>
> (ii) vermiculite?

Could he also tell us how these values compare with those obtained by other workers, using, for example, neutron scattering studies, in assessments of ease of diffusion in interlamellar water?

J. J. FRIPIAT. In phase b and for the one-layer or two-layer hydrates of Na, Ca-montmorillonites and Na-vermiculite the n.m.r. correlation times observed at 298 K were between 1 and 2×10^{-10} s whereas in liquid water it is 3×10^{-12} s.

The comparison with quasi-elastic neutron scattering (q.e.n.s.) studies is not direct since this technique gives inter-jump residence times. It may thus be expected that the activation energies observed with the two techniques will not be the same since their physical meanings are different.

However the correlation and the residence times observed in phase b, as measured by n.m.r. and q.e.n.s. are within the same order of magnitude.

Phil. Trans. R. Soc. Lond. A **311**, 301–314 (1984)   [301]

Printed in Great Britain

Thermodynamics of ion exchange in clays

By R. P. Townsend

Department of Chemistry, The City University, Northampton Square, London EC1V 0HB, U.K.

As an introduction, the general thermodynamic treatments for binary cation exchange are reviewed. The normal choice of standard states for the cations in the clay phase, and commonly used thermodynamic formulations with different scales of measurement are both discussed. The definition of standard states for the clay phase which is due to Gaines & Thomas (*J. chem. Phys.* **21**, 714 (1953)) is considered in most detail; the effect of water activity changes in the homo-ionic forms of the clay on the magnitude of the cation activity coefficients is considered. In the main part of the paper, three topics of current interest are considered. First, the current controversies regarding the appropriate choice of measurement scale for exchange reactions in clays are reviewed, together with the consequent implications that a particular choice of concentration scale have regarding the definition of ideal behaviour in the exchange phase. It is emphasized that thermodynamic formulations in terms of either the cationic mole fraction scale or the equivalent function scale are equally valid.

Second, some recent developments concerning the thermodynamics of ternary and multicomponent exchange are reviewed. The question as to whether multi-component equilibria may be predicted from binary data alone is briefly discussed, and recent rigorous thermodynamic formulations of Chu & Sposito (*J. Soil Sci. Soc. Am.* (1981)) and Fletcher & Townsend (*J. chem. Soc. Faraday Trans.* II **77**, 955; 965; 2077 (1981)) are compared. These formulations, also based on different measurement scales, are shown to be compatible and complementary. Finally, the question of site heterogeneity in clays is considered briefly, and some current attempts to interpret non-ideal behaviour in the exchange reaction in terms of this phenomenon are discussed.

Introduction

The ubiquity of clays in the environment, and the consequent importance of understanding how their properties influence the overall properties of soils, has resulted in extensive studies of their equilibrium cation exchange properties, the earliest of which precede this century. Thus despite the overwhelming importance today of organic-based resins for ion exchange applications, the basic thermodynamic formulations of ion exchange which are used so widely today are based on precepts and principles which were enunciated long ago by researchers on inorganic exchange materials, especially clays. The names of Vanselow (1932), Gapon (1933), Kielland (1935) and Gaines & Thomas (1953) still appear widely in current ion exchange literature, irrespective of the nature of the exchanger under study.

This serves to emphasize a most important basic principle concerning thermodynamic studies and one which is frequently overlooked. This principle has been reiterated recently by Sposito (1981 a, b): because thermodynamics is concerned with the measurement of changes in certain physical quantities which describe the macroscopic system under study, it is not possible thence to speculate from these thermodynamic data regarding the fundamental nature of the microscopic mechanisms which underlie the observed behaviour. Thus any properly conceived thermodynamic formulation of ion exchange should be of general application to ion exchangers, irrespective of their chemical nature.

[81]

In what sense then can one consider the thermodynamics of ion exchange in clays in particular? First, experience may show that a particular choice either of reference states or of concentration scale is specially convenient for studies on clay minerals, even though there is nothing wrong with alternative choices, and even when these alternatives may be preferred for other types of ion exchangers. Second, although it is wrong to use thermodynamic data to infer mechanisms that occur on the microscopic scale within the exchanging phases, nevertheless the characteristic behaviour of different clays is manifested in the thermodynamic data obtained for cation exchange within these materials. A recognition of this characteristic behaviour enables one to appreciate the limits to which one can then predict exchange properties from a necessarily limited quantity of basic thermodynamic data.

This paper is concerned with recent developments in these areas, and specially with the questions of choice of standard reference states, definitions of ideal behaviour, multicomponent cation exchange properties and site heterogeneity within the exchanger.

1. Basic theoretical considerations

(a) The thermodynamic equilibrium constant

By considering first an exchange involving only two different types of cations, the basic reaction is

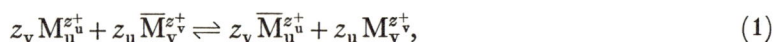

$$z_v M_u^{z_u+} + z_u \overline{M_v^{z_v+}} \rightleftharpoons z_v \overline{M_u^{z_u+}} + z_u M_v^{z_v+}, \tag{1}$$

where $M_u^{z_u+}$, $M_v^{z_v+}$ are cations of valency z_u, z_v respectively. The superscripted bar refers to the exchanger phase. The thermodynamic equilibrium constant then follows:

$$K_a = (a_v^{z_u}/a_u^{z_v})\,(\bar{a}_u^{z_v}/\bar{a}_v^{z_u}), \tag{2}$$

where $\bar{a}$ and a refer to the relative activities of the species in the clay and external solution respectively. The magnitude of K_a for a given pair of exchanging cations in a particular clay must then depend not only on the intrinsic properties of the system itself, but also on the choice of the reference states to which the relative activities $\bar{a}$, a refer. This (apparently) obvious point is overlooked frequently. Thus choosing, for example, similar criteria for ideal behaviour of the ions in *both* the exchanging phases leads to a standard free energy change of approximately zero for all exchanging pairs of cations (Helfferich 1962). Such a choice has not been favoured for studies on clays.

(b) Non-ideality in the solution phase

For the solution of electrolyte exterior to the clay exchanger, it has been normal practice in ion exchange studies to adopt the usual Henry law reference state to define ideal behaviour of the exchanging species, in which the relative activity of a species is regarded as becoming numerically equal to its concentration as its concentration tends to zero. The standard state is then the 'hypothetical ideal molal solution' of the appropriate species (Robinson & Stokes 1970), in which the ion is regarded as still behaving in the one molal solution (1 mol kg^{-1}) as it does when it is infinitely diluted with solvent. (If one prefers, one can of course use other concentration units such as mol dm^{-3}, provided the choice made is clearly specified.) To allow for non-ideality in the real solution, the activity coefficient γ is also introduced, so that for any solution concentration $a_i = \gamma_i m_i/m_i^\ominus$, where m_i, $m_i^\ominus$ are the concentrations of species i in the real solution and the hypothetical ideal molal solution respectively. The relative activity ratio for the solution phase (equation (2)) then becomes

$$(a_v^{z_u}/a_u^{z_v}) = (\gamma_v m_v/m_v^{\ominus})^{z_u}/(\gamma_u m_u/m_u^{\ominus})^{z_v}, \tag{3}$$

and for any solution concentration and for one co-anion $A^{z_{\bar{x}}}$ (valency z_x) the activity coefficient ratio is (Robinson & Stokes 1970):

$$(\gamma_v^{z_u}/\gamma_u^{z_v}) = [(\gamma_{\pm v}^{(u)})^{z_u(z_v+z_x)}/(\gamma_{\pm u}^{(v)})^{z_v(z_u+z_x)}]^{1/z_x}, \tag{4}$$

where $\gamma_{\pm vx}^{(ux)}$, $\gamma_{\pm ux}^{(vx)}$ are the mean activity coefficients of the salts $M_x A_v$ and $M_x A_u$ respectively in the *mixed electrolyte solution* at the experimental ionic strength (Dyer et al. 1981). For binary mixtures of two electrolytes with a common anion, the values of these functions may be found by using the model of Glueckauf (1949). Other model-dependent methods are available to determine these functions for multicomponent cation–anion systems (Fletcher & Townsend 1981 c) or for solutions of high ionic strength (Pitzer 1973; Scatchard 1961).

(c) Non-ideality in the clay exchanger

For clays or soils, with very few exceptions (see, for example, Babcock & Duckart 1980) the normal practice has been to follow Gaines & Thomas (1953) and make the standard reference state for each exchanging cation within the clay the appropriate homo-ionic form of the clay in equilibrium with an infinitely dilute solution of the same cation, so that the thermodynamic equilibrium constant is then a measure of the relative affinity between the clay and the two cations involved in the exchange (Helfferich 1962; Bolt 1967).

Despite the near-universal use of these reference states as standards, two different measurement scales are employed commonly by different researchers to describe concentrations of cations within the clay phase. The first of these was adopted by Vanselow (1932) and Kielland (1935) and involves the mole fraction $\overline{X}$ of ion in the exchange, so that for a binary exchange

$$\overline{X}_u = \bar{m}_u/(\bar{m}_u + \bar{m}_v), \tag{5}$$

where $\bar{m}_u$, $\bar{m}_v$ are the concentrations of the respective ions in the clay. A more precise term for this function might be 'cationic mole fraction', as it should be noted that no account is taken in (5) of adsorbed or imbibed solvent (Barrer & Townsend 1984). In contrast, Gaines & Thomas employed the equivalent cation fraction $\overline{E}$ to describe cation concentrations in the clay. This is defined as

$$\overline{E}_u = z_u \bar{m}_u/(z_u \bar{m}_u + z_v \bar{m}_v). \tag{6}$$

Again, a perhaps better name for $\overline{E}$ might be 'charge fraction', since the function is a measure of the proportion of the anionic charge on the clay which is neutralized by a given cation (and therefore of the mole fraction of clay associated with that cation).

In terms of cationic mole fractions, (2) becomes

$$K_a = (a_v^{z_u}/a_u^{z_v}) \, (\overline{X}_u^{z_v} f_u^{z_v}/\overline{X}_v^{z_u} f_v^{z_u}) = K_V(f_u^{z_v}/f_v^{z_u}), \tag{7}$$

where f_u, f_v are the activity coefficients of cations $M_u^{z_u^+}$, $M_v^{z_v^+}$ in association with their equivalents of anionic exchanger framework and K_V is a corrected selectivity quotient, sometimes referred to as the Vanselow coefficient (Sposito 1977). Argersinger et al. (1950) were first to use the

Gibbs–Duhem equation to obtain expressions which enabled the functions f_u, f_v and K_a to be evaluated from appropriate experimental data for K_V, namely

$$\ln f_u^{z_v} = -\bar{E}_v \ln K_V^* + \int_{\bar{E}_u}^1 \ln K_V \, d\bar{E}_u, \tag{8}$$

$$\ln f_v^{z_u} = \bar{E}_u \ln K_V^* - \int_0^{\bar{E}_u} \ln K_V \, d\bar{E}_u \tag{9}$$

and

$$\ln K_a = \int_0^1 \ln K_V \, d\bar{E}_u, \tag{10}$$

where the superscript * refers to the value of K_V at a particular composition $\bar{E}_u$, $\bar{E}_v$. (Note that although K_V is defined in terms of $\bar{X}_u$, $\bar{X}_v$, (8)–(10) require a knowledge of this coefficient as a function of *charge* fraction $\bar{E}_u$ also.)

By using equivalent, or charge, fractions to define the concentrations of the cations in the clay, (2) becomes

$$K_a = (a_v^{z_u}/a_u^{z_v}) \, (\bar{E}_u^{z_v} g_u^{z_v}/\bar{E}_v^{z_u} g_v^{z_u}) = K_G(g_u^{z_v}/g_v^{z_u}), \tag{11}$$

where g_u, g_v are the activity coefficients of cations M_u^{z+}, M_v^{z+} in association with their equivalents of exchanger framework and K_G is another corrected selectivity coefficient. If $z_u \neq z_v$, then $K_V \neq K_G$ and f_u, f_v will also differ in magnitude from their corresponding g_u, g_v functions (except in the case of the common standard states).

If (in order to allow comparison with (8)–(10)), the effects of sorbed solvent(s) in the exchanger are neglected, then Gaines & Thomas (1953) showed that

$$\ln g_u^{z_v} = (z_v - z_u) \, \bar{E}_v - \bar{E}_v \ln K_G^* + \int_{\bar{E}_u}^1 \ln K_G \, d\bar{E}_u, \tag{12}$$

$$\ln g_v^{z_u} = -(z_v - z_u) \, \bar{E}_u + \bar{E}_u \ln K_G^* - \int_0^{\bar{E}_u} \ln K_G \, d\bar{E}_u, \tag{13}$$

and

$$\ln K_a = (z_v - z_u) + \int_0^1 \ln K_G \, d\bar{E}_u. \tag{14}$$

Equations (8)–(10) are simpler than the alternative formulation of Gaines & Thomas, although the latter formulation has the compensating advantage of employing charge fractions only, rather than both $\bar{E}$ and $\bar{X}$.

(d) Activity of sorbed solvent

Gaines & Thomas's treatment is more comprehensive than that of Argersinger *et al.* (1950) in that it encompassed the effects of either sorbed or imbibed solvent, and of imbibed salts (Gaines & Thomas 1953). Both these factors may be significant in ion exchange studies at higher solution ionic strengths. It is here that it becomes especially important to specify that the homo-ionic forms of the clay be in equilibrium with an infinitely dilute solution of the same ion in the standard state. If the ionic strength of the electrolyte is not zero, then the activity coefficient g_u (or f_u) will *not* normally be unity even when the clay is in the homo-ionic M_u^{z+} form. Considering for example Gaines & Thomas's first case (Gaines & Thomas 1953), in which solvent only may be sorbed, then if $g_{u(I_u)}$ is the activity coefficient of ion $M_u^{z_u^+}$ in the homo-ionic M_u-exchanged clay at a solution ionic strength of I,

$$\ln (g_{u(I_u)}/g_{u(\infty)}) = z_u \int_{a_{w(\infty)}}^{a_{w(I_u)}} \left[\frac{\bar{V}}{V} - \bar{\nu}_w \right] d \ln a_w; \tag{15}$$

V is the molar volume of the sorbed solvent vapour, $\bar{V}$ and $\bar{\nu}_w$ are (per exchange *equivalent*) the volume and water content of the clay, and $g_{u(\infty)}$, $a_{w(\infty)}$ are respectively the values of g_u and the water activity when the homo-ionic clay is immersed in an infinitely dilute solution of $M_u^{z_u^+}$ ions (i.e. $g_{u(\infty)} = a_{w(\infty)} = 1$). When the bathing electrolyte solution is dilute, neglect of the correction shown in (15) is justified (Barrer & Klinowski 1974; Fletcher & Townsend 1981 *b*; Sposito 1981 *b*). For solutions of higher ionic strength, the deviation of $g_{u(I_u)}$ from unity can be significant. Thus Fletcher & Townsend (1981 *b*) found that at solution concentrations of 1 mol dm^{-3}, activity coefficients for the ions Na$^+$, K$^+$ and Li$^+$ in homo-ionic forms of zeolite A deviated by *ca.* 8 % from unity at 298 K. At the same solution strength, the deviation for Ca^{2+} was more than 30 %.

Even when the absolute magnitudes of the activity coefficients for the homo-ionic forms of the clay deviate significantly from unity, the overall effect on K_a may only be small, as the water activity terms tend to be self-cancelling. One may use the mean value theorem to effect the integrations, and assume that the volume per exchange equivalent of the clay is negligible compared with the molar volume of the vapour of the sorbed water. Then, providing the water vapour behaves as a perfect gas, the expression for K_a approximates to

$$\ln K_a = (z_v - z_u) + z_u z_v \{ \bar{\nu}_{w(b)} \ln (p/p_0)_{(b)} - \bar{\nu}_{w(a)} \ln (p/p_0)_{(a)}$$
$$- \bar{\nu}_{w(\overline{uv})} \ln [(p/p_0)_{(b)}/(p/p_0)_{(a)}] \} + \int_0^1 \ln K_G d\bar{E}_u, \tag{16}$$

where $\bar{\nu}_{w(\overline{uv})}$ is the mean water content of the mixed u, v clay exchanger, and (p/p_0) the relative pressure. Evaluations of the water terms have been carried out for binary exchanges in zeolites and resins by Barrer & Klinowski (1974) and similar ternary exchange studies have been undertaken by Fletcher & Townsend (1981 *b*). The *total* effect of the water terms on the magnitude of K_a was never found to be significant, even though the individual activity coefficient values do change markedly. Since it is activity coefficient data rather than the standard free energies which are of importance when one is mathematically modelling changes in the selectivity of a system as a function of electrolyte concentrations (Townsend *et al.* 1984; Fletcher *et al.* 1984), the greater sensitivity of activity coefficient values to water activity changes must be borne in mind if one is working with concentrated electrolyte solutions.

2. Use of different ionic concentration scales

(a) Introduction

Questions have been raised recently as to whether the equivalent fraction scale is indeed an appropriate choice for a rigorous thermodynamic formulation of ion exchange (Sposito & Mattigod 1979; Sposito 1981 *b*). Sposito & Mattigod re-examined data obtained by Maes *et al.* (1975) for exchange equilibria involving different transition metals on montmorillonite. Maes *et al.* (1975) found that the function $\ln K_G$ increased nearly linearly with the equivalent fraction of exchanging cation up to an exchange level of *ca.* 75 %. Using cationic mole fractions rather than equivalent fractions, Sposito & Mattigod (1979) showed that within experimental uncertainty, plots of $\ln K_V$ against $\bar{E}_u$ for the same exchanges were near constant for $\bar{E}_u < 0.7$. They concluded therefore that 'ideality in a cation exchanger must be studied in terms of the Vanselow selectivity coefficient', and not K_G. As a direct result of these comments, Goulding (1983) has reinterpreted some earlier thermodynamic studies on clays (see, for example, Goulding & Talibudeen 1980) using the Vanselow coefficient K_V rather than the originally used K_G function.

There are two separate important issues that must be considered here. The first is whether there is anything *intrinsically* wrong with using K_G rather than K_V to describe the selectivity (and non-ideality) of the exchange process. Sposito implies that there is something intrinsically wrong when he states that the Gaines & Thomas activity coefficients g_u, g_v 'are only formal parameters without a strict thermodynamic meaning within themselves' (Sposito 1981 b). The second issue is quite different: if both approaches are thermodynamically rigorous and compatible it may nevertheless be more *convenient* to use one measurement scale rather than the other for a particular type of exchanger (e.g. a clay).

(b) Comparison of Vanselow with Gaines & Thomas approaches

Considering first the question as to whether there is anything fundamentally wrong with the use of the equivalent fraction scale, and in particular whether the functions g_u, g_v are indeed activity coefficients rather than (*sic*) 'formal parameters' (Sposito 1981 b) the standard free energy change per equivalent of exchange for the reaction in (1) is

$$\Delta G^\ominus = [(1/z_u)\,\bar{\mu}_u^\ominus + (1/z_v)\,\mu_v^\ominus - (1/z_u)\,\mu_u^\ominus - (1/z_v)\,\bar{\mu}_v^\ominus]. \tag{17}$$

Since the reference states for $\mu_u^\ominus$, $\mu_v^\ominus$ in the solution and for $\bar{\mu}_u^\ominus$, $\bar{\mu}_v^\ominus$ in the clay are the same irrespective of whether the Vanselow or Gaines & Thomas formulations are used, then the K_a value derived from each formulation *must* be the same for a given exchange system, as must the functions $\Delta G^\ominus$, $\Delta H^\ominus$ and $\Delta S^\ominus$. However, Goulding (1983) derives an expression

$$\ln K_a' = \ln K_a'' + \int_0^1 f(\bar{E}_u)\,d\bar{E}_u - (z_v - z_u), \tag{18}$$

where K_a', K_a'' are the values found for the thermodynamic constant K_a by using respectively the Vanselow and Gaines & Thomas approaches. Goulding next writes

$$\begin{aligned}\ln K_a' &= \ln K_a'' + C_1, \\ \Delta G^{\ominus'} &= \Delta G^{\ominus''} + C_2,\end{aligned} \tag{19}$$

and thence concludes that there is a constant difference between the standard thermodynamic functions K_a, $\Delta G^\ominus$ and $\Delta S^\ominus$ for a given exchange reaction when these functions are calculated using both the Vanselow and the Gaines & Thomas formulations. However, from (7) and (11)

$$K_V = K_G(z_v^{z_v}/z_u^{z_u})\,(z_u + (z_v - z_u)\,\bar{E}_u)^{(z_u - z_v)}, \tag{20}$$

so that the integral in (18) is

$$\int_0^1 f(\bar{E}_u)\,d\bar{E}_u = \int_0^1 \ln(z_v^{z_v}/z_u^{z_u})\,d\bar{E}_u + \int_0^1 (z_u - z_v)\ln[z_u + (z_v - z_u)\,\bar{E}_u]\,d\bar{E}_u. \tag{21}$$

This function is in fact equal to $(z_v - z_u)$ (Barrer & Townsend 1984) which means that C_1 and C_2 in (19) are both zero, a conclusion that must follow from (17). Therefore the conclusions drawn by Goulding (1983) regarding K_a, $\Delta G^{\ominus}$ and $\Delta S^{\ominus}$ are erroneous. Regarding the activity coefficients g_u, g_v, these are indeed different from f_u, f_v in magnitude for a given clay composition $\bar{E}_u$, $\bar{E}_v$ (except, as again *must* follow, in the standard states when $f_{u(\infty)} = g_{u(\infty)} = 1$, etc.). However, the behaviour of g_u, g_v on the one hand and of f_u, f_v, on the other are entirely symmetric and complementary (Barrer & Townsend 1984) and there are therefore no grounds for stating that only f_u, f_v are 'true' (*sic*) activity coefficients (Sposito & Mattigod 1979). A detailed study of this matter is presented elsewhere (Barrer & Townsend 1984).

(c) Definitions of ideal exchange behaviour in a clay

The question as to whether one should use equivalent fractions rather than cationic mole fractions for thermodynamic studies is not therefore a fundamental one. The important point to remember is that when one chooses a particular concentration scale in a thermodynamic formulation, then usually one is setting up a criterion for ideal behaviour in terms of that scale. Thus using cationic mole fractions, the criterion for ideal behaviour is that the relative activity of cation i equals its cationic mole fraction for all values of $\bar{X}_i$. Similarly, for equivalent fractions the criterion is that $a_i = \bar{E}_i$ for all $\bar{E}_i$. For exchange levels of transition metal ion less than 70%, Sposito & Mattigod (1979) showed that if the cationic mole fraction scale is used, near-ideal behaviour is observed in Camp Berteau montmorillonite, whereas if equivalent fractions are used for these data, the dependence of K_G on $\bar{E}_u$ is monotonic (Maes *et al.* 1975). For exchange levels up to 70% therefore, use of $\bar{X}_u$ yields a simpler description of this system than does the use of $\bar{E}_u$. Generally, however, which of the two scales to use must be a matter of personal preference and convenience.

3. MULTICOMPONENT EXCHANGE REACTIONS

(a) Introduction

In recent years, the importance of developing adequate thermodynamic formulations for multicomponent exchange processes in clays, zeolites and resins has been recognized. Any detailed consideration of cation exchange processes in soils must involve multicomponent exchange studies, and in fact many apparently binary exchanges are in reality at least ternary. Thus (Talibudeen 1981) the well-known pH-dependent negative charge on clays (which arises from the ionization of surface hydroxyls as the pH is raised) not only leads to site heterogeneity in the exchanger, but also to participation in the exchange reaction of hydronium ions at higher pH values (Elprince *et al.* 1980). Another common cause for apparent binary exchanges being in actuality ternary has been demonstrated by some elegant studies on the Na–Cu, Na–Ca and Na–Mg exchange equilibria in Wyoming bentonite (Sposito *et al.* 1981; Sposito *et al.* 1983). Sposito *et al.* comment on the frequent observation that as a clay exchanged with alkali metal is saturated with a bivalent metal cation, both the cation exchange capacity and the selectivity of the clay for the bivalent cation increase (see, for example, Maes *et al.* 1975), and they demonstrate that these effects are due to the participation of metal complexes (e.g. $CuCl^+$) in the exchange reaction. This has of course the effect of converting what is apparently a binary (M^{2+}–Na^+) exchange into a ternary (M^{2+}–MCl^+–Na^+) one.

Attempts have been made, with varying degrees of success, to predict ternary or multicomponent exchange equilibria from experimental data obtained from the component binary exchanges using different model-based approaches (Elprince & Babcock 1975; Wiedenfeld &

Hossner 1978; Soldatov & Bychkova 1980; Barri & Rees 1980; Elprince *et al.* 1980; Chu & Sposito 1981; Fletcher *et al.* 1984). It now appears generally agreed that *a priori* prediction of multicomponent equilibria is not normally possible with binary data alone (Barri & Rees 1980; Chu & Sposito 1981; Fletcher *et al.* 1984). Direct measurements of multicomponent equilibrium compositions combined with a comprehensive and rigorous thermodynamic formulation (i.e. independent of any particular microscopic model for the system) is the best means of mathematically predicting the exchange behaviour of the clay, zeolite or resin over a range of experimental conditions.

(b) *Thermodynamic formulations of ternary ion exchange*

Systematic thermodynamic formulations for multicomponent ion exchange have been developed independently by three groups of workers. Soldatov & Bychkova (1980) developed theirs with resins particularly in mind; the model is considered elsewhere (Fletcher *et al.* 1984). The formulations of Fletcher & Townsend (1981 *a–c*) and Chu & Sposito (1981) complement each other in that the former uses the equivalent fraction scale, whereas the latter employs ionic mole fractions. Chu & Sposito (1981) develop their equations in terms of the three component binary reactions of the appropriate ternary exchange system; they obtain finally a general expression for the three binary thermodynamic equilibrium constants in terms of pseudo-binary Vanselow selectivity coefficients $^cK_{ij}^T$, where

$$K_{a,ij} = (a_j^{z_i}/a_i^{z_j})\,(\overline{X}_i^{z_j}f_i^{z_j}/\overline{X}_j^{z_i}f_j^{z_i}) = {}^cK_{ij}^T(f_i^{z_j}/f_j^{z_i}). \tag{22}$$

(Note that although $^cK_{ij}^T$ appears identical to K_V in (7) the similarity is deceptive. In (7) $\overline{X}_u + \overline{X}_v = 1$ whereas in (22) $\overline{X}_i + \overline{X}_j + \overline{X}_k = 1$. Also f_i, f_j differ from f_u, f_v in that they allow for the presence of other exchanging ions in addition to the ith and jth cations.) However, $\lim_{(\overline{X}_k \to 0)} {}^cK_{ij}^T = K_V$). Considering then three different exchanging cations $M_1^{z_1^+}$, $M_2^{z_2^+}$ and $M_3^{z_3^+}$ in the clay, the general expression for the binary equilibrium constants is (Chu & Sposito 1981):

$$\ln K_{a,ij} = \int_{\substack{\overline{E}_i=1,\overline{E}_j=0,\overline{E}_k=0 \\ \text{(any path)}}}^{0,1,0} \left[\ln\,(^cK_{ij}^T)\,\mathrm{d}\overline{E}_j - \frac{z_j}{z_i}\ln\,(^cK_{ki}^T) \right] \mathrm{d}\overline{E}_k \tag{23}$$

for the cyclic permutations

$$\begin{bmatrix} i \\ j \\ k \end{bmatrix} \equiv \begin{bmatrix} 1, & 2, & 3 \\ 2, & 3, & 1 \\ 3, & 1, & 2 \end{bmatrix}.$$

The second term in the square brackets requires comment. At first examination, since the integration is effected with respect to $\overline{E}_k$ and the limits of the integration are $\overline{E}_k = 0$ to $\overline{E}_k = 0$, it seems that this function is zero. This is not correct, however, since by definition

$$\overline{E}_i + \overline{E}_j + \overline{E}_k = 1, \quad \mathrm{d}\overline{E}_k = -\mathrm{d}\overline{E}_i - \mathrm{d}\overline{E}_j, \tag{24}$$

and therefore

$$\int_{\substack{1,0,0 \\ \text{(any path)}}}^{0,1,0} \ln\,(^cK_{ki}^T)\,\mathrm{d}\overline{E}_k = -\int_{\substack{1,0,0 \\ \text{(any path)}}}^{0,1,0} \ln\,(^cK_{ki}^T)\,\mathrm{d}\overline{E}_i - \int_{\substack{1,0,0 \\ \text{(any path)}}}^{0,1,0} \ln\,(^cK_{ki}^T)\,\mathrm{d}\overline{E}_j. \tag{25}$$

It is axiomatic that the evaluation of the function $K_{a,ij}$ in (23) be independent of the path of integration; with a limited quantity of data for the ternary system a numerical integration may be effected across the $^cK_{ij}^T$ and $^cK_{ki}^T$ surfaces between the appropriate limits. Figure 1 shows diagrammatically the numerical evaluation of the first integral in (23) with only a limited quantity of experimental data. Evaluation of the second integral in (22) involves excursions from the edge of the $^cK_{ki}^T$ surface (at which $\bar{E}_k = 0$ for all $\bar{E}_i$, $\bar{E}_j$) to values of $\bar{E}_i$, $\bar{E}_j$ for which experimental data for $^cK_{ki}^T$ are available.

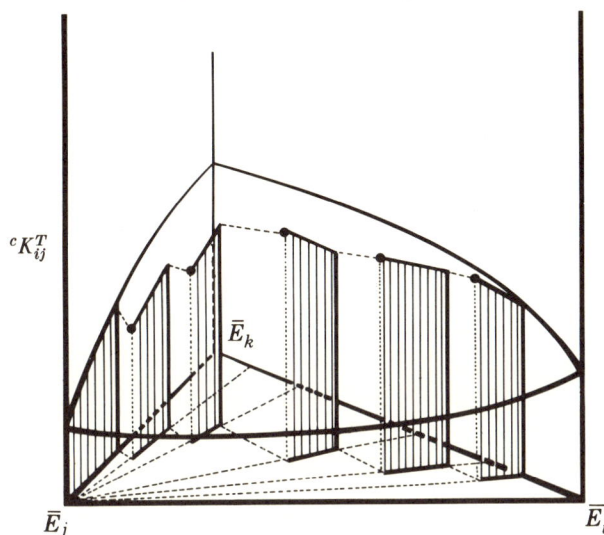

FIGURE 1. Diagrammatic representation of an evaluation by numerical integration of the first term in (23). The ternary composition of the system is depicted in triangular coordinates at the base of the perspective diagram, with the composition $\bar{E}_k = 1$ the vertex at the rear of the diagram. In the vertical direction are plotted values of $^cK_{ij}^T$ as a function of the ternary composition. The shaded areas give the value of the integral by one path across this $^cK_{ij}^T$ surface from $\bar{E}_i = 1$ to $\bar{E}_j = 1$.

The thermodynamic formulation of Fletcher & Townsend (1981 a–c) takes into account also the water activity in the exchanger, although (see §2) in agreement with earlier studies on binary systems (Barrer & Klinowski 1974), the water activity corrections are not normally significant at low electrolyte concentrations in the external solution (Fletcher & Townsend 1981 b). The general derivation begins from an opposite viewpoint to that of Chu & Sposito (1981), by directly defining ternary reaction equations and corresponding ternary equilibrium constants. Thus for three different exchanging cations $M_1^{z_1^+}$, $M_2^{z_2^+}$ and $M_3^{z_3^+}$ the ternary exchange reactions are

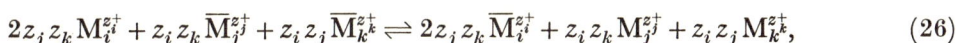

$$2z_j z_k M_i^{z_i^+} + z_i z_k \overline{M}_j^{z_j^+} + z_i z_j \overline{M}_k^{z_k^+} \rightleftharpoons 2z_j z_k \overline{M}_i^{z_i^+} + z_i z_k M_j^{z_j^+} + z_i z_j M_k^{z_k^+}, \qquad (26)$$

for the cyclic permutations

$$\begin{bmatrix} i \\ j \\ k \end{bmatrix} \equiv \begin{bmatrix} 1, & 2, & 3 \\ 2, & 3, & 1 \\ 3, & 1, & 2 \end{bmatrix}.$$

The general expression for the corresponding thermodynamic equilibrium constants is then

$$_{j,k}^{i}K_a = (a_j^{z_i z_k} a_k^{z_i z_j} / a_i^{2z_j z_k}) \, (\bar{E}_i \phi_i)^{2z_j z_k} / (\bar{E}_j \phi_j)^{z_i z_k} (\bar{E}_k \phi_k)^{z_i z_j}), \qquad (27)$$

where ϕ_i, ϕ_j, ϕ_k are activity coefficients for the cations $M_i^{z_i^+}$, etc. in the clay. Although the integration of the appropriate form of the Gibbs–Duhem equation between the prescribed

[89]

limits can be by any path, Fletcher & Townsend (1981b) chose to specify particular paths, always in terms of two of the equivalent fractions only, namely $\bar{E}_1$ and $\bar{E}_2$. This simplifies the final equations, and therefore also the evaluation of both the K_a and ϕ values. The general expression for the thermodynamic equilibrium constant is

$$\ln \left(_{j,k}^{i}K_a\right) = z_j(z_k - z_i) + z_k(z_j - z_i) + \int_0^1 I_1 \ln \ell_{3/1} \mathrm{d}\bar{E}_1 + \int_0^1 I_2 \ln \ell_{3/2} \mathrm{d}\bar{E}_2, \tag{28}$$

for the permutations

$$\begin{bmatrix} i \\ j \\ k \\ I_1 \\ I_2 \end{bmatrix} \equiv \begin{bmatrix} 1, & 2, & 3 \\ 2, & 3, & 1 \\ 3, & 1, & 2 \\ -2, & 1, & 1 \\ 1, & -2, & 1 \end{bmatrix}$$

and the activity coefficients are respectively

$$\ln \phi_1^{z_2 z_3} = \int_1^{\bar{E}_1} \ln \chi_{3/1} \mathrm{d}\bar{E}_1 + \int_0^{\bar{E}_2} \ln \chi_{3/2} \mathrm{d}\bar{E}_2 \tag{29}$$

$$\ln \phi_2^{z_1 z_3} = \int_0^{\bar{E}_1} \ln \chi_{3/1} \mathrm{d}\bar{E}_1 + \int_1^{\bar{E}_2} \ln \chi_{3/2} \mathrm{d}\bar{E}_2 \tag{30}$$

$$\ln \phi_3^{z_1 z_2} = \int_0^{\bar{E}_1} \ln \chi_{3/1} \mathrm{d}\bar{E}_1 + \int_0^{\bar{E}_2} \ln \chi_{3/2} \mathrm{d}\bar{E}_2 \tag{31}$$

where

$$\begin{aligned} \ell_{3/1} &= \bar{E}_3 a_1 / \bar{E}_1 a_3, \\ \ell_{3/2} &= \bar{E}_3 a_2 / \bar{E}_2 a_3, \end{aligned} \tag{32}$$

$$\begin{aligned} \chi_{3/1} &= (\ell_{3/1}/\ell_{3/1}^*) \exp \left[-z_2(z_3 - z_1) \right], \\ \chi_{3/2} &= (\ell_{3/2}/\ell_{3/2}^*) \exp \left[-z_1(z_3 - z_2) \right]; \end{aligned} \tag{33}$$

$\ell_{3/1}^*$ and $\ell_{3/2}^*$ are the values of $\ell_{3/1}$, $\ell_{3/2}$ at the composition for which values of ϕ_1, ϕ_2 and ϕ_3 are required. Provided sufficient accurate experimental data on the ternary exchange equilibrium are available, it is comparatively easy to evaluate ϕ_1, ϕ_2 and ϕ_3 from (29)–(31). The dependence of $\ln \chi_{3/1}$ and of $\ln \chi_{3/2}$ on $\bar{E}_1$, $\bar{E}_2$ are expressed as polynomials:

$$\begin{aligned} \ln \chi_{3/1} &= \sum_{a=0}^{e} \alpha_a (\bar{E}_1)^a + \sum_{b=1}^{f} \beta_b (\bar{E}_2)^b, \\ \ln \chi_{3/2} &= \sum_{c=0}^{g} \gamma_c (\bar{E}_1)^c + \sum_{d=1}^{h} \delta_d (\bar{E}_2)^d, \end{aligned} \tag{34}$$

where α_a, β_b, γ_c, δ_d are appropriate coefficients of $\bar{E}_1$, $\bar{E}_2$. Then on taking (29) as an example,

$$\ln \phi_1^{z_2 z_3} = \sum_{a=0}^{e} \frac{\alpha_a}{a+1} (\bar{E}_1)^{(a+1)} + \sum_{b=1}^{f} \beta_b \bar{E}_1 (\bar{E}_2)^b + \sum_{c=0}^{g} \gamma_c \bar{E}_2 (\bar{E}_1)^c + \sum_{d=1}^{h} \frac{\delta_d}{d+1} (\bar{E}_2)^{(d+1)}. \tag{35}$$

The integration procedures corresponding to (35) are shown in figure 2. Further details of these procedures, and examples of experimentally determined ϕ data, are given elsewhere (Fletcher *et al.* 1984; Franklin & Townsend 1984).

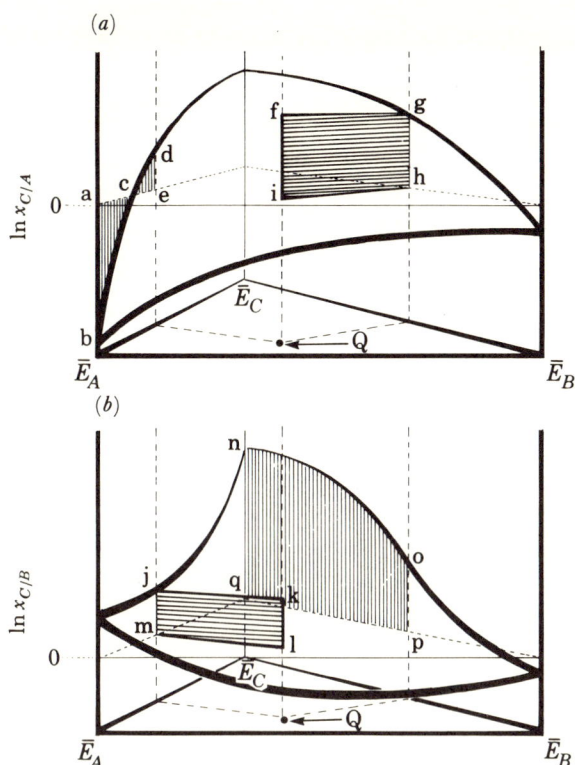

FIGURE 2. Representations with perspective diagrams of the integration procedures required to evaluate $z_2 z_3 \ln \phi_1$ at a composition Q by using (29) and (33). Figure 2a corresponds to the first integral in (29), or the first two summation terms in (35). Figure 2b corresponds to the second integral in (29) or the last two summations in (35). The value of $z_2 z_3 \ln \phi_1$ at the composition Q is seen to be equal to area [(cde) − (abc) + (fghi) + (jklm) + (nopq)].

(c) Compatibility of different thermodynamic formulations for ternary ion exchange

Thermodynamic rigour requires that the treatments of Chu & Sposito (1981) and Fletcher & Townsend (1981a–c) be fully compatible. The use of different concentration scales, and the choice of different integration paths makes a comprehensive proof of compatibility somewhat involved. However, for the case of exchanges involving univalent ions only $\bar{E}_i = \bar{X}_i$, etc., and if $M_i = M_1, M_j = M_2, M_k = M_3$ then (23) becomes

$$\ln K_{a,12} = \int_{\substack{\bar{E}_1=1, \bar{E}_2=0, \bar{E}_3=0 \\ (\text{any path})}}^{0,1,0} [\ln (^cK_{12}^T) \, \mathrm{d}\bar{E}_2 - \ln (^cK_{31}^T) \, \mathrm{d}\bar{E}_3]. \tag{36}$$

Applying the thermodynamic closure rule (Chu & Sposito 1981) together with (24) gives

$$\ln K_{23} + \ln K_{31} = \int_{\bar{E}_1=1, \bar{E}_2=0, \bar{E}_3=0}^{0,1,0} [\ln (^cK_{23}^T) \, \mathrm{d}\bar{E}_2 - \ln (^cK_{31}^T) \, \mathrm{d}\bar{E}_1]. \tag{37}$$

Also, from (22) and (27)

$$\tfrac{1}{3} \ln \{{}_{2,3}^1 K_a / {}_{3,1}^2 K_a\} = \ln K_{23} + \ln K_{31} \tag{38}$$

and from (28)

$$\tfrac{1}{3} \ln \{{}_{2,3}^1 K_a / {}_{3,1}^2 K_a\} = \int_0^1 \ln k_{3/2} \mathrm{d}\bar{E}_2 - \int_0^1 \ln k_{3/1} \mathrm{d}\bar{E}_1 \tag{39}$$

A comparison of (37), (38) and (39) shows that the right side of (37) and (39) should be the same if the treatments are compatible. From the definitions in (22) and (32)

$$\left.\begin{array}{l} k_{3/1} = \bar{E}_3 a_1 / \bar{E}_1 a_3 = {}^c K_{13}^T = ({}^c K_{31}^T)^{-1}, \\ k_{3/2} = \bar{E}_3 a_2 / \bar{E}_2 a_3 = {}^c K_{23}^T, \end{array}\right\} \tag{40}$$

and (37) and (39) are seen to be identical. The complicated relations between $\bar{X}_{i,j,k}$ and $\bar{E}_{i,j,k}$ when $z_i \neq z_j \neq z_k$ make proof of compatibility much more involved in these cases.

Two thermodynamically rigorous and mutually compatible procedures for determining thermodynamic functions are therefore available for ternary exchange studies in clays. These two formulations correspond in choice of scales of measurement with the Vanselow and Gaines & Thomas treatments for binary cation exchange.

4. SITE HETEROGENEITY IN CLAYS

(a) Introduction

Site heterogeneity in exchangers is a well-known phenomenon, which occurs not only in clays but also in resins (Soldatov & Bychkova 1971) and in zeolites (Barrer & Klinowski 1977, 1979; Fletcher et al. 1984). The nature of site heterogeneity in clays is, however, different from that found in the other most common forms of inorganic cation exchanger, the zeolites. In zeolites, the site heterogeneity arises because the exchanger normally contains several crystallographically distinct yet intimately mixed sets of sub-lattices for cations, the sites on which are at least partly occupied by exchange ions. In clays the heterogeneous cation sites are not necessarily intimately mixed. Thus in addition to those cation sites that are randomly spaced throughout the expansible interlayer spaces (Talibudeen 1981) and that are an *intrinsic* property of the clay, a significant quantity of sites (the number dependent on average particle size) may arise from faults, dislocations and exposed edges within the crystallites (Talibudeen 1981; Brouwer et al. 1983). pH-dependent site heterogeneity may be present also (Talibudeen 1981; Elprince et al. 1980).

(b) Site heterogeneity and thermodynamic functions

Site heterogeneity in an exchanger, if present, is indeed likely to manifest itself in the variation with the exchange composition of functions such as the differential enthalpy and entropy of exchange, or in the variation with composition of the corrected selectivity coefficient and associated activity coefficients. However, as emphasized in the introduction section, to reverse this 'cause and effect' phenomenon, and instead use thermodynamic measurements alone to infer details of a particular heterogeneous site model for the exchanger, is a highly questionable procedure. If other independent experimental measurements that provide information on the structure and site heterogeneity within the clay are also applied to the material (e.g. X-ray, electron microscopy, neutron diffraction, magic angle spinning n.m.r.), only then is it strictly permissible to interpret thermodynamic data in terms of site heterogeneity.

Site heterogeneity is undoubtedly responsible for the very complicated dependence on exchange composition that is often seen for activity coefficients and corrected selectivity coefficients, and is likely to be one reason why it is difficult to predict multicomponent equilibrium compositions from binary data alone. Failures of the 'sub-regular' model of Hardy

[92]

(1953) when it is applied to binary and ternary cation exchanges were attributed to site heterogeneity by Elprince *et al.* (1980) and Fletcher *et al.* (1984). The former case is an example where site heterogeneity appears to have arisen from the pH-dependent contribution to the cation exchange capacity of the clay. The authors emphasize rightly that while this is the likely cause for the failure 'the sub-regular model, being thermodynamic, neither assumes nor yields any information about the mechanism of cation exchange', and therefore 'the reasons for the failure of the model ... cannot be established precisely' (Elprince *et al.* 1980).

Sharp and large changes with composition in the corrected selectivity coefficient have also been interpreted in terms of site heterogeneity. Thus observed changes in K_G for a series of different exchanges in illite clay have been interpreted in terms of exchange into three kinds of sites within the clay (Brouwer *et al.* 1983). Earlier studies had indicated the presence of these three sets of sites (Bolt *et al.* 1963). In another recent study, Goulding & Talibudeen (1980) have shown that site heterogeneity can be used to characterize soils and clays and even to distinguish between materials having the same generic name. These characterizations are drawn from the behaviour with exchange composition of the differential enthalpy and entropy functions.

Concluding remarks

Thermodynamic formulations of ion exchange that are in current use today for all manner of ion exchange materials owe much to early studies on clays. The Gaines & Thomas (1953) convention for thermodynamic reference states in the exchanger phase has been almost universally adopted for thermodynamic studies on clays, yet two alternative scales of measurement for cation concentrations in the exchange are in common use. Despite recent criticism of the equivalent fraction scale, thermodynamic formulations based on either this or on the cationic mole fraction scale are equally acceptable and mutually compatible; the question of which scale to adopt is one of personal preference.

Recently, thermodynamic formulations have been developed independently by different workers to describe ternary or multicomponent exchanges. These treatments enable the direct measurement of activity coefficients for multicomponent cation systems within clays, and also the prediction of selectivity behaviour over a range of conditions. The different formulations are based on the same two scales of measurement which have been used for binary exchanges, and as in the binary case, the two treatments are both thermodynamically rigorous and mutually compatible. Direct measurement of activity coefficients for ternary systems is now possible, and is necessary owing to the failure of attempts to predict the values of these functions from binary data alone. A major cause of this failure must be due to the presence of site heterogeneity within clay exchangers, even though the nature and the degree of this heterogeneity is different to that found in zeolites and resins.

References

Argersinger, W. J., Davidson, A. W. & Bonner, O. D. 1950 *Trans. Kansas Acad. Sci.* 53, 404.
Babcock, K. L. & Duckart, E. C. 1980 *Soil Sci.* 130, 64.
Barrer, R. M. & Klinowski, J. 1974 *J. chem. Soc. Faraday Trans.* I 70, 2080.
Barrer, R. M. & Klinowski, J. 1977 *Phil. Trans. R. Soc. Lond.* A 285, 247.
Barrer, R. M. & Klinowski, J. 1979 *J. chem. Soc. Faraday Trans.* I 75, 247.
Barrer, R. M. & Townsend, R. P. 1984 *J. chem. Soc. Faraday Trans.* II. (In the press.)
Barri, S. A. I. & Rees, L. V. C. 1980 *J. Chromatogr.* 201, 21.

Bolt, G. H. 1967 *Neth. J. agric. Sci.* **15**, 81.

Bolt, G. H., Sumner, M. E. & Kamphorst, A. 1963 *Proc. Soil Sci. Soc. Am.* **27**, 294.

Brouwer, E., Baeyens, B., Maes, A. & Cremers, A. 1983 *J. phys. Chem.* **87**, 1213.

Chu, S.-Y. & Sposito, G. 1981 *J. Soil Sci. Soc. Am.* **45**, 1084.

Dyer, A., Enamy, H. & Townsend, R. P. 1981 *Sep. Sci. Technol.* **16**, 173.

Elprince, A. M. & Babcock, K. L. 1975 *Soil Sci.* **120**, 332.

Elprince, A. M., Vanselow, A. P. & Sposito, G. 1980 *J. Soil Sci. Soc. Am.* **44**, 964.

Fletcher, P., Franklin, K. R. & Townsend, R. P. 1984 *Phil. Trans. R. Soc. Lond.* A (In the press.)

Fletcher, P. & Townsend, R. P. 1981*a J. chem. Soc. Faraday Trans.* II **77**, 955.

Fletcher, P. & Townsend, R. P. 1981*b J. chem. Soc. Faraday Trans.* II **77**, 965.

Fletcher, P. & Townsend, R. P. 1981*c J. chem. Soc. Faraday Trans.* II **77**, 2077.

Franklin, K. R. & Townsend, R. P. 1984 (In preparation.)

Gaines, G. L. & Thomas, H. C. 1953 *J. chem. Phys.* **21**, 714.

Gapon, Y. N. 1933 *J. gen. Chem., U.S.S.R.* **3**, 144.

Glueckauf, E. 1949 *Nature, Lond.* **163**, 414.

Goulding, K. W. T. 1983 *J. Soil Sci.* **34**, 69.

Goulding, K. W. T. & Talibudeen, O. 1980 *J. Colloid Interface Sci.* **78**, 15.

Hardy, H. K. 1953 *Acta metall.* **1**, 202.

Helfferich, F. 1962 *Ion exchange.* London: McGraw-Hill.

Kielland, J. 1935 *J. Soc. chem. Ind., Lond.* **54**, 232T.

Maes, A., Peigneur, P. & Cremers, A. 1975 *Proc. Int. Clay Conference*, Wilmetle, Illinois: Applied Publishing Ltd. 1975, p. 319.

Pitzer, K. S. 1973 *J. phys. Chem.* **77**, 268.

Robinson, R. A. & Stokes, R. H. 1970 *Electrolyte solutions.* London: Butterworths.

Scatchard, G. 1961 *J. Am. chem. Soc.* **83**, 2636.

Soldatov, V. S. & Bychkova, V. A. 1971 *Russ. J. phys. Chem.* **45**, 707.

Soldatov, V. S. & Bychkova, V. A. 1980 *Sep. Sci. Tehnol.* **15**, 89.

Sposito, G. 1977 *J. Soil Sci. Soc. Am.* **41**, 1205.

Sposito, G. 1981*a The thermodynamics of soil solutions*, p. 2. Oxford University Press.

Sposito, G. 1981*b Chemistry in the soil environment* (ed. M. Stelly). *Am. Soc. Agron. Spec. Publ.* **40**, 13.

Sposito, G., Holtzclaw, K. M., Charlet, L., Jouany, C. & Page, A. L. 1983 *J. Soil Sci. Soc. Am.* **47**, 51.

Sposito, G., Holtzclaw, K. M., Johnston, C. T. & Levesque-Madore, C. S. 1981 *J. Soil Sci. Soc. Am.* **45**, 1079.

Sposito, G. & Mattigod, S. V. 1979 *Clays Clay Mineral.* **27**, 125.

Talibudeen, O. 1981 *The chemistry of soil processes* (ed. D. J. Greenland & M. B. H. Hayes), p. 115. New York: John Wiley.

Townsend, R. P., Loizidou, M. & Fletcher, P. 1984 *Proc. 6th Int. Conf. Zeolites, Nevada* 1983. (In the press.)

Vanselow, A. P. 1932 *Soil Sci.* **33**, 95.

Wiedenfeld, R. P. & Hossner, L. R. 1978 *J. Soil Sci. Soc. Am.* **42**, 709.

Phil. Trans. R. Soc. Lond. A **311**, 315–332 (1984)   [315]

Printed in Great Britain

Clay–organic interactions

By G. Lagaly

Institut für anorganische Chemie der Universität Kiel,
Olshausenstraße 40, 2300 Kiel, F.R.G.

Clay minerals interact with organic materials by adsorption, intercalation and cation exchange.

Basic principles of intercalation reactions were obtained with kaolinite which intercalates a limited number of neutral organic compounds.

The interaction of neutral organic compounds with mica-type layer silicates (2/1 clay minerals) is of quite different type. As illustrated for the interaction with nuclein bases, the adsorption can be strikingly dependent on the layer charge and the concentration of salts and co-adsorption phenomena can occur.

Various organic materials are bound by cation exchange. Besides some other examples, the reaction with alkylammonium ions is of interest because of widespread practical applications. From a more scientific point of view, the interactions of alkylammonium ions with clays provide models for studying surfactant aggregations on solid surfaces and possible conformational changes in aggregates of long chain compounds (mono- and bimolecular films, as in biomembranes).

Negatively charged organic ions can also be bound by clays. The main mechanisms are binding by positive edge charges or exchanging structural OH-groups.

INTRODUCTION

The alteration of clay properties by adsorbed organic materials has been known from the earliest use of clays by man about 7000 B.C. The Greeks profited by the interaction of clays with vegetable tannins. They probably used tannins besides potash for peptization of clays because they needed a very careful fractionation to obtain distinct clay fractions for red–black decors (Hofmann 1962). The likely use of urea for preparing very thin-walled porcelain during the Sung period in China (see below) is a further striking example of the practical application of clay–organic interactions.

INTERCALATION INTO KAOLINITE MINERALS

Type of guest molecules

Kaolinite (and also dickite, nacrite and halloysite) intercalate distinct organic compounds (Wada 1962; Weiss 1962). The guest molecules enter the interlayer spaces and prise apart the silicate layers. The reactive guest molecules were classified by Weiss *et al.* (1966) as follows:

(i) Compounds forming strong hydrogen bonds to the silicate layers: urea, formamide, acetamide, hydrazine. Suitable organic molecules have to have acceptor *and* donator sites for hydrogen bonds, for example, formamide:

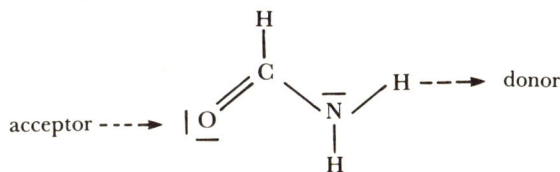

316   G. LAGALY

(ii) Compounds with pronounced betaine-like character with the possibility of strong dipole interactions with the silicate layers, for example, dimethyl sulphoxide (DMSO):

$$\begin{array}{c} CH_3 \\ \diagdown \\ | \, S = O \\ \diagup \\ CH_3 \end{array} \quad \longleftrightarrow \quad \begin{array}{c} CH_3 \\ \diagdown \\ | \, S \overset{+}{} - \overline{O} | \overset{-}{} \\ \diagup \\ CH_3 \end{array}$$

(iii) Alkali salts (K^+, NH_4^+, Rb^+, Cs^+, not Li^+, Na^+) of short chain fatty acids, in particular acetic and propionic acid.

TABLE 1. PREPARATION OF KAOLINITE INTERCALATION COMPOUNDS

(Weiss et al. 1966; Weiss et al. 1970; Weiss & Orth 1973.)

guest compound	reaction time and temperature	basal spacing 10^{-10} M
hydrazine hydrate	1d, 60 °C	10.4
urea†	8d, 60–110 °C	10.7
formamide	4d, 60 °C	10.1
N-methylformamide	2d, 60 °C	10.8
acetamide	30d, 100 °C	10.9
dimethyl sulphoxide	30h, 50 °C	11.2
pyridine-N-oxide†	12d, 50 °C	12.6
imidazole†	3d, 60 °C	11.4
ammonium acetate† (pH = 8–9)	20d, 20 °C	14.1
potassium acetate† (pH = 8)	1d, 65 °C	14.0
N,N-dimethyl formamide	8d, 50 °C‡	12.0
N,N-dimethylurea	8d, 50 °C‡	10.9; 12.2
pyridine	20d, 50 °C‡	12.0
2-picoline-N-oxide	20d, 25 °C‡	12.9
alkali chlorides, bromides and iodides	5d, 25 °C‡	11.1–11.7

† Concentrated aqueous solution.
‡ From ammonium acetate kaolinite.

A large variety of organic molecules are intercalated by the displacement procedure. For example, ammonium acetate is first intercalated, then displaced by alkylamines. In a similar way ammonium acetate can be displaced by ammonium propionate and then ammonium propionate by glycol. Diaminohexane instead of glycol combines with the acid anion to the corresponding amide in the interlayer space (Seto et al. 1978a, b).

A different mechanism is the 'Schleppreaktion' (the entraining reaction, Weiss et al. 1966). Non-reacting compounds are entrained between the layers by reactive guest molecules.

In table 1 are listed some experimental conditions for preparing kaolinite-intercalation products.

Structure of kaolinite intercalates

The structure of kaolinite intercalates is now established by several X-ray diffraction studies (DMSO, NMFA, imidazole, pyridine-N-oxide, picoline-N-oxide, Weiss et al. 1966; Weiss & Orth 1973; Weiss et al. 1973). The first three-dimensional crystal structure analysis was made of a NMFA–dickite complex (Adams 1978a). A neutron powder diffraction study allowed insight into the hydrogen bonding scheme in the formamide–kaolinite complex (Adams et al. 1976).

Crystallographically ordered structures are only expected in cases where there are functional groups capable of hydrogen bond formation to the silicate oxygen atoms. Dipole and van der Waals interactions may not be strong enough to maintain a well-ordered structure with sharp (*hkl*)-reflections.

Kinetics of intercalation

The degree of reaction increases with time in an S-shaped curve starting after a more or less pronounced induction period. According to recent neutron scattering experiments (Weiss *et al.* 1981) the adsorption on the external surfaces causes a reorientation of the OH-groups or a migration of protons during the induction period. This initiates the elastic deformation of the layers and the opening of interlayer space. In the next step the molecules enter the interlayer space. The reaction rate obeys the Avrami–Erofeev equation (two-dimensional phase boundary reaction, Fenoll & Weiss 1969) and can eventually change to a two-dimensional diffusion controlled reaction.

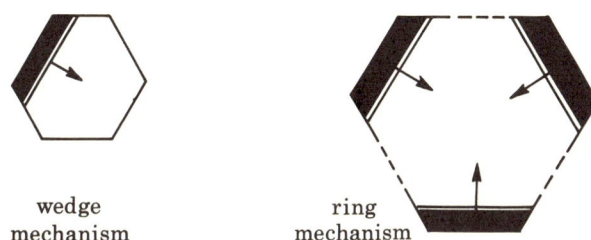

wedge
mechanism

ring
mechanism

FIGURE 1. Intercalation mechanism for small and large particles according to A. Weiss.

The reaction rate as a function of particle-size has a maximum for 3.8–5.0 μm particles (Weiss *et al.* 1970). The penetration of the guest molecules between the layers requires an elastic deformation of the silicate layers. Owing to the polarity of the kaolinite layer the decisive steps of deformation proceed in one direction. This kind of deformation leads to different reaction mechanisms for small and large particles. In the case of small particles the guest molecules enter the interlayer space only at one site. The neighbouring zones are blocked, because their opening would strongly increase the energy of deformation. Large particles can react at different sites, so that the reaction rate exceeds that of somewhat smaller crystals (figure 1). The movement of the reaction front from the edge to the interior is stopped by larger defects. The reaction rate was thus found to increase with increasing crystallization index (Weiss *et al.* 1970).

The reaction rate, surprisingly, depends on the ratio of DMSO to kaolinite. A maximum appears at a very low ratio which almost corresponds to a monomolecular covering of the surface by DMSO. The formation of the liquid structure (association of the DMSO molecules) apparently retards the entry of DMSO between the layers. The tendency for 'self-preservation' of the liquid structure is evident.

The desorption of intercalated guest molecules apparently is more complex than originally thought. In an attempt to follow the thermal decomposition of DMSO–kaolinite, Adams & Waltl (1980) found a two-dimensional contracting-circle type of mechanism to be consistent with the thermogravimetric result provided that the nucleation process is exponential rather than instantaneous. The high activation energy of 105 kJ mol^{-1} DMSO, however, is inconsistent with the assumption (as in the case of NMFA–kaolinite, Adams 1978*b*) that the rate-determinant step is the desorption of single organic molecules from the silicate layer.

[97]

Applications

Intercalation of long chain alkylamines by displacement reactions drastically changes the crystal shape. Without any mechanical treatment the plates are split into thinner layers which roll up forming halloysite-like tubes or grooves and troughs. The driving force is the misfit between the dimensions of the octahedral and tetrahedral sheet which is suppressed in thicker crystals by the superposition of the layers (Weiss & Russow 1963; Poyato-Ferrera *et al.* 1977).

The disaggregation of the kaolinite platelets by intercalation strongly increases the dry strength. Very likely the Chinese ceramists manufacturing thin-walled porcelain ('egg-shell porcelain', particular in the Sung period) took advantage of the increased dry strength of urea-treated kaolins (Weiss 1963a).

The maximum degree of reaction of many kaolins remains below 100 %, even after very long reaction times. Weiss and coworkers (Range *et al.* 1970) concluded that kaolins generally are mixtures of three or four types which differ by their chemical reactivity. Based on these observations a method was evaluated for estimating the composition of kaolins from the degree of reaction with DMSO and urea (Lagaly 1981a). Kaolins are indeed commonly and intimately intermixed in microdimensions (Keller & Haenni 1978). The experiments are still outstanding to prove this concept. It also seems likely that a relation exists between the technical applicability and the chemical reactivity of the kaolins (Fernandez-Gonzales *et al.* 1976).

Finally, the reactivity of kaolins towards DMSO and hydrazine can be used to identify even small amounts of kaolinites and halloysites (Range *et al.* 1970; Wilson & Tait 1977; Jackson & Abdel-Kader 1978). The procedure is also recommended for disintegrating flint clays and related kaolinite containing rocks (Weiss & Range 1970).

SMECTITES WITH NEUTRAL ORGANIC MOLECULES

Interlamellar solvates

The interlayer water molecules can be replaced by a variety of neutral organic molecules. A part of the organic molecules more or less strongly solvate the interlayer cations, the others are associated between the solvation shells. Like the solvation of ions in solution the bonding strength between the interlayer cations and the solvent molecules covers the range from loose aggregates to complex bonding.

Many organic molecules displace the water molecules from the interlayer cations more or less quantitatively (short chain alcohols, ethylene glycol, carbohydrates; Brindley 1966; Bissada *et al.* 1967; Dowdy & Mortland 1967; Stul & Uytterhoeven 1975; Chassin 1976). These interlayer solvates can thus be prepared by reacting air-dried smectites with the organic liquids. Other guest compounds such as nitriles have to be introduced in the interlayers by reaction with the dried smectites (Serratosa 1968; Yamanaka *et al.* 1975).

Widely applied is the use of ethylene glycol and glycerol for identifying smectites and vermiculites and for estimating the interlamellar surface area (Brindley 1966; Lagaly 1981a).

A basic question is that of the localization of the interlayer cations. If the interaction with the organic molecules is strong, the cations are completely solvated and solvent molecules are between the cations and the surface oxygen atoms. For weak interactions the cations may reside at the surface of the silicate layers so that only a few of their coordination sites are occupied by the solvent molecules. Berkheiser & Mortland (1975) obtained information about solvent-dependent cation positions from the intensity of the high field Fe(III) electron spin resonances ($g = 3.6$). The intensity varies with the degree to which the ion is removed from the smectite surface by solvation and is related to the Gutmann donor number of the solvent. For Na-smectite a Gutmann number of 14 is required for complete solvation. On the other hand, multi-valent cations can retain water molecules in the presence of solvents with a low Gutmann number and are precluded from close contacts to the surface.

The problem of the interlamellar structure can also be discussed from another point of view. Liquids of highly polar molecules show a pronounced tendency to maintain the molecular association on the solid surface. The 'self-preservation' tendency competes with the forces that are introduced by the field arising from the surface charges and the gegen ions and that tend to destroy the association (Lagaly & Witter 1982). An instructive example is reported by Annabi-Bergaya et al. (1981). Methanol molecules on smectite surfaces tend to maintain the typical zig-zag chain aggregation with strong hydrogen bonds between the molecules. Lithium as interlayer cation does not disturb this continuous network of adsorbed species, rather it fits in the holes between the aggregated methanol molecules. Calcium ions, owing to their stronger electrical field, disturb the association completely. In the case of sodium and barium ions the chains are broken into smaller fragments.

Long chain compounds

Interlamellar long chain compounds exhibit a strong tendency to aggregation in bimolecular films. The formation of these structures in most cases requires the propping-open procedure of Brindley & Ray (1964). These authors also observed a phase transition with increasing temperature (see also Pfirrmann et al. 1973). One or more temperature dependent phase transitions are very characteristic of interlamellar bimolecular films and are also observed for interlamellar triglycerides (Roloff & Weiss 1966), fatty acids (Brindley & Moll 1965) and primary amines (Brindley 1965; G. Lagaly and G. Söffing, unpublished). The transitions result from trans-gauche isomerization of the alkyl chains (see below).

Besides high-spacing structures long chain compounds often form complexes with very low spacings in which the alkyl chains are parallel to the silicate surface. This arrangement favours the interactions with the interlayer cations (for fatty acids: Yariv & Shoval 1982).

It stands to reason that interlamellar long chain compounds, in particular glycerides and fatty acids, are possible precursors of petroleum.

Interlamellar complexes

Many complex-forming ligands are strongly coordinated to the interlayer cations and form complexes identical with or similar to those in solution. Examples of ligands are diacetyl dioxime (Weiss & Hofmann 1951), oxime (Yamamoto et al. 1969), thiourea (Pleysier & Cremers 1975), acetylacetone (Parfitt & Mortland 1968), bipyridyl (Taylor et al. 1978), porphyrines (Cady & Pinnavaia 1978).

Recently special rhodium phosphonium interlayer complexes like $Rh(PPh_3)_2^+$ were developed as a new class of selective heterogeneous catalysts. A typical authentic metal complex catalysed reaction is the hydrogenation of terminal olefins (Pinnavaia 1983; Farzaneh & Pinnavaia 1983).

Resolution of optically active compounds by stereoselective adsorption was reported by Yamagishi (1983). Several Co-III-chelates were resolved at least partially into two configurational isomers, using the montmorillonite (Δ-tris (1,10-phenanthroline)nickel-II complex as column material).

An interesting group of complexes was obtained when arenes were reacted with Cu^{2+}-montmorillonite (table 2). In the yellow benzene complex the interlayer water has been partly displaced by benzene molecules which interact by their π-electrons with the Cu^{2+} ions (type I complexes). Complete removal of water (for instance in a desiccator over P_4O_{10}) leads to the

TABLE 2. AROMATIC COMPLEXES OF Cu^{2+}-MONTMORILLONITE AND

Ag^+, Fe^{3+}, VO_2^+-HECTORITE

type I			
(ring planar, aromaticity retained)	Cu^{2+}: benzene	phenol	
	toluene‡	anisole†	
	xylenes	phenyl ether	
	mesitylene	benzyle methyl ether	
	Fe^{3+}: toluene‡		
	Ag^+:	phenol	
		anisole	
type II			
(ring distorted, aromaticity lost)	Cu^2: benzene	anisole†	biphenyl
	thiophene	butyl phenyl ether	naphthalene anthracene
	Fe^{3+}: benzene	anisole†‡	
	thiophene		
	VO_2^+: thiophene	anisole	

† Oxidation to 4.4′-dimethoxy biphenyl.   ‡ Polymeric materials observed.

red type II complexes in which the aromaticity of the ligands is lost. The organic radical cations $C_6H_6^+$, formed by one electron transfer from the benzene molecules to the Cu^{2+} ions are probably coupling themselves (Pinnavaia & Mortland 1971; Rupert 1973; Pinnavaia *et al.* 1974).

Reaction with bases

Aliphatic and aromatic amines do not coordinate alkali- and alkaline-earth ions in aqueous solutions. This may also be true for interlamellar solvates. These bases are indeed predominantly bound to the cations by H_2O bridges (pyridine: Farmer & Mortland 1966; aniline, cyclohexylamine: Heller & Yariv 1970; ethylene diamine: Laura & Cloos 1972, 1975):

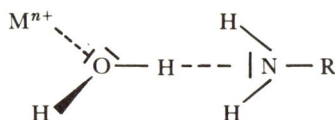

$$M^{n+} \cdots O - H --- | N - R$$

The type of coordination (directly or by H_2O-bridges) should depend on the Pearson–Schwarzenbach hardness and softness of the cations. Pyridine is directly coordinated to Cu^{2+} but bound by H_2O bridges to Mg^{2+} and Ca^{2+}-interlayer cations. A direct bonding, however, seems also to be possible, even for alkali ions, or at least is tentatively suggested in recent papers (Adams & Breen 1982).

The interlayer adsorption of bases, in particular amines, from aqueous solutions is accompanied by a cation exchange (Vansant & Uytterhoeven 1973). The protonated species

$$\text{(pyridine)} N + H_2O \rightleftharpoons \text{(pyridine)} N^{\pm} H + OH^-$$

[100]

exchange interlayer cations. The interlayer ratio between the protonated and unprotonated base increases with decreasing solution pH but certainly deviates from that in solution (Karickhoff & Bailey 1976; Feldkamp & White 1979).

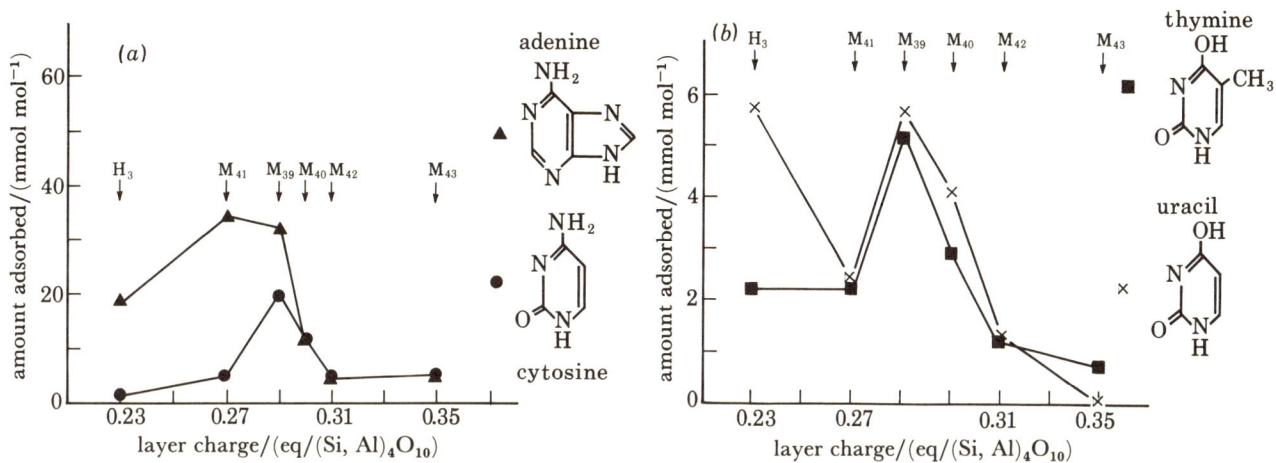

FIGURE 2. Adsorption of nuclein bases from 10^{-3} M solutions (pH = 5) on various smectites as a function of layer charge.

TABLE 3. ADSORPTION OF NUCLEIN BASES (FROM 1×10^{-3} M AQUEOUS SOLUTIONS)

montmorillonite	bases	ratio of bases adsorption from mixtures	adsorption from pure solution
Bavaria M 39	thymine/adenine	0.49	0.17
	uracil/adenine	0.24	0.19
	thymine/cytosine	0.29	0.28
	uracil/cytosine	0.17	0.31
Wyoming M 41	thymine/adenine	0.49	0.07
	uracil/adenine	0.38	0.08
	thymine/cytosine	0.63	0.46
	uracil/cytosine	0.35	0.54

The total adsorption as a function of pH often shows a maximum because protons compete with the acid form, or acid–base pairs such as

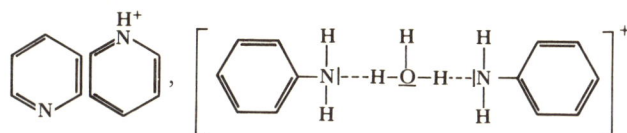

are predominantly adsorbed (Yariv & Heller-Kallai 1975).

The adsorption of nuclein bases exhibit interesting features. Not only does it depend on the type of clay mineral (Lailach et al. 1968; Thompson & Brindley 1969) but also on the kind of montmorillonite and its layer charge (figure 2, Samii 1981). The adsorption from 10^{-3} M solution is maximal at a layer charge of 0.27–0.29 eq/(Si, Al)$_4$O$_{10}$. Adenine and cytosine are adsorbed in markedly higher amounts than thymine and uracil. At layer charges above 0.30 eq/(Si, Al)$_4$O$_{10}$ the adsorption is generally very low.

The co-adsorption of nuclein bases described by Lailach & Brindley (1969) again depends on the kind of montmorillonite (table 3, Samii 1981). The presence of adenine increases the adsorption of thymine and uracil, in particular with montmorillonite from Wyoming, but the adsorption of uracil is decreased by cytosine. From a theoretical point of view the association between two different bases in solution by hydrogen bonds can decrease or increase their ratio

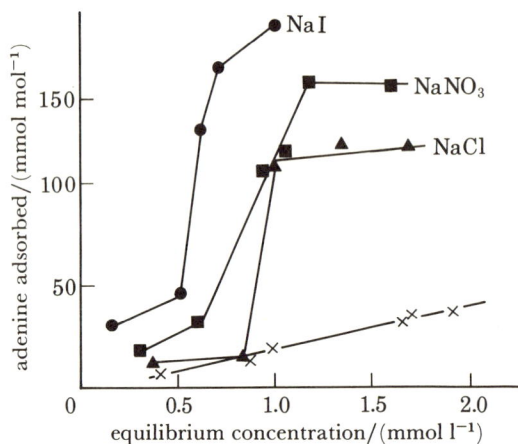

FIGURE 3. Salt effects on the adsorption of adenine on montmorillonite (from Wyoming M 41), salt concentration 0.2 M, pH = 5. ×, Without salt.

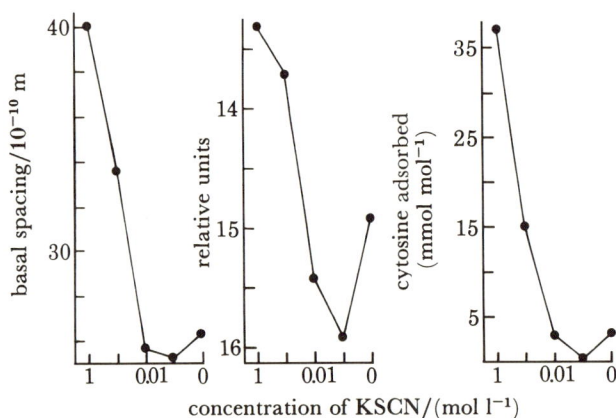

FIGURE 4. Salt dependent properties: basal spacings of tetra-decylammonium beidellite in salt solutions, water vapour pressure, cytosine adsorption on Ca-montmorillonite (from 10^{-3} M cytosine solution, pH = 5, montmorillonite: 'Greenbond' Wyoming, M 40).

in the adsorbed layer. One gets the impression that the base-pairing as in DNA plays a role, that is, besides adenine molecules pairs adenine–thymine are adsorbed. Consequently, preliminary adsorption of adenine or cytosine changes the adsorption properties of the smectite towards thymine and uracil.

The adsorption of nuclein bases is sensitively influenced by salts (figure 3). In particular the adsorption of adenine is greatly enhanced by sodium salts. (The influence of the Ca^{2+}/Na^+ exchange on the smectite is moderate.) The effect of the anions often follows the Hofmeister series indicating the adsorption to be influenced by the water structure. The directing influence of the water structure is also evident from figure 4. The adsorption of cytosine from KSCN solutions has a small minimum at 0.01 M KSCN, then sharply increases with the KSCN concentration. In a similar way the structure-breaking effect of the SCN^- anion influences the water vapour pressure (figure 4) and the swelling of tetradecylammonium beidellite in water (figure 4). One arrives to the same conclusion as stated by Fripiat and coworkers (Annabi-Bergaya *et al.* 1981) that 'the intermolecular forces between adsorbed species have been generally underestimated in studies of sorption processes by smectites'.

TABLE 4. EXCHANGE OF INTERLAYER CATIONS BY CATIONIC SURFACTANTS

ξ layer charge (eq/(Si, Al)$_4$O$_{10}$)

surfactant ion	formula $R = C_nH_{2n+1}$	montmoril-lonites $\xi = 0.3-0.4$	beidellite $\xi = 0.4-0.5$	reactivity vermiculites $\xi \leqslant 0.7$	vermiculites $\xi \geqslant 0.7$	mica
alkylammonium	RNH_3^+	+	+	+	+	slowly
trimethyl alkyl-ammonium	$RN(CH_3)_3^+$	+	+	+	−	−
dimethyl dialkyl-ammonium	$R_2N(CH_3)_2^+$	+	+	+	−	−
N-alkyl pyridinium	$R-\overset{+}{N}\langle\bigcirc\rangle$	+	+	±	−	−

The discussion of the interaction of clays with other biologically important bases and related compounds would be very fascinating but is outside the scope of this paper. For completeness some references may be listed: amino acids, peptides (Weiss 1963 *b*; Fripiat *et al.* 1960; Rausell-Colom & Salvador 1971; Raussel-Colom & Fornes 1974; Slade *et al.* 1976; Raupach & Janik 1976; Siffert & Kessaissia 1978); codeine (Rupprecht *et al.* 1975); porphyrins (van Damme *et al.* 1978; see also Theng 1974 and Mortland 1970).

The present concern about pollution control will activate studies of biocide–clay interactions. The principal interaction mechanisms are known (Mortland 1970; Weber & Weed 1974; White 1976). The protonation of the molecules and their binding as interlayer cations decrease the immediate efficacy and the pollution of the environment. On the other hand, if the protonated and adsorbed species resist hydrolysis and microbial decomposition, the compound shows biological activity when it is desorbed. The bound residues pose a possible threat for the environment as they can be released over a long time.

INTERACTIONS WITH CATIONIC SURFACTANTS

Cation exchange

Cationic surfactants (table 4) are exchanged for the inorganic interlayer cations from aqueous solutions at pH 5–7. Suitable surfactant concentrations are, for example, for alkyl-ammonium ions $C_nH_{2n+1}NH_3^+$: 0.5–2 M for $n < 10$; 0.1 M for $n = 10–15$; 0.05 M for $n > 15$. Depending on concentration, pH, kind of surfactant, alkyl chain length, and layer charge the cation exchange is accompanied by intercalation of ionic pairs: surfactant cation plus gegen ion. The total amount of interlayer surfactant in contact with the equilibrium surfactant solution can thus greatly exceed the cation exchange capacity (c.e.c.). With alkylammonium ions $C_nH_{2n+1}NH_3^+$ alkylamine molecules $C_nH_{2n+1}NH_2$ are also adsorbed, in particular at pH > 6. Most of the ionic pairs or the alkylamine molecules are removed by short washing; the fully quantitative removal (without showing any change in the X-ray diagram) requires intense washing of the samples.

Smectites bind very different kinds of surfactants. Selectivity arises with increasing layer charge (table 4). Highly charged vermiculites (the reaction of higher charged vermiculites is promoted when the vermiculite is in the Li form (Slade *et al.* 1978)) are expected to select distinct surfactants from surfactant mixtures. Micas and illites react very slowly with alkylammonium ions. A nearly complete reaction requires months or even years depending on the particle size (Weiss *et al.* 1956). Micas may first be treated with $BaCl_2$-solution at about 120 °C to exchange the K^+ ions by the Ba^{2+} ions (Reichenbach & Rich 1968). The Ba^{2+} ions are then easily exchanged by alkylammonium ions (Beneke & Lagaly 1982).

Application of alkylammonium exchanges

The reaction with alkylammonium ions is the simplest possibility for identifying even very small amounts of smectites and vermiculites in bentonites, kaolins, rocks and soils. It further provides the most reliable method for determining the layer charge and charge distribution of smectites, vermiculites and regular interstratified materials. Combined with carbon content determination (by combustion) the alkylammonium ion exchange further allows a rapid and reliable quantitative estimation of the smectite content in bentonites. The estimation is based on the assumption that the interlayer c.e.c. (derived from the layer charge) of most smectites is about 80 % of the total c.e.c. (derived from the carbon content) (Lagaly 1981 *a*).

The modification of the clay surface by surfactants changes the adsorption properties towards organic gases and liquids (Barrer & Millington 1967; Slabaugh & Hanson 1969; Stul *et al.* 1983 *a, b*; chromatographic behaviour: Barrer & Hampton 1957; Taramasso *et al.* 1971; McAtee & Robbins 1980).

The concept of pillared clays was first used by Barrer & MacLeod (1955) to induce interlayer porosity in montmorillonites by an exchange with tetraalkylammonium ions. Other organic cations that have been used include bicyclic ammonium cations, such as triethylene diammonium ions $HN^+(CH_2CH_2)_3N^+H$ and metal chelate cations. The pillared clays offer new possibilities for heterogeneous catalysis, but sufficient temperature stability often requires pillars other than organic cations. The first pillared clays with poly-condensed Al-ions as pillars and high thermal stability were prepared by Brindley & Sempels (1977). Other inorganic pillared clays can be prepared by hydrolysis of silicon acetylacetonate and niobium and tantalum metal cluster cations (Pinnavaia 1983).

The increased hydrophobicity with increasing loading by surfactants and its effect on the adsorption from binary solutions is illustrated by the following example. The excess isotherm for the binary mixture methanol–benzene on hexadecyl pyridinium montmorillonite (figure 5) is a type IV isotherm that results from the superposition of the individual methanol and benzene isotherms, with distinct plateaux. Both compounds are adsorbed but the amount of benzene almost linearly increases with the extent of surface modification by hexadecylpyridium ions (Dekany *et al.* 1978).

The binding of surfactants by cation exchange on clay surfaces provides a simple possibility for studying hydrophobic interactions from a different point of view. The hydrophobic interactions, particularly effective in biological systems, is based on a distinct association of polar molecules, especially water molecules, in the contact area to alkyl chains or low dielectric

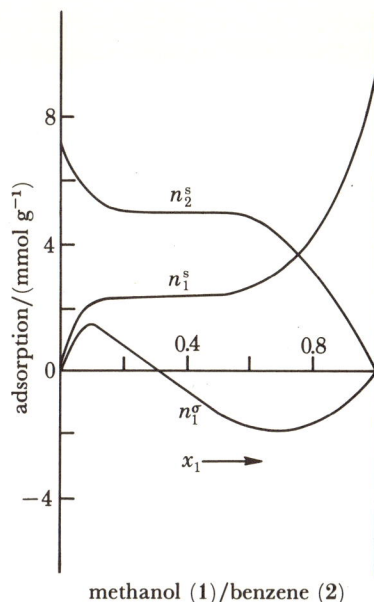

FIGURE 5. Adsorption of methanol (n_1^s) and benzene (n_2^s) from the binary mixture (X_1: mol fraction methanol) on hexadecylpyridinium montmorillonite: n_1^σ: surface excess (composite isotherm), n_1^s, n_2^s: individual adsorption isotherms of methanol (1) and benzene (2) (Dekany *et al.* 1975).

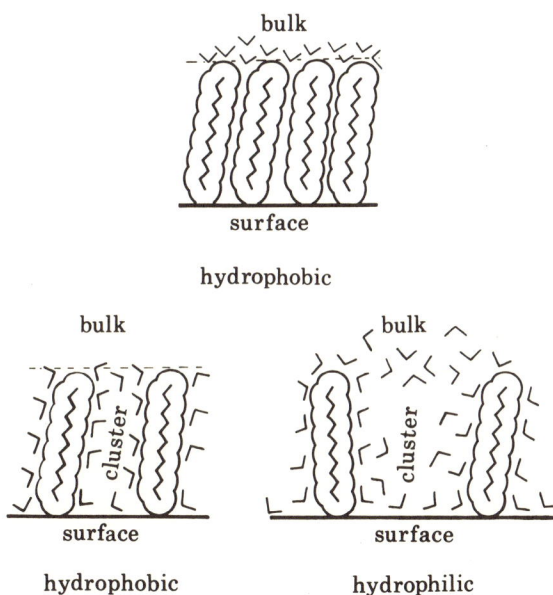

FIGURE 6. Surface hydrophobicity and hydrophilicity.

materials. Polar molecules between alkyl chains on the surface of clay minerals indeed associate to clusters (Lagaly & Witter 1982). The cluster formation of water is even enhanced by influence of the silicate surface. (The uncharged silicate surface is in principle hydrophobic. The hydrophilicity of the clay surface originates from the surface charges and the corresponding gegen ions.) A model was deduced which may explain whether an alkyl chain primed surface is hydrophobic or hydrophilic (figure 6, Lagaly *et al.* 1983). Hydrophobicity does not require the chains to be close-packed, but their distance should not exceed a critical value. Under this condition almost all water molecules form clusters between the chains. A discontinuity may be created to the bulk water phase. If the chains are primed more distantly from each other, the discontinuity disappears because the bulk water phase and disordered water molecules penetrate between the water clusters around the chains and the discontinuity disappears.

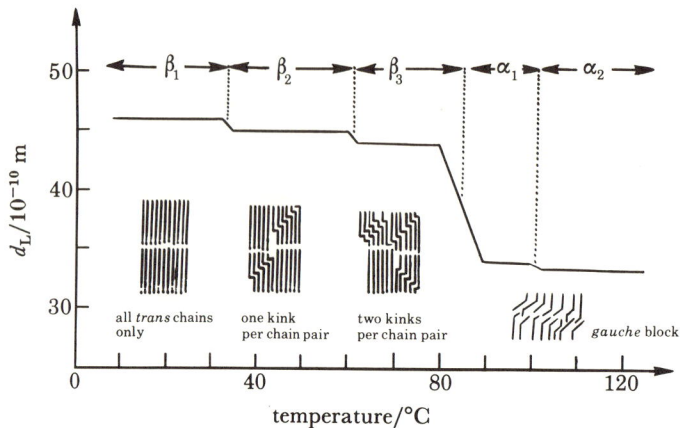

FIGURE 7. Basal spacing changes of alkylammonium Alkanol complexes (schematic) and the corresponding kink block and *gauche* block structures.

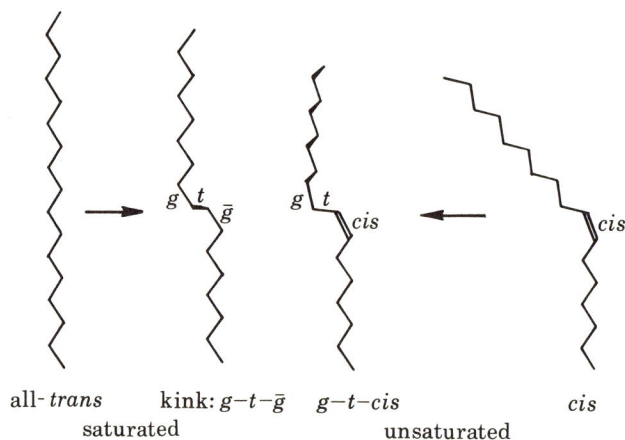

FIGURE 8. Formation of kinks (*gauche*(+)–*trans*–*gauche*(−) conformation) in saturated chains and kink-like *gauche*–*trans*–*cis* conformations in unsaturated chains with *cis*-double bonds.

The liquid clusters on the alkyl chain bearing surface are sensitively influenced by structure-breaking and structure-making ions (figure 4). The structure-breaking power of ions can increase (for example, tetradecylammonium beidellite in water) or decrease the basal spacing (tetradecylammonium vermiculite in DMSO, tetradecylammonium beidellite in ethanol, Lagaly *et al.* 1983). Consequently, the viscosity of organic dispersions of alkylammonium montmorillonites can be strongly influenced by inorganic salts (Sander & Lagaly 1983).

Technical uses

For technical uses organic-activated bentonites are usually prepared from diluted dispersions of sodium bentonite ('alkaline activated bentonite') by addition of alkylammonium salts, for instance dialkyldimethylammonium salts. Depending on the application the salts can also be incorporated in bentonite pastes.

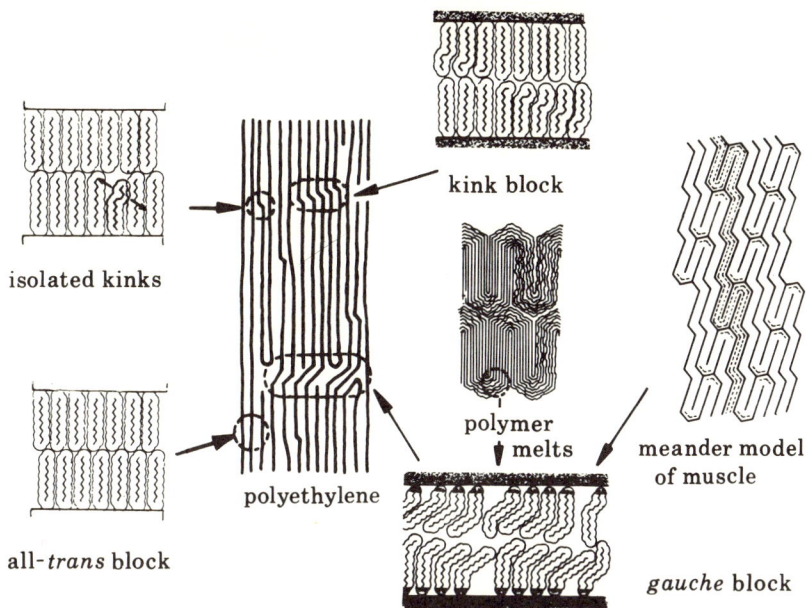

FIGURE 9. Structural elements of crystalline and melted polymers related to the alkylammonium Alkanol silicate complexes.

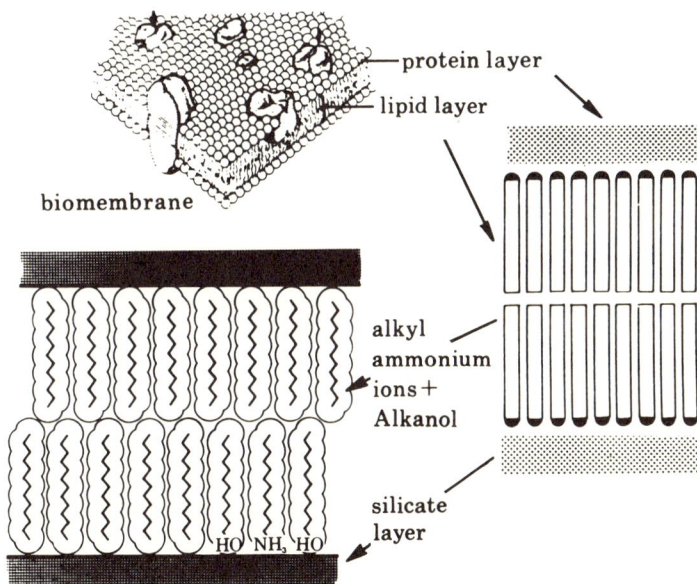

FIGURE 10. Alkylammonium Alkanol silicate complexes as models of the chain aggregation in biomembrane lipid layers.

Hydrophobic bentonites are used purely as thickeners or to produce or improve thixotropic properties (for instance of paints and waxes). Other uses are in the field of gelling agents for lubricants, as binders for water-free foundry sands and as additives in tar, asphalts and bituminous emulsions or in mastics, lutes, putties, waxes, ointments and cosmetic preparation. As fillers and strengthening agents in plastics polymeric compounds are preferred over monomeric surfactants (Lagaly 1984).

[107]

Alkylammonium Alkanol complexes

The study of interlamellar bimolecular films of alkylammonium ions and long chain alcohol molecules contributed to the present knowledge on conformational changes of lamellar-aggregated alkyl chains. The films are easily prepared by adding long chain alcohols to the alkylammonium derivatives. They undergo a series of phase transitions with rising temperature (Lagaly 1981 b). The phase changes are recognized by the stepwise decrease of the basal spacing (figure 7) and are caused by the cooperative formation of kinks in the alkyl chains. The basic reaction is the rotational isomerization of *trans* C–C bonds into *gauche* C–C bonds (figure 8).

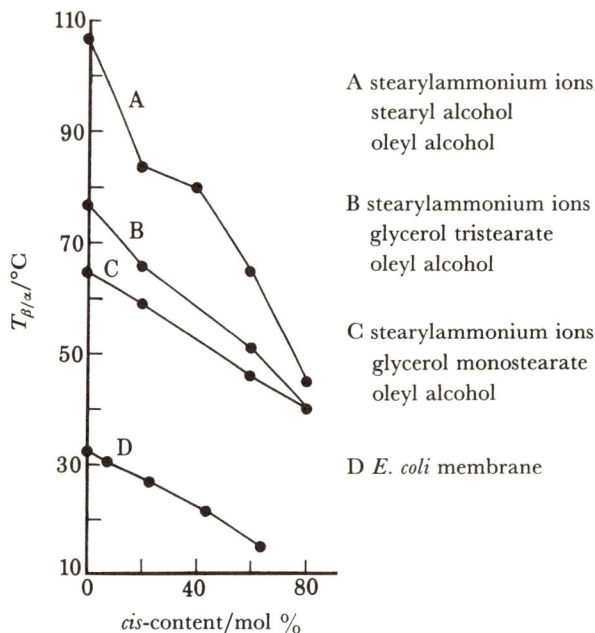

A stearylammonium ions
    stearyl alcohol
    oleyl alcohol

B stearylammonium ions
    glycerol tristearate
    oleyl alcohol

C stearylammonium ions
    glycerol monostearate
    oleyl alcohol

D *E. coli* membrane

FIGURE 11. Transition temperatures of interlamellar bimolecular films (β–α transition) and of *Escherichia coli* membranes (Overath *et al.* 1971) as a function of unsaturation.

The sequence of kink blocks finally transforms to a sequence of *gauche* blocks. The corresponding changes of entropy and enthalpy make evident that the phase transitions in lamellar aggregations of alkyl chains are sensitively related to the interchain interaction parameters (Lagaly 1976, 1981 a; Pechhold *et al.* 1976; Baur 1975).

Figures 9 and 10 illustrate how the bimolecular alkylammonium–alcohol films are modelling the chain aggregation in polymers and in the lipid layer of PLP-biomembranes.

In partially crystalline polymers kinks and kink blocks occur as defects. Gauche blocks are elements of the meander structure postulated for amorphous and melted polymers and the filaments in muscles (Pechhold & Grossmann 1979; Pechhold *et al.* 1973).

Certainly, alkylammonium–alcohol silicate complexes are not models of biomembranes but some properties of the chain aggregation in the lipid layer can be studied. One of the problems rencountered in the studies of PLP-membranes was the stuctural effect of the high amounts of unsaturated alkyl chains with *cis*-configuration. It could be shown (Lagaly *et al.* 1977) that the *cis*-compounds adopt kink-like *cis-trans-gauche*-conformations (figure 8) and in this way are incorporated in the bimolecular film. The temperature of transition $T_{\beta/\alpha}$ from kink blocks to *gauche* blocks decreases with increasing content of *cis*-chains (figure 11). The decrease is similar to that occurring in *Escherichia coli* membranes and similarity increases when stearyl alcohol is replaced by glycerol monostearate.

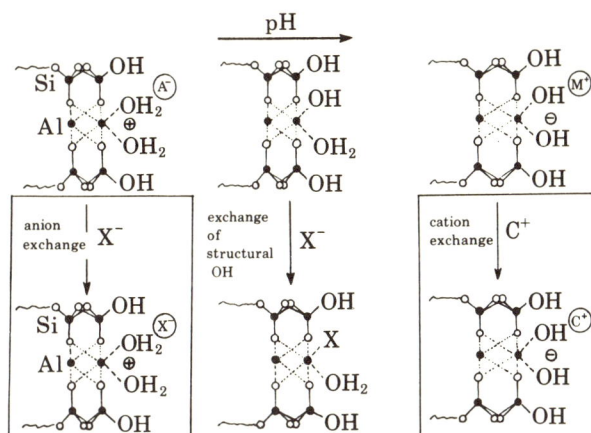

FIGURE 12. pH-dependent reactions on the edge surfaces.

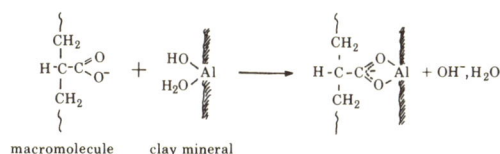

binding carboxylate groups
on the edge surface

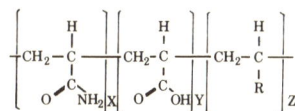

hydrolysed polyacrylamide

FIGURE 13. Binding of polyacrylamide by exchange of structural OH-groups on the edge surfaces (Siffert & Espinasse 1980).

A hypothetical model for information transfer within the lipid layer by cooperative phase transitions was suggested by Lagaly *et al.* (1977). It seems that *gauche*-conformations play a role as orientation defects on the protein–lipid contacts within the lipid layer and on splay deformations in ripple membrane structures.

INTERACTIONS WITH EDGE SURFACES

The interaction of organic compounds with the edge surface is presently not understood in full detail. The difficulties arise from the pH-dependent variation of the surface structure (figure 12). The charge of the surface changes from positive to negative within the pH range 4–5 (Rand *et al.* 1980) and consequently anion exchange changes to cation exchange.

Cationic and anionic surfactants are thus adsorbed in different ranges of pH (Hower 1970).

The corresponding changes of the zeta-potential and the type of particle–particle interactions (peptization, coagulation, sediment volumes) of kaolinite were studied by Welzen *et al.* (1981).

Besides the binding as gegen ions organic compounds can also be bound by exchanging structural OH-groups (Siffert & Espinasse 1980). The exchange of structural OH-groups likely becomes effective near the point of zero charge (p.z.c.), that is, when the charge density of the surface is low. The expected pH-dependency with adsorption maxima around the p.z.c. were indeed observed for the adsorption of the nucleotide ATP on smectites (Herrmann & Lagaly 1984).

Exchange of structural OH-groups by chelated carboxylate groups seems to be the main mechanism for binding polyanions like partially hydrolysed polyacrylamide on the edge surface (figure 13, Espinasse & Siffert 1979). Polyanions are widely used as flocculation agents and are more effective than polycations and neutral macromolecules.

The interaction of clays with polymers shows very special features and is discussed in a separate paper (Lagaly 1984).

REFERENCES

Adams, J. M. 1978*a* *Acta Crystallogr.* **35**, 1984–1988.
Adams, J. M. 1978*b* *Clays Clay Miner.* **26**, 169–172.
Adams, J. M. & Breen, Ch. 1982 *J. Colloid Interface Sci.* **89**, 272–289.
Adams, J. M., Reid, I. P., Thomas, J. M. & Walters, M. J. 1976 *Clays Clay Miner.* **24**, 267–269.
Adams, J. M. & Waltl, G. 1980 *Clays Clay Miner.* **28**, 130–134.
Annabi-Bergaya, F., Cruz, M. I., Gatineau, L. & Fripiat, J. J. 1981 *Clay Miner.* **16**, 115–122.
Barrer, R. M. & Hampton, M. G. 1957 *Trans. Faraday Soc.* **53**, 1462–1475.
Barrer, R. M. & MacLeod, D. M. 1955 *Trans. Faraday Soc.* **51**, 1290–1300.
Barrer, R. M. & Millington, A. D. 1967 *J. Colloid Interface Sci.* **25**, 359–372.
Baur, H. 1975 *Prog. Colloid Polymer. Sci.* **58**, 1–18.
Beneke, K. & Lagaly, G. 1982 *Clay Miner.* **17**, 175–183.
Berkheiser, V. & Mortland, M. M. 1975 *Clays Clay Miner.* **23**, 404–410.
Bissada, K. K., Johns, W. D. & Cheng, F. S. 1967 *Clay Miner.* **7**, 155–166.
Brindley, G. W. 1965 *Clay Miner.* **6**, 91–96.
Brindley, G. W. 1966 *Clay Miner.* **6**, 237–259.
Brindley, G. W. & Moll, W. F. 1965 *Am. Miner.* **50**, 1355–1370.
Brindley, G. W. & Ray, S. 1964 *Am. Miner.* **49**, 106–115.
Brindley, G. W. & Semples, R. E. 1977 *Clay Miner.* **12**, 229–237.
Cady, S. & Pinnavaia, T. J. 1978 *Inorg. Chem.* **17**, 1501–1507.
Chassin, P. 1976 *Clay Miner.* **11**, 23–29.
Damme, van H., Crespin, M., Obrecht, F., Cruz, M. I. & Fripiat, J. J. 1978 *J. Colloid Interface Sci.* **66**, 43–54.
Dékány, I., Nagy, L. G. & Schay, G. 1978 *J. Colloid Interface Sci.* **66**, 197–199.
Dékány, I., Szántó, F., Nagy, L. G. & Fóti, G. 1975 *J. Colloid Interface Sci.* **50**, 265–271.
Dowdy, R. H. & Mortland, M. M. 1967 *Clays Clay Miner. 15th natn. Conf.* pp. 259–271.
Espinasse, P. & Siffert, B. 1979 *Clays Clay Miner.* **27**, 279–284.
Farmer, V. C. & Mortland, M. M. 1966 *J. chem. Soc.* A, 344–351.
Farzaneh, F. & Pinnavaia, Th. J. 1983 *Inorg. Chem.* **22**, 2216–2220.
Feldkamp, J. R. & White, J. L. 1979 *J. Colloid Interface Sci.* **69**, 97–106.
Fenoll, H.-A. P. & Weiss, A. 1969 *Quimica, Riode. J.* **65**, 769–790.
Fernandez-Gonzales, M., Weiss, A. & Lagaly, G. 1976 *Keram. Z.* **28**, 55–58.
Fripiat, J. J., Cloos, P., Calicis, B. & Makay, K. 1966 *Proc. int. Clay Conf. Jerusalem 1966*, vol. 1, pp. 223–245. Jerusalem: Israel University Press.
Heller, L. & Yariv, S. 1970 *Israel J. Chem.* **8**, 391–397; 935–945.
Herrmann, H. & Lagaly, G. 1984 *Proc. European Clay Conf. Prague 1983.* (In the press.)
Hofmann, U. 1962 *Angew. Chem.* **74**, 397–442.
Hower, W. F. 1970 *Clays Clay Miner.* **18**, 97–105.
Jackson, M. L. & Abdel-Kader, F. H. 1978 *Clays Clay Miner.* **26**, 81–87.
Karickhoff, S. W. & Bailey, G. W. 1976 *Clays Clay Miner.* **24**, 170–176.
Keller, W. D. & Haenni, R. P. 1978 *Clays Clay Miner.* **26**, 384–396.
Lagaly, G. 1976 *Angew. Chem. int. Ed. Engl.* **15**, 575–586.

Lagaly, G. 1981a *Clay Miner.* **16**, 1–24.
Lagaly, G. 1981b *Naturwissenschaften* **68**, 82–88.
Lagaly, G. 1984 (In preparation.)
Lagaly, G., Weiss, A. & Stuke, E. 1977 *Biochem. Biophys. Acta* **470**, 331–341.
Lagaly, G. & Witter, R. 1982 *Ber. BunsenGes. phys. Chem.* **86**, 74–80.
Lagaly, G., Witter, R. & Sander, H. 1983 In *Adsorption from solution* (ed. R. H. Ottewill, C. H. Rochester, A. L. Smith), pp. 65–77. London: Academic Press.
Lailach, G. E. & Brindley, G. W. 1969 *Clays Clay Miner.* **17**, 95–100.
Lailach, G. E., Thompson, T. D. & Brindley, G. W. 1968 *Clays Clay Miner.* **16**, 285–293.
Laura, R. D. & Cloos, P. 1972 *Clays Clay Miner.* **20**, 259–270.
Laura, R. D. & Cloos, P. 1975 *Clays Clay Miner.* **23**, 61–69; 343–348; 417–423.
McAtee, J. L. & Robbins, R. C. 1980 *Clays Clay Miner.* **28**, 61–64.
Mortland, M. M. 1970 *Adv. Agron.* **22**, 75–117.
Parfitt, R. L. & Mortland, M. M. 1968 *Soil Sci. Soc. Am. Proc.* **32**, 355–363.
Pechhold, W., von Soden, R. & Kimmich, R. 1973 *Kolloid Z. Z. Polymere Sci.* **251**, 829–842.
Pechhold, W., Liska, E., Grossmann, H. P. & Hägele, P. C. 1976 *J. pure appl. Chem.* **46**, 127–134.
Pechhold, W. R. & Grossmann, H. P. 1979 *Discuss. Faraday Soc.* **68**, 58–77.
Pfirrmann, G., Lagaly, G. & Weiss, A. 1973 *Clays Clay Miner.* **21**, 239–247.
Pinnavaia, J. T. 1983 *Science, N.Y.* **220**, 365–371.
Pinnavaia, Th. J. & Mortland, M. M. 1971 *J. phys. Chem.* **75**, 3957–3962.
Pinnavaia, Th. J., Hall, P. L., Cady, Sh. S. & Mortland, M. M. 1974 *J. phys. Chem.* **78**, 994–999.
Pleysier, J. & Cremers, A. 1975 *J. chem. Soc. Farad. Trans.* **71**, 256–264.
Poyato-Ferrera, J., Becker, H. O. & Weiss, A. 1977 *Proc. 3rd European Clay Conf. Oslo 1977*, pp. 148–149.
Rand, B., Pekené, E., Goodwin, J. W. & Smith, R. W. 1980 *J. chem. Soc. Farad. Trans.* I **76**, 225-235.
Range, K. J., Range, A. & Weiss, A. 1970 *Proc. int. Clay Conf. Tokyo 1969*, vol. 1, pp. 3–13. Jerusalem: Israel University Press.
Raupach, M. & Janik, L. J. 1976 *Clays Clay Miner.* **24**, 127–133.
Rausell-Colom, J. A. & Fornes, V. 1974 *Am. Miner.* **59**, 790–798.
Rausell-Colom, J. A. & Salvador, P. S. 1971 *Clay Miner.* **9**, 139–149; 193–208.
Reichenbach, H. Graf von & Rich, C. I. 1968 *Trans. 9th int. Congr. Soil Sci.* **1**, 709–719.
Roloff, G. & Weiss, A. 1966 *Proc. int. Clay Conf. Jerusalem 1966*, vol. 1, pp. 263–275. Jerusalem: Israel University Press.
Rupert, J. P. 1973 *J. phys. Chem.* **77**, 784–790.
Rupprecht, H., Stanislaus, F. & Lagaly, G. 1975 *Colloid Polymer. Sci.* **253**, 773–780.
Samii, A. M. 1981 Thesis, University of Kiel.
Sander, H. & Lagaly, G. 1983 *Keram. Z.* **35**, 584–587.
Serratosa, J. M. 1968 *Am. Miner.* **53**, 1244–1251.
Seto, H., Cruz, M. I. & Fripiat, J. J. 1978a *Am. Miner.* **63**, 572–583.
Seto, H., Cruz-Cumplido, M. I. & Fripiat, J. J. 1978b *Clay Miner.* **13**, 309–323.
Siffert, B. & Espinasse, P. 1980 *Clays Clay Miner.* **28**, 381–387.
Siffert, B. & Kessaissia, S. 1978 *Clay Miner.* **13**, 255–270.
Slabaugh, W. H. & Hanson, D. B. 1969 *J. Colloid Interface Sci.* **29**, 460–463.
Slade, P. G., Raupach, M. & Emerson, W. W. 1978 *Clays Clay Miner.* **26**, 125–134.
Slade, P. G., Telleria, M. I. & Radoslowich, E. W. 1976 *Clays Clay Miner.* **24**, 134–141.
Stul, M. S. & Uytterhoeven, J. B. 1975 *J. chem. Soc. Farad. Trans.* I, **6**, 1396–1401.
Stul, M., van Leemput, L., Rutsaert, M. & Uytterhoeven, J. B. 1983a *J. Colloid Interface Sci.* **92**, 222–231.
Stul, M., van Leemput, L., Leplat, L. & Uytterhoeven, J. B. 1983b *J. Colloid Interface Sci.* **94**, 154–165.
Taramasso, M., Lagaly, G. & Weiss, A. 1971 *Kolloid Z. Z. Polymere* **245**, 508–518.
Taylor, M. F., Mortland, M. M. & Pinnavaia, T. J. 1978 *Clays Clay Miner.* **26**, 318–326.
Theng, B. K. G. 1974 *The chemistry of clay–organic reactions.* London: Adam Hilger.
Thompson, T. D. & Brindley, G. W. 1969 *Am. Miner.* **54**, 858–868.
Vansant, E. F. & Uytterhoeven, J. B. 1973 *Clay Miner.* **10**, 61–69.
Wada, K. 1962 *Am. Miner.* **46**, 78–91.
Weber, J. B. & Weed, S. B. 1974 In *Pesticides in soil and water* (ed. W. D. Guenzi), p. 223. Madison, Wisconsin: Soil Science Society of America.
Weiss, A. 1962 *Angew. Chem.* **73**, 736–737.
Weiss, A. 1963a *Angew. Chem.* **75**, 755–762.
Weiss, A. 1963b *Angew. Chem.* **75**, 113–122.
Weiss, A., Becker, A. O., Orth, H., Mai, G., Lechner, H. & Range, K. J. 1970 *Proc. int. Clay Conf. Tokyo 1969*, vol. 2, pp. 180–184. Jerusalem: Israel University Press.
Weiss, A., Choy, J. H., Meyer, H. & Becker, H. O. 1981 *Proc. int. Clay Conf. Bologna, Pavia 1981, Abstracts*, p. 331.
Weiss, A. & Hofmann, U. 1951 *Z. Naturf.* **6b**, 405–409.

Weiss, A., Mehler, A. & Hofmann, U. 1956 *Z. Naturf.* **11***b*, 435–438.
Weiss, A. & Orth, H. 1973 *Z. Naturf.* **28***b*, 252–254.
Weiss, A. & Range, K. J. 1970 *Proc. int. Clay Conf. Tokyo 1969*, vol. 2, pp. 185–186. Jerusalem: Israel University Press.
Weiss, A. & Russow, J. 1963 *Proc. int. Clay Conf. Stockholm 1963*, pp. 69–74. London: Pergamon Press.
Weiss, A., Ruthard, R. & Orth, H. 1973 *Z. Naturf.* **28***b*, 446–449.
Weiss, A., Thielepape, W. & Orth, H. 1966 *Proc. int. Clay Conf. Jerusalem*, vol. 1, pp. 277–293. Jerusalem: Israel University Press.
Welzen, J. T. A. M., Stein, H. N., Stevels, J. M. & Siskens, C. A. M. 1981 *J. Colloid Interface Sci.* **81**, 455–467.
White, J. L. 1976 *Proc. int. Clay Conf. Mexico 1975*, pp. 391–398. Wilmette, U.S.A.: Applied Publishing.
Wilson, M. J. & Tait, J. M. 1977 *Clay Miner.* **12**, 59–66.
Yamagashi, A. 1983 *J. chem. Soc. Dalton Trans.* pp. 679–681.
Yamanaka, S., Kanamura, F. & Koizumi, M. 1975 *Phys. Chem.* **79**, 1285–1288.
Yamamoto, D., Wakasugi, N. & Ono, K. 1970 *Proc. int. Clay Conf. Tokyo 1969*, vol. 1, pp. 735–739. See also: vol. 2, pp. 187–188. Jerusalem: Israel University Press.
Yariv, S. & Heller-Kallai, L. 1975 *Clay Miner.* **10**, 479–481.
Yariv, S. & Shoval, S. 1982 *Israel J. Chem.* **22**, 259–265.

Discussion

R. M. BARRER, F.R.S. (*Chemistry Department, Imperial College, London SW 7, U.K.*). In connection with the uptake of polymers by smectites does Professor Lagaly know whether it has proved possible first to disperse the lamellae of the mineral in water and then, by quickly adding a polymer solution, to catch the lamellae in their dispersed state and so to augment both rate and amount of uptake?

G. LAGALY. This is indeed possible and has recently been proved by an interesting paper of N. Larsson and B. Siffert (*J. Colloid Interface Sci.* **93**, 424–431 (1983)). Lysoszyme molecules are adsorbed by this disaggregation–reaggregation mechanism without any diffusion in lateral direction. So, a complete saturation is achieved which very likely would not be attained by lateral diffusion.

K. GOULDING (*Soils and Plant Nutrition Department, Rothamsted Experimental Station, Harpenden, Hertfordshire, AL5 2JQ, U.K.*). Are the intercalated charged organic molecules easily replaced by inorganic cations?

G. LAGALY. They cannot be replaced. Even uncharged molecules are often displaced with difficulty, particularly if they are polymeric molecules. There are very few studies concerning the desorption of intercalated organic materials.

Phil. Trans. R. Soc. Lond. A **311**, 333–352 (1984)   [333]

Printed in Great Britain

Sorption and molecular sieve properties of clays and their importance as catalysts

By R. M. Barrer, F.R.S.

Chemistry Department, Imperial College, London SW7 2AY, U.K.

Kandites, smectites and vermiculites all form interlamellar inclusion complexes. A comparison has been made between sorption in zeolites, clay minerals and clathrates. In smectites and vermiculites sorption isotherms reflect in contour and behaviour three situations. Kandites show only the first of these, in which the water-free crystals are used and the interlamellar cations, if any, are small inorganic ions (for example Na⁺, Ca²⁺). The guest molecules are polar and penetration of the host occurs with swelling after a threshold pressure or activity is reached. The isotherms may show one or more steps and hysteresis between sorption and desorption cycles is normal. In the second situation all the interlamellar space is filled by long chain organic cations. If a potential guest is imbibed there is further expansion of the host and isotherms sometimes resemble in contour those obtained when benzene is imbibed by rubber. Cohesive energy densities of the interlamellar regions relative to those of the potential guest species strongly influence the interlamellar uptake and the selectivity. In the third situation the clay lamellae are propped apart permanently by cations that do not fill all interlamellar space and are chosen to give vertical free distances between sheets and horizontal free distances between adjacent cations which are of molecular dimensions. The resultant porous crystals behave like zeolite molecular sieves both in the contours of isotherms for the interlamellar part of the sorption and in the molecule sieving effects that have been obtained.

In addition suitably modified or natural clay minerals can catalyse many kinds of reaction. These include cracking, isomerization, dimerization, and oligomerization, redox reactions and hydrogen transfer, hydration and dehydration, esterification, lactonization, etherification, conversion of primary to secondary amines and reaction between guest and interlamellar organic cations. Some comments on and examples of certain of these processes are given.

1. Introduction

Interlamellar inclusion of guest species is a notable characteristic of 1:1 and 2:1 sheet silicates (kandites, smectites and vermiculites). Inclusion is facilitated because the forces binding the sheets to one another are not chemical but physical, that is, dispersion, close range repulsion, hydrogen bonding, and especially for smectites, vermiculites and micas, electrostatic (Jenkins & Hartman 1982) interactions. For kandites the ideal composition is $Al_2Si_2O_5(OH)_4$. For the 2:1 sheet silicates the negative charge on the lamellae varies between zero for an ideal pyrophyllite, $Al_2Si_4O_{10}(OH)_2$, to a maximum of about 500 meq 100 g⁻¹ for margarite micas. The charge densities of a number of 2:1 sheet silicates have been tabulated by Weiss (1958a) as the area per unit charge. Some of these values are given in table 1. Swelling in water can, in the Na-clay minerals of charge density corresponding with areas of 47 to at least 100 Å²† per unit charge, result in complete dispersion of smectite lamellae. Swelling is zero for the most highly charged mica lamellae (margarite) and also for uncharged lamellae of pyrophyllite and talc.

† 1 Å = 0.1 nm = 10⁻¹⁰ m.

Table 1 also shows that for a given lamellar charge density the charge on the interlamellar cation can modify the swelling behaviour. Thus the Ca-forms of the clay minerals swell in water to a different extent to the Na-forms at the same charge density. When crystals that swell in water are exposed to water vapour at various relative humidities (r.h.) and the $d(001)$ spacings at equilibrium are plotted against r.h. there tend to be plateaux at uptakes which have been interpreted as representing completion of one- or two-layer hydrates between each pair of lamellae (figure 1) (Gillery 1959). In the same connection the isotherms tend to show steps, as illustrated for water in Mg-vermiculite (figure 2) (van Olphen 1969). As with the 2:1 sheet silicate pyrophyllite, water does not intercalate between the 1:1 sheets in kandites, the lamellae of which ideally carry no negative charge so that there would be no interlamellar cations. Halloysite, $Al_2(Si_2O_5)(OH)_4 \cdot 2H_2O$, in which there is a single layer of water molecules between each pair of lamellae, is an exception. However, when this water is removed by heat and evacuation the process is irreversible: no water can subsequently be intercalated.

TABLE 1. CHARGE DENSITY AND SWELLING OF CLAY MINERALS IN DISTILLED WATER (WEISS 1958a)

mineral	area per unit charge Å²	swelling for interlamellar cations Å Na⁺	Ca²⁺
margarite	12	0	0
muscovite	24	1.9	2.8
biotite	24	1.9	2.8
lepidolite	24	1.9	2.8
seladonite	27	2.4	2.8
glauconite	31	3.8	2.8
trioctahedral illite	36	5.1	4.3
vermiculite	37	5.1	4.3
beidellite I	41	5.4	4.9
nontronite	46	∞	9.2
beidellite II	57	∞	9.2
montmorillonite I	60	∞	9.2
montmorillonite II	75	∞	9.6
hectorite	100	∞	10.6
pyrophyllite	∞	0	0
talc	∞	0	0

2. INCLUSION BY ZEOLITES, CLAY MINERALS AND CLATHRATES: A COMPARISON

Inclusion by zeolites, clay minerals and clathrates can be of technical importance, and a comparison of some typical features is of particular interest (table 2). In table 2 under the heading of clay minerals, A and B denote different types of sorbent. Thus type A sorbents can intercalate guest molecules only when they expand as uptake proceeds. Type B sorbents on the other hand have been expanded and made permanently porous by exchange using cations of chosen shapes and sizes as props or pillars (§ 4).

Isotherms for interlamellar sorption in clay mineral sorbents of type A often show steps, as in figure 2, and as often happens in clathrates (Allison & Barrer 1968). On the other hand isotherms in type B sorbents are continuous (§ 4). Interlamellar sorption can be regarded as a type of solution, just as for sorption in zeolites or clathration. At an equilibrium pressure, p, of

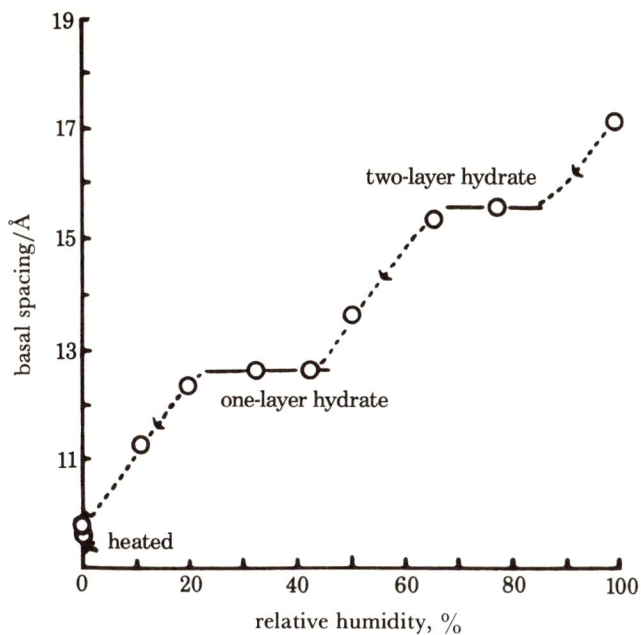

FIGURE 1. Relation between $d(001)$ and percentage relative humidity for water in a natural Na-montmorillonite (Gillery 1959).

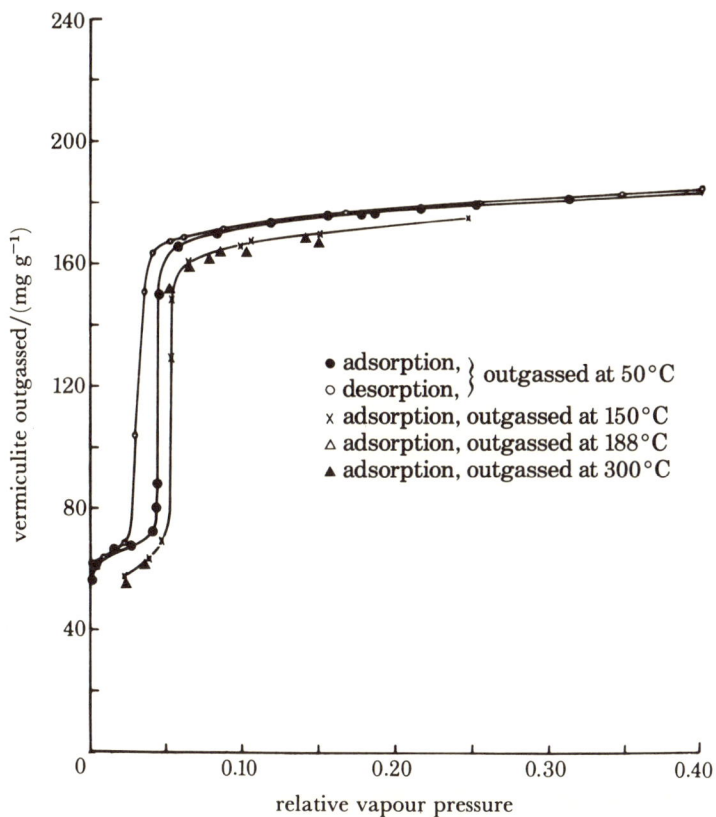

FIGURE 2. The stepwise sorption isotherm of water at 50 °C in vermiculite outgassed at various temperatures (van Olphen 1969).

[115]

the guest the lowering in chemical potential of the host crystal, $\Delta\mu$, owing to uptake of the guest is (Barrer 1960)

$$\Delta\mu = (\mu - \mu^0) = \int_0^p \frac{V\,dp}{(1-X_G)} - RT \int_0^p \frac{X_G\,dp}{(1-X_G)p} \tag{1}$$

where μ and μ^0 are chemical potentials of the lattice-forming units of the host crystal containing the guest molecules and when empty respectively; X_G is the mole fraction of the guest species in the host–guest solution; and V is the volume of one mixed mole of the solution. The gaseous phase of the guest is assumed ideal. For pressures normally involved the first integral is small. The second integral can be evaluated by graphical integration once the isotherm is known.

TABLE 2. COMPARISON OF ASPECTS OF INCLUSION OF GUEST MOLECULES IN
ZEOLITES, CLAY MINERALS AND CLATHRATES

zeolites	clay mineral sorbents	clathrates
porosity permanent and stable; pores and channels of molecular dimensions	(A) porosity absent until $d(001)$ expanded by guest (B) permanent porosity created before sorption	porosity created by recrystallization of host in presence of an adequate vapour pressure of guest
isotherms normally continuous and of Type I†	(A) isotherms often stepped (B) isotherms continuous and of Type I†	isotherms continuous and of Type I† over region of vapour pressure of guest in which clathration occurs; no uptake below this region
guest molecules distributed spatially according to numerous channel and cavity patterns	guest molecules in layers	guest molecules normally isolated in cavities or channels

guest molecules oriented in some examples; 'lock and key' relations between guest and host sometimes shown

guest removable with minimal unit cell change	(A) guest removable with decrease in $d(001)$ (B) guest removable with minimal change in $d(001)$	guest removal causes recrystallization of host to the compact non-porous original form
external area small compared with intracrystalline area	external surface can be more important than for zeolites	external area small; uptake inside crystals only
ion exchange can modify sorption often profoundly	ion exchange can modify sorption, often profoundly (but kandites and pyrophyllite have no inter-layer exchange capacity)	ion exchange properties are absent

† In the Brunauer classification (Brunauer 1944).

Equation (1) refers to host crystals already opened to accept the guest molecules without perceptible changes in lattice structure (as for zeolites and the permanently porous clay mineral sorbents of type B). In clathration, or when the clay minerals have first to expand to accommodate the guest molecules (sorbents of type A) one must add to $\Delta\mu$ an extra free energy term $\Delta\mu_{\alpha\beta}$ which represents the increase in chemical potential when the denser or α-phase of the host is changed to the empty porous or β-phase, free of the guest species. For inclusion of guest molecules at the pressure p the sum

$$\Delta\mu + \Delta\mu_{\alpha\beta}$$

[116]

must be negative. This means that p must be such that

$$-RT \int_0^p \frac{X_G \, dp}{(1-X_G)p} \quad \text{exceeds} \quad \Delta\mu_{\alpha\beta} + \int_0^p \frac{V \, dp}{(1-X_G)}.$$

There can therefore sometimes be a measurable threshold pressure below which clathration or inclusion in clay mineral sorbents of type A does not occur. This behaviour is shown for pyridine and water in Na-montmorillonite in figure 3 (Barrer & McLeod 1954). The isotherms represent the sum of adsorption on external surfaces plus inclusion between the clay lamellae. Because of this and because of imperfections in the small host crystals the steps are not sharp. The water isotherm indicates more than one stage and reflects the stages in expansion seen in figure 1. That is, after completion of an interlamellar monolayer of water, there is a further expansion to accommodate a second layer, followed by still further expansion to accommodate still thicker layers. Quantitative modelling of this behaviour provides an interesting challenge.

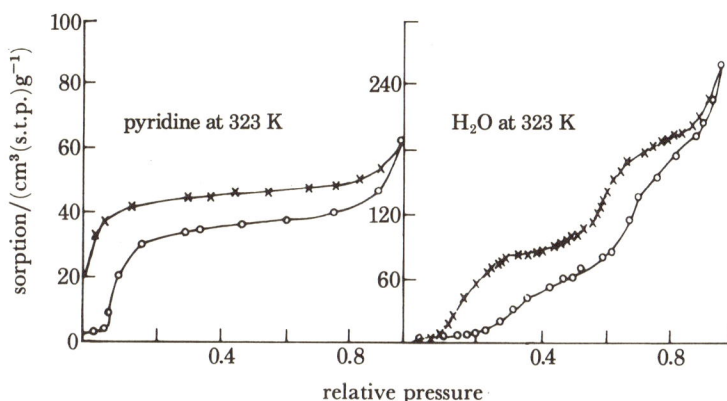

FIGURE 3. Sorption of pyridine (left) and water (right) in natural Na-rich montmorillonite, showing threshold pressure, isotherm steps and hysteresis (Barrer & McLeod 1954).

The onset of inclusion at a critical pressure sometimes observed for type A sorbents is further modified because the expanded host lattice has to nucleate from the parent phase at the edges of the clay lamellae. This introduces two new free energy terms: Δg_σ due to interfacial free energy; and Δg_s due to misfit and hence strain between the nucleus and the parent lattice. Δg_σ and Δg_s are both positive and so tend to delay the onset of interlamellar penetration beyond the true equilibrium threshold pressure. Conversely on desorption a Δg_σ and Δg_s again arise, this time delaying the nucleation of the collapsed guest-free phase of the host lattice when the vapour pressure of the guest falls below the equilibrium threshold value. As a result hysteresis can occur between sorption and desorption cycles (figure 3).

3. CLAY MINERAL SORBENTS OF TYPE A

As already noted, type A sorbents cannot include guest molecules without expansion of the $d(001)$ spacing. There are however two subdivisions of these sorbents which behave very differently.

Type A1. This group comprises smectites and vermiculites having exchangeable inorganic cations such as Li+, Na+, K+, Mg2+ or Ca2+ between the lamellae; and kandites which ideally are free of interlamellar cations.

Type A2. These sorbents comprise organo-clay minerals in which, by ion-exchange, the inorganic cations are replaced by alkylammonium ions with long chain alkyl groups sufficiently large to fill all the available interlamellar space. Even organo-mica of this type can be made by exchange of K^+, although the process occurs with extreme slowness (Weiss 1958*b*). In this organo-mica the area available per chain is about 24 $Å^2$ (table 1), which is about the cross-sectional area of a chain normal to the basal surfaces. As the charge density decreases (vermiculites and smectites) the area available per chain increases (table 1), chain orientation can be less steep (vermiculites) while in smectites the chains may often lie parallel with the basal surfaces, as evidenced by the $d(001)$ spacings. For all sorbents of type A2 however the chains must fill the interlamellar region.

TABLE 3. SOME ONE- AND TWO-LAYER COMPLEXES OF ORGANIC MOLECULES
IN Ca-MONTMORILLONITE (BRINDLEY & RAY 1958)

guest	$d(001)/Å$ one layer	$d(001)/Å$ two layer	one layer spacing $- 9.5$ Å	probable orientation of chain zigzag to basal surfaces
acetylacetone	13.0_0	17.1	3.5_0	‖
α-methoxyacetylacetone	13.0_0	16.9_5	3.5_0	‖
acetoaceticethylester	13.1_0	17.1_5	3.6_0	‖
nonanetrione-2,5,8	13.1_0	17.3_5	3.6_0	‖
hexanedione-2,5	13.0_5	16.8_5	3.5_5	‖
β, β'-oxydipropionitrile	13.1_5	15.7_5	3.6_5	‖
β-ethoxypropionitrile	13.2_5	—	3.7_5	‖ (?)
bis-(2-ethoxyethyl)-ether	13.4_0	16.7_5	3.9_0	⊥
bis-(2-methoxyethyl)-ether	13.2_5	17.0_0	3.7_5	⊥
ethyleneglycoldiglycidether	13.4_0	17.6_0	3.9_0	⊥
triethyleneglycol	13.3_0	17.3_5	3.8_0	⊥
diethyleneglycol	13.3_0	15.7_5	3.8_0	⊥
triethyleneglycoldiacetate	13.3_5	16.5_0	3.8_5	⊥
diethyleneglycoldiacetate	13.1_5	16.5_5	3.6_5	‖ (?)
hexanediol-1,6	13.5_5	17.4_5	4.0_5	⊥
pentanediol-1,5	13.6_5	17.4_5	4.1_5	⊥
2,3-hexadiynediol-1,6	13.0_5	16.1_0	3.5_5	‖

3.1. *Inclusion by sorbents of type A1*

Intercalation by well outgassed smectites and vermiculites of type A1 normally requires that the guest molecules should be polar, because of the high polarity of the sorbents. Some characteristics of this group of sorbents are already outlined in § 1 and 2. The $d(001)$ spacings for some organic guest species in Ca-montmorillonite (Brindley & Ray 1958) serve to illustrate other inclusion complexes (table 3). The spacings have been interpreted as owing to one or two monolayers of the organic molecules between the clay lamellae. The molecules all lie along the basal surfaces, but the zigzag of the chain can lie flat on the basal surfaces (giving monolayer thicknesses *ca.* 3.5_0 to 3.6_5 Å) or perpendicular to these surfaces (monolayer thicknesses *ca.* 3.7_5 to 4.1_5 Å).

Glycol- and glycerol-smectite complexes have double layers with $d(001)$ *ca.* 17.4 Å and *ca.* 17.7 Å respectively. Each also intercalates in vermiculites as layers one and sometimes two molecules thick. Nitriles, amines, diamines, glycine and its peptides and even proteins and other polymers are among the species which can be intercalated in sorbents of type A1.

Kaolinite forms intercalation complexes (Barrer 1978) involving the very polar or ionic species urea, formamide, hydrazine, acetamide and salts of organic acids (Li-, Na-, K-, Rb-Cs-, NH_4- and N_2H_5- acetates; K-, Rb- and Cs-propionates; and K-cyanoacetate). The guest often forms a monolayer ca. 2.9–3.8 Å thick. However, K-, Rb-, Cs- and NH_4-acetates and K-cyanoacetate give thicker layers from ca. 5.8–10 Å. Kaolin demonstrates other features of interest. Thus, as shown in figure 4 (Weiss et al. 1963) the kinetics of uptake of formamide have the sigmoid form expected for penetration only after nucleation of the expanded phase (§ 2).

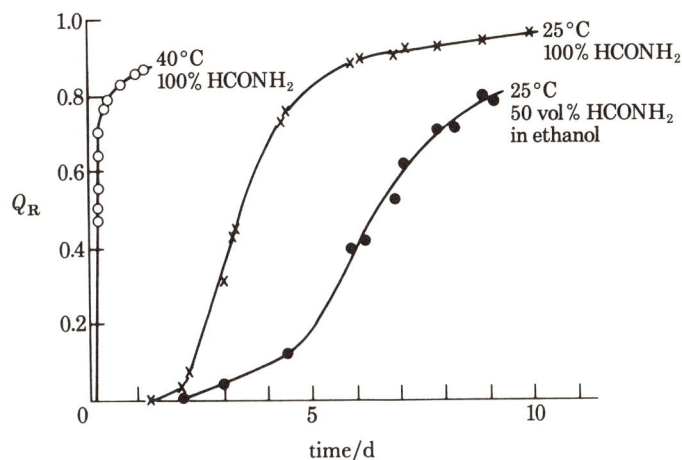

FIGURE 4. Degree of intercalation, Q_R, as a function of time and temperature for formamide in kaolinite. The figure illustrates the induction period for the nucleation of the expanded phase (Weiss 1963).

TABLE 4. $d(001)$ VALUES FOR COMPLEX FORMED WITH THE ASSISTANCE OF 24 % HYDRAZINE IN WATER (WEISS ET AL. 1963)

potential guest	$d(001)/Å$ in aqueous hydrazine solution of guest	after hydrazine solution replaced by aqueous solution of guest
K-acetate	10.4	14.0
Na-acetate	10.4	10.0_6
K-oxalate	10.4	10.2_8
K-glycollate	10.7	12.4_4
K-alaninate	10.7	12.4_5
K-lysinate	10.6	14.8_8
K-lactate	10.6	11.1_8
glycerine	10.4	10.5_2
n-octylamine	31.7_1	31.7_1
benzidine†	10.6	10.8

† Benzidine in 24% hydrazine in aqueous alcohol.

The uptake is also pH and temperature sensitive. NH_4-acetate for example was intercalated only in the pH range 7–10.5. Between 0 and 25 °C the rate of uptake increased rapidly but above 40 °C no intercalation occurred. If the 14.0 Å complex was first prepared at low temperature and was then heated while still in the acetate solution to 65 °C the c-repeat distance increased to 17.1_5 Å. The repeat distance in the parent kaolinite is 7.2 Å so that the interlamellar layer of NH_4-acetate has risen from 6.8 Å to 9.9_5 Å in thickness.

The number of compounds that can be intercalated in kaolin can be increased by including in the aqueous phase not only the potential guest but also a second species such as hydrazine

[119]

340 R. M. BARRER

which can intercalate directly and independently. Examples are given in table 4 in which complexes first form from solutions of various species in 24% aqueous hydrazine. When the hydrazine component is now removed the aqueous guests swelled the kaolinite as shown in the final column of table 4. The spacing *ca.* 10.7 Å is found for hydrazine acetate solutions in absence of other guests.

3.2. *Organo-clay sorbents of type A2*

As noted in § 3 in these sorbents the interlamellar space is fully occupied by long chain organic cations. For alkyl chains of carbon number up to about 10 the chains can lie along the basal surface but for still longer chains the thickness of the organic layer rises from the monolayer value of *ca.* 3.5 Å to about 7 or 8 Å, which could represent bi-layers of ions possibly with some excess interlamellar alkylammonium salt. In partially ion-exchanged N-montmorillonites there is clear evidence of randomly interstratified layers containing either the organic or the inorganic ion but not both, when the entering ions were $NH_3^+CH_3$ or $N(CH_3)_4^+$ (Barrer & Brummer 1963). This behaviour could be general for partial exchanges with alkylammonium ions.

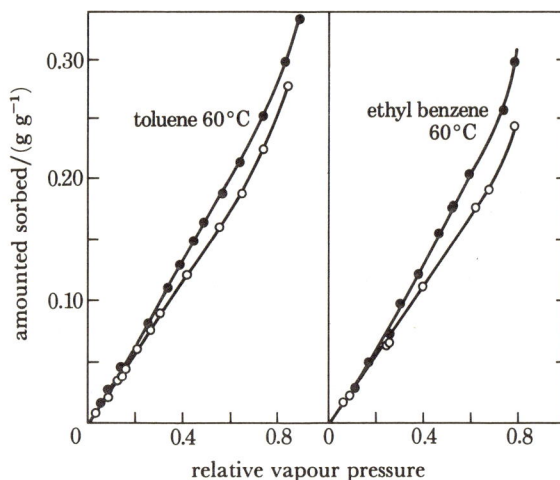

FIGURE 5. Isotherms for toluene and ethyl benzene obtained with Bentone 34 (dimethyldioctadecylammonium montmorillonite) (Barrer & Kelsey 1961 *b*). O, Adsorption; ●, desorption.

Whatever the packing and orientation of the alkyl chains in fully exchanged organo-clays these chains provide beds of interlamellar material of low cohesive energy density. The organo-clay sorbents of type A2 now imbibe not only polar but also non-polar guest species, by mixing of these species with the alkyl chains of the organic cations. Isotherms for toluene and ethyl benzene in dimethyldioctadecylammonium montmorillonite (Bentone 34) are shown in figure 5 (Barrer & Kelsey 1961 *b*), which represent total sorptions on external surfaces and between the lamellae. Where interlamellar uptake is large, as for toluene and ethylbenzene, the isotherm contours recall those when benzene and other hydrocarbons are imbibed by rubber (Gee 1942), and are greatly changed in shape from those in type A1 sorbents (figures 2 and 3). The isotherm contours support the view that sorption involves mixing of hydrocarbon and alkyl chains. Such mixing involves further swelling of the clay sorbent and also the extent of uptake at a given relative pressure, p/p_0, of the guest is a function of its cohesive energy density (c.e.d.). These features are illustrated in table 5. The uptake at a given relative pressure and temperature should

[120]

be a maximum for a c.e.d. which matches most nearly that of the interlamellar region as modified by the influence of the adjacent lamellae. The largest uptake was observed with pyridine. The very limited swelling for n-heptane and cyclohexane indicates that interlamellar uptake is slight, so that much of the observed total sorption is here on external sufaces of the sorbent. Interlamellar uptake is a very selective function of $\delta = \rho^{\frac{1}{2}}$, where ρ is the c.e.d. and is not closely related to molecular size and shape differences as in the case of molecule sieving by zeolites.

TABLE 5. CORRELATION OF SORPTION BY BENTONE 34 (ALCOHOL EXTRACTED) WITH COHESIVE ENERGY DENSITY (C.E.D., ρ) OF GUEST SPECIES (BARRER & KELSEY 1961a)

guest	$\delta = \rho^{\frac{1}{2}}$, at 25 °C $(cal\ddagger\ cm^{-3})^{\frac{1}{2}}$	T °C	sorption at T and $p/p_0 = 0.2$ mmol g^{-1}	sorption at T and $p/p_0 = 0.6$ mmol g^{-1}	swelling at T and $p/p_0 = 0.6$ A°
iso-butane	6.25	-30	0.06_2	0.21	—
n-butane	6.7	-30	0.07_7	0.22	—
iso-octane	6.85	45	0.11	0.31	—
n-heptane	7.45	45	0.11	0.34	0.4
cyclohexane	8.20	45	0.16	0.48	0.4
ethylbenzene	8.80	60	0.52	1.60	—
toluene	8.90	45	0.65	1.93	11.8
benzene	9.15	45	0.81	2.33	11.2
dioxane	10.0	60	0.92	2.07	—
pyridine	10.7	60	1.79	3.08	—
nitromethane†	12.6	45	1.38	2.54	—

† Purity not known.
‡ Cal = calorie. 1 calorie = 4.187 J.

The selectivity shown by Bentone 34 towards different potential guests has led to its successful use in a variety of chromatographic separations. These include separations of mixtures of ethylbenzene, p-xylene, o-xylene and m-xylene (van Rysselberge & van der Stricht 1962); of aromatic and aliphatic compounds (White 1957); of toluidines and cresols (White 1959); of benzene and its di- and trichloro derivates (Cowen & Hartwell 1961); and of o-, m- and p-isomers of terphenyl, phenoxyphenyl and nitrophenol (Kiselev et al. 1972).

4. SORBENTS OF TYPE B

In sorbents of type B the exchange ions expand the clay mineral but do not fill all the interlamellar space, so that there are pathways among the ions. The ions must be large enough to hold the clay lamellae apart by distances comparable with the dimensions of the guest. Provided lateral free distances between adjacent ions are also of molecular dimensions permanent interlamellar porosity is available, just as with zeolites, for selective uptake of guests on the basis of size and shape differences of the molecules of competing guest species. This behaviour was first envisaged and studied in the 1950s in the author's laboratory (Barrer & McLeod 1954, 1955; Barrer & Hampton 1957; Barrer & Reay 1957). The exchange ions introduced were $CH_3NH_3^+$, $(CH_3)_2NH_2^+$, $(CH_3)_3NH^+$, $(CH_3)_4N^+$ and $(C_2H_5)_4N^+$. This work was developed and extended using the above and other cations until the mid 1970s (Barrer & Perry 1961a, b; Barrer & Kelsey 1961a, b; Barrer & Brummer 1963; Barrer & Millington 1967; Barrer & Jones 1971; Craven 1976), and has recently been further extended in several laboratories

(Mortland & Berkheiser 1976; Shabtai *et al.* 1976; Vaughan & Lussier 1980; Ocelli *et al.* 1981; Pinnavaia *et al.* 1983). It has been shown that sensitive molecule sieving is possible with these new kinds of sieve (Craven 1976) in full parallel with zeolites. By changing the size and shape of the entering exchange ions, by selecting or making smectites and vermiculites of differing exchange capacities and by altering the charge of the entering exchange ions one may vary and control both lateral and vertical free distances in the resultant interlamellar cavities and channels. Access to these tailor-made cavities and channels involves diffusion of guests in two dimensions in each interlamellar region.

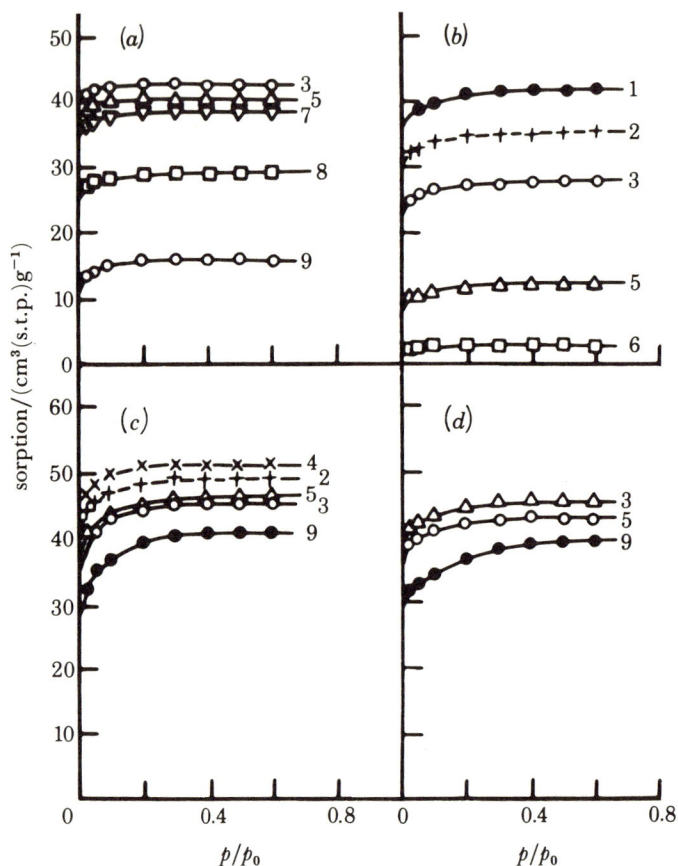

FIGURE 6. Estimated interlamellar sorptions of N_2 and Ar at 78 K (Barrer & Millington 1967). (a) N_2 in alkyldiammonium montmorillonites; (b) N_2 in alkylammonium montmorillonites; (c) N_2 in alkyldiammonium hectorites; and (d) Ar in alkyldiammonium hectorites. The numbers by the curves are the carbon numbers of the organic ions.

A sorption isotherm for these expanded, permanently porous clay minerals is a composite of that on external surfaces of the crystals and that between the clay sheets. The total isotherm can therefore be of Type I or Type II in Brunauer's classification. If the isotherm of N_2 at 78 K is measured on the parent non-porous Na-smectite and then on a series of its expanded porous forms in which the exchange ions are, for example, alkyl- or dialkylammonium ions and the isotherm for the non-porous Na-smectite is subtracted from those for the porous forms the difference could be expected to approximate to the interlamellar parts of the sorption. Such difference isotherms obtained with organo-smectites are shown in figure 6 (Barrer & Millington 1967) for N_2 and Ar at 78K. Just as with zeolites they are all of Type I. As the carbon number

of the exchange ion increases (indicated by the numbers on the curves) the saturation interlamellar uptake decreases. This is as expected for alkyl chains parallel with basal surfaces of
lamellae because more and more interlamellar free volume is occupied by the chains. Eventually, as discussed in § 3.2, permanent intracrystalline porosity becomes negligible and the
sorbents become fully of type A2.

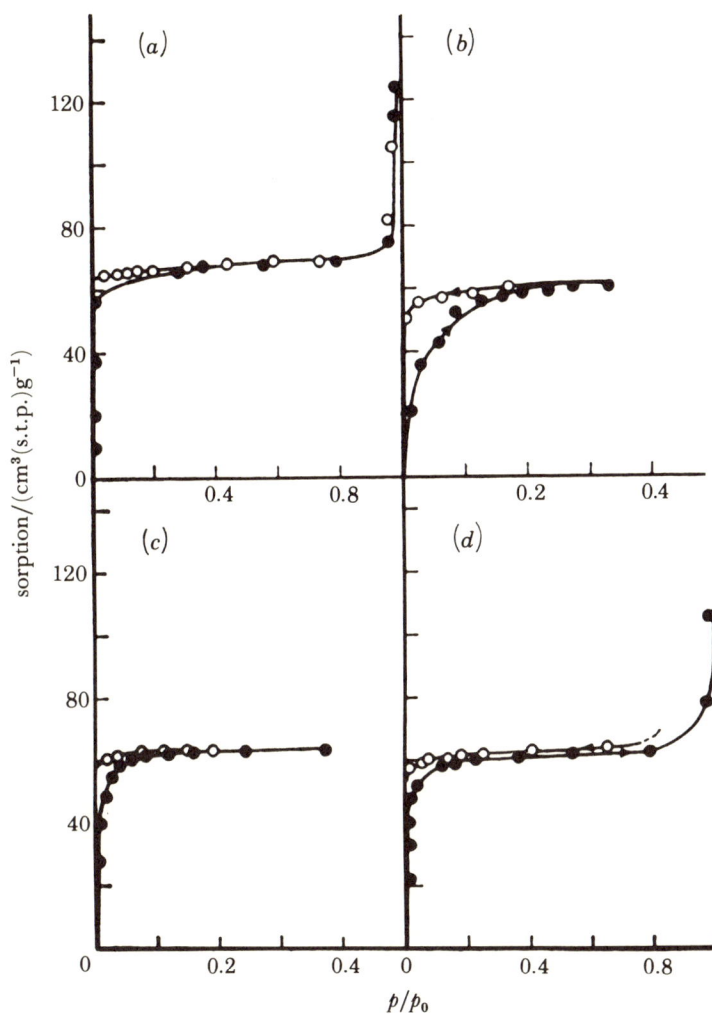

FIGURE 7. Isotherms of O_2, N_2 and Ar in en-FH90 (en = ethylenediammonium) (Barrer & Jones 1971). (a) O_2 at
    78 K, allowing 20 min per isotherm point; (b) N_2 at 78 K, allowing 15 min per point; (c) O_2 at 90 K, 20 min
    per point; and (d) Ar at 78 K, 30 min per point.

As an alternative to estimating intracrystalline sorption isotherms of smectite sorbents of
type B by the above difference method much larger crystals of fluorhectorites were synthesized
by sintering and melt procedures (Barrer & Jones 1971). For these as for zeolites sorption on
external surfaces becomes very small. Isotherms of O_2, N_2 and Ar at 78 or 90 K are shown in
figure 7 for ethylammonium fluorhectorite of exchange capacity 90 meq 100 g^{-1} (termed
en-FH90). Except at the highest relative pressures all are of Type I.

Examples of interlamellar saturation capacities for N_2 at 78 K are given in column 7 of

[123]

table 6 for some alkylammonium and alkyldiammonium smectites. The calculated interlamellar free areas were obtained as follows. The N_2 isotherms on the Na-montmorillonite or hectorite gave external areas appropriate to average values of 26 and 9 lamellae per crystal. Crystals containing these numbers of lamellae have total interlamellar areas of 270×10^3 and 250×10^3 m² per gram unit cell (g.u.c.). From these areas one must subtract areas occupied by cations after allowing for the fractions of the total cations on external areas. These fractions were estimated as 4 % for montmorillonite and 11 % for hectorite. Areas per cation were estimated from their lengths multiplied by 4.6 Å, corresponding with ions lying along basal surfaces (table 6, column 3). The total interlamellar area less that occupied by the interlamellar ions then gives the calculated interlamellar free area of table 6 column 4. The $d(001)$ repeat distance of table 6 column 5 less 9.4 Å was taken as the vertical free distance. The free area multiplied by this distance gives a free volume, not tabulated, and from the free volume and the molar volume of N_2 at 78 K (34.8 cm³ mol⁻¹) the calculated saturation capacities of column 6 were obtained. The ratio of the observed saturation (column 7) to that calculated is given in column 8.

At 78 K these ratios show an interesting trend. For the alkyldiammonium hectorites observed and calculated sorption capacities are not very dissimilar. However for the alkylammonium montmorillonites the ratios decline strongly from about 1 for $CH_3NH_3^+$ to 0.12 for $C_6H_{13}NH_3^+$. It seems likely that at 78 K chains are immobile along basal surfaces and that juxtaposition of the CH_3 ends of alkyl chains can lead to nearly closed loops of chains from which N_2 is excluded. Such behaviour is not possible for an ion as small as $CH_3NH_3^+$ but is more and more likely as the chains increase in length. The above effect should occur less often with alkyldiammonium smectites, where both ends of the ions are -NH_3^+ and so would repel other ends. Moreover, because these ions carry two charges they are only half as numerous as the alkylammonium ions. For both reasons the ratios of observed to calculated saturation values should decline less rapidly with carbon number for the alkyldiammonium smectites.

The extreme sensitivity of low temperature access to the interlamellar region of some permanently porous clays is shown in figure 8 (Craven 1976) for $Co^{III}(en)_3$-fluorhectorite of exchange capacity 150 meq 100 g⁻¹ (Barrer & Jones 1971), designated as $Co^{III}(en)_3$-FH150. Here 'en' denotes ethylene diamine. The van der Waals dimensions of O_2 and N_2 are respectively 2.8×3.9 and 3.0×4.1 Å so that at 78 K these small dimensional differences serve to differentiate O_2 and N_2 in a remarkable way. These differences are reflected in the sorption rates, which were at ca. 77 K extremely fast in $Co^{III}(en)_3$-FH150 for H_2, D_2, Ne and O_2 but very slow for N_2, Ar and CH_4 (Craven 1976). One can produce in clay mineral sorbents of type B rate and selectivity differences based on molecular shape and size differences just as remarkable as those found with zeolite molecular sieves.

Selectivites in $N(CH_3)_4$-montmorillonite are illustrated in figure 9 (Barrer & Perry 1961b) for the pairs benzene-n-heptane, benzene-cyclohexane and cyclohexane-n-heptane. These are equilibrium isotherms. $N(CH_3)_4$-montmorillonite gave excellent chromatographic separations of n-heptane from isooctane and from cyclohexane (Barrer & Hampton 1957). Sorption in ethylenediammonium fluorhectorite (en-FH90) suggested that n-paraffins and, or, aromatics could be separated from paraffins with a quaternary carbon or from cyclohexane. En-FH150 behaved in a significantly different way. It sorbed benzene but not toluene, m-xylene or mesitylene under the conditions employed. Like en-FH90 it sorbed n-paraffins but did not intercalate paraffins with a quaternary carbon atom (Barrer & Jones 1971). $Co^{III}(en)_3$- and $Cr^{III}(en)_3$-montmorillonite gave good separations of C_1 to C_5 hydrocarbons on very short

Table 6. Observed and calculated interlamellar saturation capacities (cm^3 (at s.t.p.) g^{-1}) for N_2 at 78 K in some organo-clays (Barrer & Millington 1967)

cation	gram unit cell mass	area of ion $\overline{Å^2}$	calculated free area $\overline{m^2 g^{-1}}$	$\dfrac{d(001)}{Å}$	calculated saturation	observed saturation	$\dfrac{\text{observed saturation}}{\text{calculated saturation}}$
(a) alkylammonium montmorillonites (c.e.c. 85 meq 100 g^{-1})							
$CH_3\overset{+}{N}H_3$	746	21.6	294	11.6	$41._7$	$41._5$	0.99
$C_2H_5\overset{+}{N}H_3$	755	29.8	214	12.8	$46._9$	35	0.74
$C_3H_7\overset{+}{N}H_3$	765	35.6	184	13.1	$43._7$	$27._5$	0.63
$C_5H_{11}\overset{+}{N}H_3$	783	47.1	126	13.4	$32._1$	$12._5$	0.39
$C_6H_{13}\overset{+}{N}H_3$	792	52.9	98	13.4	$25._0$	3	0.12
(b) alkyldiammonium montmorillonites (c.e.c. 85 meq 100 g^{-1})							
$\overset{+}{N}H_3(CH_2)_3\overset{+}{N}H_3$	750	40.6	262	12.9	$59._1$	43	0.73
$\overset{+}{N}H_3(CH_2)_5\overset{+}{N}H_3$	759	52.2	231	13.3	$57._8$	40	0.69
$\overset{+}{N}H_3(CH_2)_7\overset{+}{N}H_3$	769	63.7	201	13.3	$50._2$	38	0.76
$\overset{+}{N}H_3(CH_2)_8\overset{+}{N}H_3$	773	69.5	187	13.3	$46._9$	29	0.62
$\overset{+}{N}H_3(CH_2)_9\overset{+}{N}H_3$	778	75.3	172	13.3	$43._0$	16	0.37
(c) alkyldiammonium hectorites (c.e.c. 91 meq 100 g^{-1})							
$\overset{+}{N}H_3(CH_2)_2\overset{+}{N}H_3$	769	34.8	240	12.2	$43._2$	49	1.14
$\overset{+}{N}H_3(CH_2)_3\overset{+}{N}H_3$	774	40.6	224	13.0	$51._8$	44	0.85
$\overset{+}{N}H_3(CH_2)_3\overset{+}{N}H_3$	779	46.4	208	[13.3]†	$51._9$	51	0.98
$\overset{+}{N}H_3(CH_2)_5\overset{+}{N}H_3$	783	52.2	192	13.3	$48._1$	49	1.02
$\overset{+}{N}H_3(CH_2)_9\overset{+}{N}H_3$	802	75.3	128	13.3	$32._0$	41	1.28

† Assumed value.

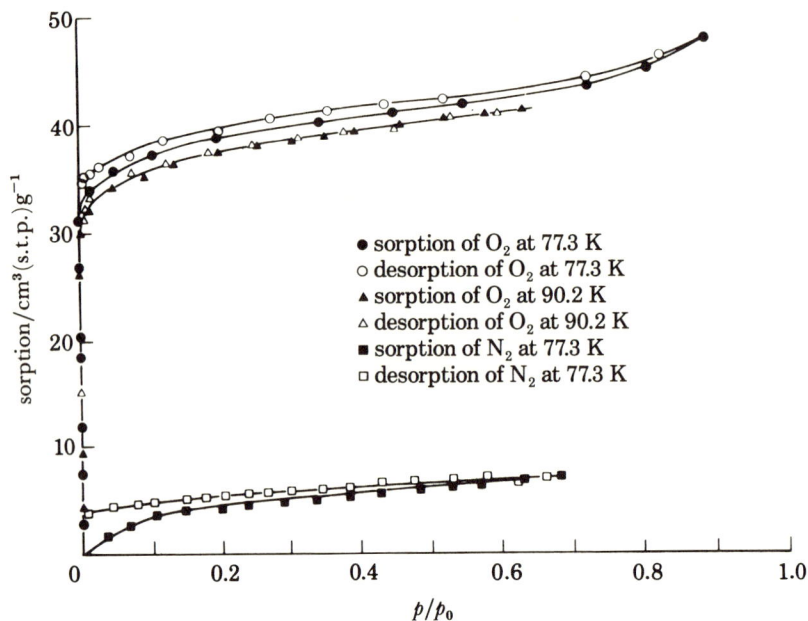

FIGURE 8. Isotherms for O_2 and N_2 at about 78 K and 90 K in $Co^{III}(en)_3$-FH150, showing very large difference in the uptake (Craven 1976; Barrer 1978).

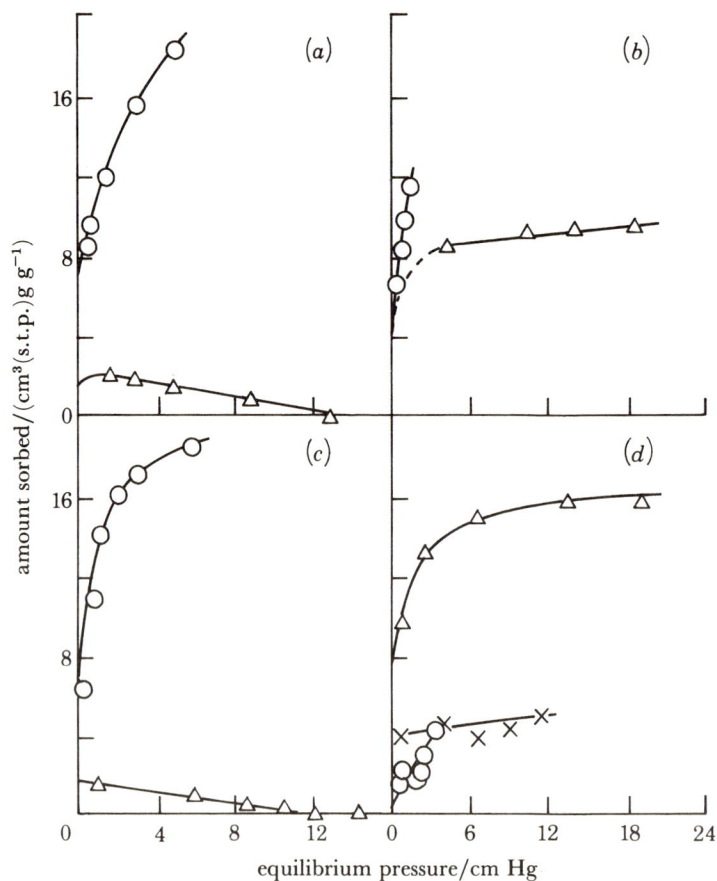

FIGURE 9. (a) and (b). Isotherms in $N(CH_3)_4$-montmorillonite for benzene ($\bigcirc$) and n-heptane ($\triangle$) from two mixtures with (a) 0.67 and (b) 0.30 mole fraction of benzene. (c) Isotherms for benzene ($\bigcirc$) and cyclohexane ($\triangle$) from a mixture with 0.67 mole fraction of benzene. (d) Cyclohexane isotherms from several mixtures with n-heptane with mole fractions 0.85 ($\bigcirc$) 0.56 ($\times$) and 0.16 ($\triangle$) of n-heptane (Barrer & Perry 1961 b). 1 mmHg = 133.322 Pa.

[126]

chromatographic columns at relatively high temperatures (Thielmann & McAtee 1975). Montmorillonite which had intercalated Λ-$[Ru(phen)_3]^{2+}$ (where phen is 1,10-phenanthroline) gave $d(001) = 18.0$ Å and so had a free distance of $(18.0 - 9.4) = 8.6$ Å. When dl-2,3-dihydro-2-methyl-5,6-diphenylpyrazine was sorbed into a short column of this complex and then progressively eluted with 1/2 (by volume) methanol water solvent a partial resolution of the dl-compound was achieved (Yamagishi 1983).

Permanently porous clay mineral sorbents are likely to attract growing interest. The globular ion H$\overset{+}{N}$⟨ ⟩$\overset{+}{N}$H exchanges with ions in montmorillonite and vermiculite (Mortland & Berkheiser 1976) and gives sorbents with free distances between lamellae of *ca.* 5.4 Å. These sorbents took up N_2, C_2H_6 and 2,4-dimethylpentane and had monolayer equivalent areas of 280 and 144 $m^2\,g^{-1}$ for the montmorillonite and vermiculite sorbents respectively. For some uses, however, organic and metal chelate cations in clay minerals have the disadvantage of thermal instability at high temperatures. Catalytic applications in particular could be limited by this instability to low temperature processes. It is however possible to produce rather stable permanently porous sorbents from smectites by exchanging into them the cations present in chlorhydrol solutions (Vaughan & Lussier 1980). These ions are thought to be $[Al_{13}O_4(OH)_{24}(H_2O)_{12}]^{7+}$ and the resulting 'pillared' clays have an interlamellar free distance of about 9.6 Å. When the pillared clay mineral is heated with loss of zeolitic and hydroxyl water this difference is still about 9.4 Å. The product thus obtained sorbed *n*-butane, cyclohexane, CCl_4 and 1,3,5-trimethylbenzene but not 1,2,3,5-tetramethylbenzene or perfluorotributylamine. In another study (Ocelli *et al.* 1981) of a bentonite in which Na^+ was also replaced by oxy-aluminium cations the expanded clay mineral was reported to have a pore volume of 0.16 $cm^3\,g^{-1}$, a surface area of 250–300 $m^2\,g^{-1}$ and to be hydrothermally stable to 1000 °F. The micropore space was easily accessible to C_6–C_{10} *n*-paraffins and to substituted aromatics such as 1,3,5-trimethylbenzene.

Montmorillonite and vermiculite have also been exchanged by $[SiA_3]^+$ where A is acetonylacetate. The exchange forms were hydrolysed at pH *ca.* 9 and heated in air at 100–600°. The product was reported as stable at high temperature and with silica groups separating the lamellae of the clay minerals (Pinnavaia *et al.* 1983).

5. INTERLAMELLAR REACTIONS

Reactions which may occur in interlamellar spaces in clay minerals include the kinds given in table 7. The examples given are merely typical and the table could be much extended, both as regards the types of process catalysed and the number of examples of each type.

The substitution of Si^{4+} by Al^{3+} and of Al^{3+} by Mg^{2+} creates in smectites and vermiculites Lewis base sites, while in the interlamellar region there may be Lewis acid sites. Treatment with acids creates Bronsted acid sites. Accordingly catalytic properties in clay minerals are to be expected. An early use of acid clays was in catalytic cracking of petroleum but in this area such catalysts as well as alumina-silica cracking catalysts have been largely superseded by zeolite-based catalysts. Clay catalysts can also be used in dimerization of unsaturated fatty acids to dicarboxylic acids. Figure 10 (Weiss 1981) shows oligomerization reactions of oleic acid using a $(CH_3)_4$N-montmorillonite catalyst, as a function of charge density measured as $e/(Si, Al)_4O_{10}$ where e is the charge associated with the unit $(Al, Si)_4O_{10}$. The figure shows the influence of

[127]

348 R. M. BARRER

TABLE 7. SOME TYPES OF REACTION PROMOTED BY CLAYS OR MODIFIED CLAYS

reaction type	example
cracking hydrocarbons	processing of petroleum
dimerization and oligomerization	oleic acid to di- and tri-carboxylic acids (Weiss 1981). Cyclohexene to oligomers (Barrer & Perry 1961 a)
isomerization	*cis-trans* isomerization in oleic acid (Weiss 1981)
hydrogen transfer	oleic to di-unsaturated acid and stearic acid (Weiss 1981)
dehydration	NH_4-acetate in kaolinite → acetamide (Weiss 1981). NH_4-salts of aminoacids → peptides, in kaolinite and montmorillonite (Weiss 1981; Paecht-Horowitz *et al.* 1970)
oxidation	leuco-bases oxidized in copying papers in air or with other oxidant (Barrett *et al.* 1951)
esterification	ethylacetate from ethylene and acetic acid (Ballantine *et al.* 1981 a)
lactonization	cyclooctene carboxylic acid → 1,4 and 1,3 lactones (Thomas 1982; Adams *et al.* 1982)
etherification	di(alk-1-yl)ether from alk-1-ene (Thomas 1982). Di(alk-1-yl)ethers from alkan-1-ols (Thomas 1982)
reduction	1-hexene $\xrightarrow{H_2}$ hexane with catalyst of Rh-phosphine-Na-hectorite complex (Raythatha & Pinnavaia 1983)
substitution	acetonitrile + $\overset{+}{N}H_3(CH_2)_n\overset{+}{N}H_3$ → $CH_3\overset{\overset{NH}{\|}}{C}$—$\overset{+}{N}H_2(CH_2)_n\overset{+}{N}H_3$ in smectite (Barrer & Millington 1967)
primary to secondary amines	$2R$—CH_2NH_2 → R—CH_2—NH—CH_2—R + NH_3 (Ballantine *et al.* 1981 b)

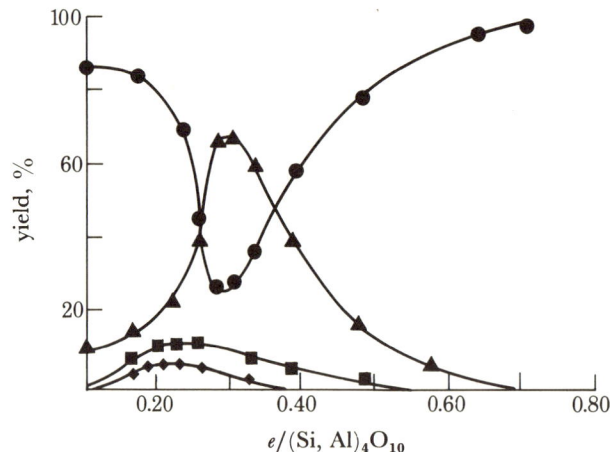

FIGURE 10. Oligomerization of oleic acid to di-, tri- and oligocarboxylic acids. $(CH_3)_4$N-montmorillonite is the catalyst and yield of each product is plotted against the charge density measured as the charge, e, per $(Si, Al)_4O_{20}$ unit. Starting material: oleic acid 98%; 0.5 gram catalyst per 100 ml of oleic acid (Weiss 1981). ● Oleic + stearic (C_{18}); ▲ dicarboxylic acids (C_{36}); ■ tricarboxylic acids (C_{54}); ◆ oligocarboxylic acids (≥ C_{72}).

[128]

charge density upon the behaviour. For $e > 0.6$ oligomerization was replaced by *cis–trans* isomerization and by hydrogen transfer to give di-unsaturated C_{18} acids and stearic acid (a redox-type process).

Technical development of copying papers illustrates the use of clay catalysts in oxidation. The front of the paper is filled with a montmorillonite containing some acid centres which catalyses oxidation of leucobases. The back of the paper has a coating containing leucobase and the pressure of writing forces contact between catalyst and leucobase. The oxidizing agent can be air (Barrett *et al.* 1951).

With reference to figure 10 further examples of the influence of charge density are of interest. Aniline with high charge density smectite readily oxidizes to a black 'polymer'; at medium charge densities it yields a red or blue 'oligomer'; and at low charge densities the aniline is not changed (Weiss 1963). Again, in the interlayer region of high charge density smectite liganded Co^{3+} is stable, while in low charge density smectite it is reduced to Co^{2+} (Weiss 1981):

$$2[CoL_n]^{3+} + 2H_2O \rightleftharpoons 2nL + 2Co^{2+} + 2H^+ + H_2O_2.$$

The peroxide at once decomposes to O_2 and H_2O. As a final example one may consider the isomerization:

$$[Cr(H_2O)_5Cl]^{2+} + H_2O \rightleftharpoons [Cr(H_2O)_6]^{3+} + Cl^-,$$

which is influenced by the charge density of the clay mineral. The trivalent interlayer cation is stable at high charge density but at low charge density the divalent cation is much preferred (Weiss 1981).

In a review of some reactions catalysed by clay minerals Thomas (1982) suggests that considerable parallels may exist between clay mineral and zeolite catalysts, an example being the scheme considered valid for several reactions catalysed by zeolites (Knözinger & Kohne 1966):

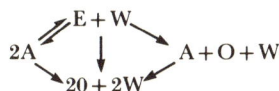

$$2A \underset{\searrow 2O + 2W \nearrow}{\overset{\nearrow E + W \searrow}{\underset{\downarrow}{}}} A + O + W$$

In this scheme A, E, O and W denote respectively alcohol, ether, olefine and water. The diagram summarizes reaction pathways without indicating mechanisms.

Mechanisms may in some processes be given a rational formulation in terms of carbenium ion intermediates, the proton being donated by Bronsted acid centres in the clay catalyst. As an example one may consider the lactonization reactions of cyclo-octene-5-carboxylic acid (Adams *et al.* 1982). Five lactones were produced, four of these being 1,4-cyclo-octanecarbolactone, 1,3-cyclo-octanecarbolactone; 3-methyl-3-cycloheptanecarbolacone and 4-methyl-4-cyclo-heptanecarbolactone. The carbenium ion mechanism for production of the last of these is representative (Thomas 1982):

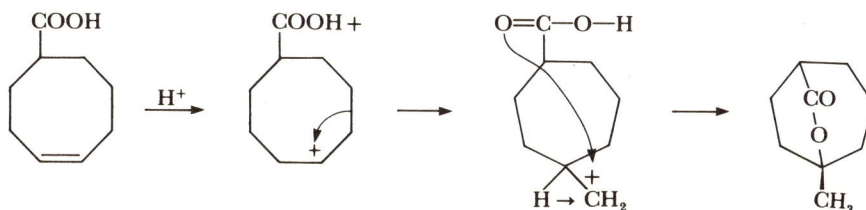

However, mechanisms of acid clay catalyses cannot always be expressed unambiguously in

[129]

terms of carbenium ion intermediates. For example butyl ether formation from butene has been formulated by either of the pathways (i) Ballantine *et al.* 1981*b* or (ii) Adams *et al.* 1971:

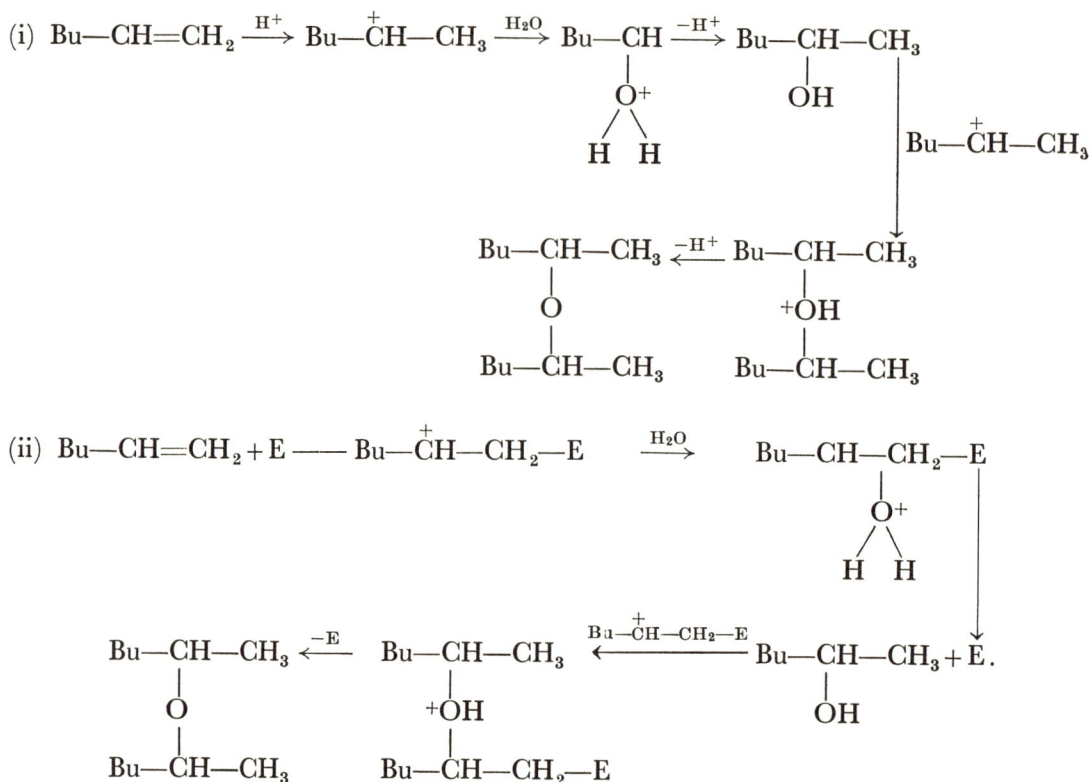

(i) $Bu-CH=CH_2 \xrightarrow{H^+} Bu-\overset{+}{C}H-CH_3 \xrightarrow{H_2O} Bu-CH \xrightarrow{-H^+} Bu-CH-CH_3$

with $Bu-CH$ bearing $\overset{+}{O}$ with H H below; $Bu-CH-CH_3$ bearing OH; and $Bu-\overset{+}{C}H-CH_3$

$Bu-CH-CH_3 \underset{}{\overset{-H^+}{\rightleftharpoons}} Bu-CH-CH_3$

$Bu-CH-CH_3$ with O and below $Bu-CH-CH_3$; $Bu-CH-CH_3$ with ^+OH and below $Bu-CH-CH_3$

(ii) $Bu-CH=CH_2 + E \longrightarrow Bu-\overset{+}{C}H-CH_2-E \xrightarrow{H_2O} Bu-CH-CH_2-E$

with $\overset{+}{O}$ H H branch;

$Bu-CH-CH_3 \xleftarrow{-E} Bu-CH-CH_3 \xleftarrow{Bu-\overset{+}{C}H-CH_2-E} Bu-CH-CH_3 + E.$

$Bu-CH-CH_3$ with O below $Bu-CH-CH_3$; $Bu-CH-CH_3$ with ^+OH below $Bu-CH-CH_2-E$; $Bu-CH-CH_3$ with OH

Mechanism (ii) involves the Lewis acid site E, i.e. an electron accepting site in the catalyst. In both mechanisms Bu denotes butyl.

Finally it is noted that the external surfaces of clay minerals, which are often of significant extent, may catalyse reactions so that independent evidence is required that the seat of reaction is in the interlamellar region. An example of catalysis involving the external surface only is provided by the polymerization of vinylethers using kaolinite at room temperature (R. M. Barrer and A. T. T. Oei, unpublished). Here there is no expectation of intercalation of the ether but the reaction occurred very vigorously as soon as the ether and kaolinite made contact.

6. Conclusion

This account of interlamellar sorption and catalysis by clay minerals demonstrates the notable richness in the phenomena observed and shows a still largely undeveloped potential for selective and molecular sieve separations, and catalyses which may as with zeolites be shape selective. The future of clay minerals in chemistry and chemical engineering looks very promising.

References

Adams, J. M., Ballantine, J. A., Graham, S. H., Laub, R. J., Purnell, J. H., Reid, P. I., Sharman, W. Y. M. & Thomas, J. M. 1971 *J. Catal.* **58**, 238.

Adams, J. M., Davies, S. E., Graham, S. H. & Thomas, J. M. 1982 *J. Catal.* **78**, 197.

Allison, S. A. & Barrer, R. M. 1968 *Trans. Faraday Soc.* **64**, 549.

Ballantine, J. A., Purnell, J. H., Rayanakorn, M., Thomas, J. M. & Williams, K. J. 1981*a J. chem. Soc. chem. Commun.* pp. 8–9.

Ballantine, J. A., Davies, M., Purnell, J. H., Rayanakorn, M., Thomas, J. M. & Williams, K. J. 1981*b J. chem. Soc. chem. Commun.* p. 427.

Barrer, R. M. 1960 *J. Phys. Chem. Solids* **16**, 84.

Barrer, R. M. 1978 *Zeolites and clay minerals as sorbents and molecular sieves* pp. 416–20. London: Academic Press.

Barrer, R. M. & Brummer, K. 1963 *Trans. Faraday Soc.* **59**, 959.

Barrer, R. M. & Hampton, M. G. 1957 *Trans. Faraday Soc.* **53**, 1462.

Barrer, R. M. & Jones, D. L. 1971 *J. chem. Soc.* A, 2595.

Barrer, R. M. & Kelsey, K. 1961*a Trans. Faraday Soc.* **57**, 452.

Barrer, R. M. & Kelsey, K. 1961*b Trans. Faraday Soc.* **57**, 625.

Barrer, R. M. & McLeod, D. M. 1954 *Trans. Faraday Soc.* **50**, 980.

Barrer, R. M. & McLeod, D. M. 1955 *Trans. Faraday Soc.* **51**, 1290.

Barrer, R. M. & Millington, A. D. 1967 *J. Colloid Interface Sci.* **25**, 359.

Barrer, R. M. & Perry, G. S. 1961*a J. chem. Soc.* 842.

Barrer, R. M. & Perry, G. S. 1961*b J. chem. Soc.* 850.

Barrer, R. M. & Reay, J. S. S. 1957 *Trans. Faraday Soc.* **53**, 1253.

Barrett, K., Green, R. & Sandberg, R. W. 1951 *U.S.P.* 2, 550, 469.

Brindley, G. W. & Ray, S. 1958 *Am. Miner.* **49**, 106.

Brunauer, S. 1944 *The adsorption of gases and vapours*, p. 150. Oxford: University Press.

Cowan, C. T. & Hartwell, J. M. 1961 *Nature, Lond.* **190**, 712.

Craven, R. J. B. 1976 Ph.D. Thesis, London University.

Gee, G. 1942 *Trans. Faraday Soc.* **38**, 418.

Gillery, F. H. 1959 *Am. Miner.* **44**, 806.

Jenkins, H. D. B. & Hartman, P. 1982 *Phil. Trans. R. Soc. Lond.* A **304**, 397.

Kiselev, A. V., Lebedeva, N. P., Frolov, I. I. & Yashin, Ya. I. 1972 *Chromatographia* **5**, 341.

Knözinger, H. & Kohne, R. 1966 *J. Catal.* **5**, 264.

Mortland, M. M. & Berkheiser, V. 1976 *Clays Clay Miner.* **24**, 60.

Ocelli, M. L., Hwu, F. & Hightower, J. W. 1981 *Am. chem. Soc., Div. Pet. Chem.* **26**, 672.

Olphen, H. van 1969 *Proc. int. Clay Conf., Tokyo,* **1**, 649.

Paecht-Horowitz, M., Berger, J. & Katchalsky, A. 1970 *Nature, Lond.* **228**, 636.

Pinnavaia, J., Mortland, M. M. & Endo, T. 1983 *U.S.P.* 4, 367, 163.

Raythatha, R. & Pinnavaia, J. 1983 *J. Catal.* **80**, 47.

Rysselberge, J. van & Stricht, M. van der 1962 *Nature Lond.* **193**, 1281.

Shabtai, J., Frydman, N. & Lazar, R. 1976 *6th int. Cong. Catalysis, London,* 6–12 July, Paper B5.

Thielmann, V. J. & McAtee, J. L. 1975 *J. Chromatog.* **105**, 115.

Thomas, J. M. 1982 *Intercalation chemistry* (ed. M. S. Whittingham and A. J. Jacobson), p. 55. London: Academic Press.

Vaughan, D. E. W. & Lussier, R. J. 1980 *Proc. 5th int. Conf. Zeolites* (ed. L. V. C. Rees) p. 94. London: Heyden & Sons.

Weiss, A. 1958*a Chem. Ber.* **91**, 497.

Weiss, A. 1958*b Zeit. anorg. allg. Chemie* **297**, 17.

Weiss, A. 1963 *Clays Clay Miner.* **10**, 191.

Weiss, A. 1981 *Angew. Chem. int. Ed. Engl.* **20**, 850.

Weiss, A., Thielpape, W., Goring, G., Ritter, W. & Schafer, H. 1963 *Int. Clay Conf.* **14** p. 287. Oxford: Pergamon Press.

White, D. 1957 *Nature, Lond.* **179**, 1075.

White, D. 1959 *Nature, Lond.* **184**, 1795.

Yamagishi, A. 1983 *J. chem. Soc. chem. Commun.* p. 9.

Discussion

P. B. Tinker (*Rothamsted Experimental Station, Harpenden, Hertfordshire, U.K.*). Have you used water as guest molecule in the smectites with intercalated organic compounds? Natural clays must often contain some degree of intercalation; could you say anything about the effect this would have on water sorption? Is there any possibility that organic compounds could be lost from the clay if it expands on hydration?

R. M. Barrer. We have investigated only competitive sorption of two organic guest species, rather than water and an organic species. When one guest is used to displace a previously intercalated second guest the amount displaced will depend upon the relative vapour pressure (or solution concentration) of each guest, and the nature of the two guest species and hence the energy and type of bonding of each to the clay surfaces. In clay sorbents containing only small inorganic cations if the molecule first intercalated was large and polar (for example, a sugar, a peptide or a protein) displacement by water would be more difficult to achieve than for a small polar molecule such as methanol or ethanol. However, additional sorption of water could occur.

Phil. Trans. R. Soc. Lond. A **311**, 353–368 (1984)   [353]
Printed in Great Britain

Some properties of clay–water dispersions

By S. D. Lubetkin, S. R. Middleton† and R. H. Ottewill, F.R.S.

School of Chemistry, University of Bristol, Bristol BS8 1TS, U.K.

The bulk properties of clay–water dispersions, particularly with montmorillonites, depend on the very large surface area of the materials and their plate-like form. Owing to isomorphous substitution in the clay lattice the plates are charged and hence interact electrostatically with each other. The form of the interaction was tested using macroscopic clay surfaces in the form of cleaved mica and the results compared with those obtained by compression of concentrated aqueous dispersions of various montmorillonites in a homoionic form. The results obtained with mica and lithium montmorillonite were comparable suggesting the latter disperses as single plates. When sodium, potassium and caesium were used as the counter-ions for montmorillonite, differences were observed which were attributed to face–face association. The consequences of isomorphous substitution in either the tetrahedral or the octahedral layer were also examined.

1. Introduction

Naturally occurring clays owe many of their properties to the isomorphous substitution that occurs in the tetrahedral and octahedral layers during their geological formation (van Olphen 1963). The defects that this replacement leaves in the clay lattice lead to the formation of an electric charge at the clay–water interface and hence to an electrostatic potential difference between the clay surface and the solution phase, the so-called electrical double layer. Consequently, at distances of the order of a few tens of nanometres electrostatic interaction occurs between the clay plates and the electrostatic forces generated play a significant role in determining the properties of clay–water dispersions. The quantitative investigation of these electrostatic forces is therefore important in order to understand how they are influenced both by the nature of the clay and the electrolyte concentration of the dispersion medium.

The most direct method for the examination of electrostatic forces is to use macroscopic surfaces with an intervening layer of liquid between them, and in the present work, dedicated to the examination of clay–clay surface interactions, we have used cleaved mica in a specially constructed apparatus designed for the direct measurement of surface forces. With clay–water dispersions the pressure generated in a clay dispersion has been measured directly as a function of the distance of separation between the clay plates.

A direct comparison of the results obtained suggests that for well-dispersed clays such as homoionic lithium montmorillonite, the pressure measurements on the dispersions are directly comparable with the macroscopic measurements on mica. Evidence was obtained for significant electrostatic effects and with the dispersions this was found to depend on the nature of the isomorphous substitution. In the presence of counter-ions other than lithium, for example, sodium, potassium and caesium, a significant reduction in interaction occurred which appeared to be due to the formation of associated face-to-face units of clay particles.

† Present address: Pulp and Paper Research Institute of Canada, 570 St Johns Boulevard, Pointe Claire, Quebec, Canada H9R 3J9.

2. EXPERIMENTAL

Materials

The distilled water used was doubly distilled from an all-Pyrex apparatus. All salts used were Analar grade materials.

The bentonite used was a sample of montmorillonite no. 26 from Clay Spur, Wyoming, as prepared for the American Petroleum Institute Clay Minerals Standards Project no. 49. It was converted into the homoionic form of lithium, sodium, potassium or caesium montmorillonite using the procedure previously described for the preparation of sodium montmorillonite (Barclay & Ottewill 1970; Callaghan & Ottewill 1974). Electron microscope examination of samples of lithium montmorillonite using the technique suggested by Greene *et al.* (1974) showed that it was dispersed essentially as single plates. The width and breadth of the plates however, varied over the range 0.2–0.7 μm. Hence although it appeared to be monodisperse with respect to plate thickness, the sample was polydisperse with respect to its face dimensions.

The sample of Cypern montmorillonite was kindly donated by Professor G. Lagaly of the University of Kiel as a dispersion of lithium clay in an electrolyte solution. It was reprocessed in concentrated lithium chloride solution to ensure conversion into the homoionic lithium form.

The sample of beidellite was kindly given to us by Professor A. Weiss of the University of Munich. The material was received in the dry state and was pinkish-white in colour. It was converted into the homoionic form using the procedure used previously for the preparation of sodium montmorillonite (Barclay & Ottewill 1970).

The mica used was from a sample of green mica. Lyons *et al.* (1981) have given a composition for this material as

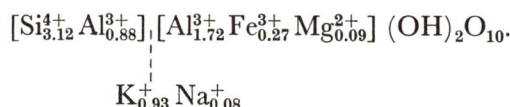

$$[Si^{4+}_{3.12}Al^{3+}_{0.88}]_{|}[Al^{3+}_{1.72}Fe^{3+}_{0.27}Mg^{2+}_{0.09}](OH)_2O_{10}.$$

$$K^+_{0.93}Na^+_{0.08}$$

Force measurements – macroscopic surfaces

These measurements were carried out using cleaved mica surfaces. The sheets of mica obtained were *ca.* 2 μm thick and were lightly silvered on the back surface; the latter was attached with adhesive to a hemi-cylindrical quartz former. Following the procedures of Tabor & Winterton (1969) and Israelachvili & Tabor (1972, 1973) this enabled a multiple beam interferometer (Tolansky 1948) to be used to determine the distance between the mica surfaces once they were immersed in electrolyte solution. The hemi-cyclindrical surfaces were used in a crossed conformation. One of the formers, the uppermost, was attached to a transducer arm and the other, the lower, was attached to a micrometer with a very fine adjustment. Both mica surfaces were immersed in the electrolyte solution which was controlled at a temperature of 25 °C. The force of interaction was thus measured directly via the transducer and the distance of separation between the mica surfaces, in the electrolyte solution, by the multiple beam interferometer. A full description of this equipment will appear elsewhere (Lubetkin & Ottewill 1984).

Pressure measurements – clay dispersions

The apparatus used for the compression of the clay dispersions and for measuring its volume under a constant applied pressure has been described previously (Barclay & Ottewill 1970). In this equipment the clay particles were confined between a rubber membrane and a filter so arranged that the electrolyte expelled by the application of pressure was maintained in equilibrium with that of the dispersion. Under these conditions the total volume of the system was also maintained constant but the ratio of the volume of the clay dispersion to that of the electrolyte changed. A hydraulic system was used for the application of pressure and this allowed constant pressure to be accurately maintained. The stainless steel pressure cell, containing the clay dispersion, was immersed in a water bath maintained at a constant temperature of 25 °C.

3. Theory

The precise measurement of the force of interaction between surfaces, either for macroscopic surfaces or particle surfaces, requires that the internal force of interaction, F_I, should be balanced by an external applied force, F_E; if the force is applied per unit area of surface this gives the pressures P_I and P_E. It is required that P_E should be measurable by a physical technique, in the present case the transducer for macroscopic surfaces and the gauge pressure in the hydraulic system for the clay dispersions. The condition of equilibrium is illustrated schematically in figure 1 and for this condition we can write

$$P_I + P_E = 0.$$

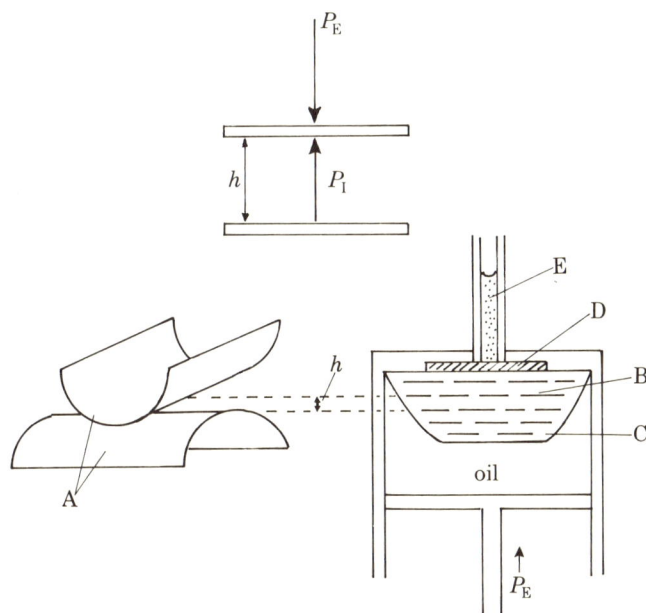

FIGURE 1. Schematic illustration of the balance of forces between plates and the apparatus used. A, mica surfaces; B, clay plates; C, rubber membrane; D, filter; E, dispersion medium.

The pressure P_I can be subdivided into various components and hence written in the form

$$P_I = P_{el} + P_A + P_B + P_{st} + P_{so},$$

where P_{el} = electrostatic pressure, P_A = the van der Waals attractive pressure, P_B = the Born repulsion, P_{st} = a steric effect due to adsorbed molecules and P_{so} = a solvation pressure.

In this work we concentrated on long range interactions and chose conditions such that P_{el} is the dominant term. This is justified by the fact that P_B and P_{so} appear to be very short

[135]

range terms and calculation shows P_A to be very small for the conditions used. In the absence of added macromolecules we will assume that $P_{st} = 0$ also.

It was shown by Langmuir (1938) that the electrostatic repulsive pressure for two flat plates both having the same surface potential ψ_s was given by,

$$P_{el} = 2n_0 kT(\cosh u - 1), \tag{1}$$

where n_0 = the number of ions per unit volume (m³) in the bulk electrolyte phase, k = the Boltzmann constant, T = absolute temperature, and u for 1:1 electrolytes is given by

$$u = e\psi_d/kT, \tag{2}$$

where e = the fundamental unit of charge and ψ_d = the electrostatic potential at a distance midway between the plates, and at a distance d from the surface of each plate. The surface to surface distance between the plates, h, is thus given by

$$h = 2d. \tag{3}$$

For the interaction of diffuse electrical double layers for flat surfaces at constant surface potential, ψ_s, the distance h is given by (Verwey & Overbeek 1948).

$$h = -\frac{2}{\kappa} \int_z^u \frac{dy}{[2(\cosh y - \cosh u)]^{\frac{1}{2}}} \tag{4}$$

with $z = e\psi_s/kT$ and $y = e\psi/kT$, the general variable. κ for 1:1 electrolytes is given by,

$$\kappa^2 = 2n_0 e^2/\epsilon_r \epsilon_0 kT \tag{5}$$

with ϵ_r = the relative permittivity of the solution phase and ϵ_0 that of free space.

For long range interaction defined by the condition that $h > 2/\kappa$, we find the equation,

$$P_{el} = 64 n_0 kT \tanh^2 \left(\frac{e\psi_s}{kT}\right) \exp(-\kappa h) \tag{6}$$

or

$$\ln P_{el} = \ln\left[64 n_0 kT \tanh^2\left(\frac{e\psi_s}{kT}\right)\right] - \kappa h \tag{7}$$

so that for long distances of separation it can be anticipated that curves of $\ln P_{el}$ against h should be linear. Furthermore the negative slope should yield a value of κ and the intercept at $h = 0$ a value of the surface potential ψ_s.

In this work the surface potential ψ_s has been taken as the diffuse layer potential, and hence it has been assumed that it is close to the Stern potential. Detailed arguments about the inner part of the double layer will not be considered.

The geometry of crossed cylinders used in the case of the mica experiments is equivalent in interaction terms to interaction between a flat surface and a sphere of radius R. This gives, again for the condition $h > 2/\kappa$, the equation,

$$\ln\left(\frac{F_{el}}{2\pi R}\right) = \ln\left[\frac{64 n_0 kT}{\kappa} \cdot \tanh^2\left(\frac{e\psi_s}{kT}\right)\right] - \kappa h, \tag{8}$$

where F_{el} = the electrostatic force between the sphere and the plate. Thus (8) can be compared with (7). R was taken as the radius of the hemi-cylinder.

The significance of h in the two systems is illustrated in figure 1.

4. Results

Mica surfaces

Figure 2 shows the curves of lg $(F/2\pi R)$ against h obtained using mica surfaces in solutions of potassium chloride at concentrations of 10^{-2} and 10^{-3} mol dm^{-3}. F was taken as the force of repulsion measured directly using the transducer arrangement. As can be seen from the figure the range of the electrostatic forces is much greater in 10^{-3} mol dm^{-3} potassium chloride than in 10^{-2} mol dm^{-3} potassium chloride. For the range of distances shown in figure 2 good linear plots of lg $(F/2\pi R)$ against h were obtained. In 10^{-3} mol dm^{-3} potassium chloride solution an experimental value of 9.6 nm was obtained for κ using (8) which compares favourably with the theoretical calculation of 9.62 nm from (5). A value of 84 mV was obtained for $|\psi_s|$ from the intercept at $h = 0$ and (8). In 10^{-2} mol dm^{-3} potassium chloride solutions an

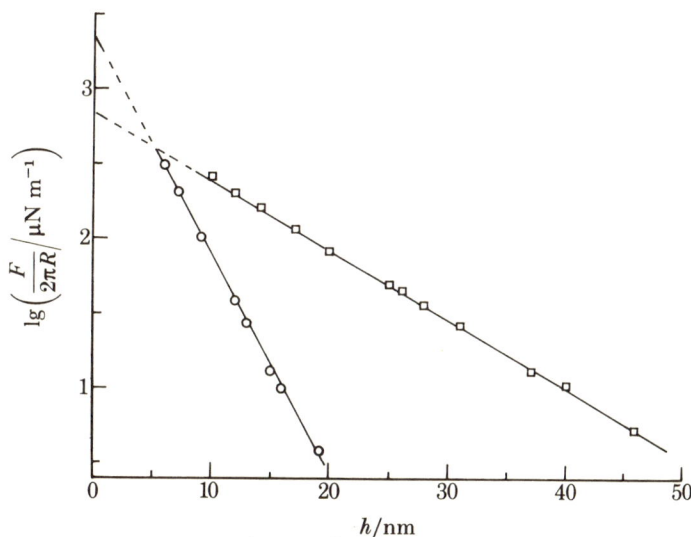

FIGURE 2. Lg $(F/2\pi R)$ against h for mica surfaces in, —○—, 10^{-2} mol dm^{-3} potassium chloride; —□—, 10^{-3} mol dm^{-3} potassium chloride.

experimental value of 3.5 nm was obtained for κ compared with the theoretical value of 3.04 nm. The value of $|\psi_s|$ obtained was 79 mV. These values of $|\psi_s|$ are comparable with the values of ζ-potential obtained by Lyons et al. (1981) using the streaming potential technique on freshly cleaved sheets of green mica.

These results show that at long distances, the forces of electrostatic repulsion appear to be the dominant forces and that these can be explained using the theory of the electrical double layer for the condition that $h > 2/\kappa$.

Wyoming bentonite

The sample used was found by analysis of the sodium form and using the procedure of Marshall (1949) to have the composition,

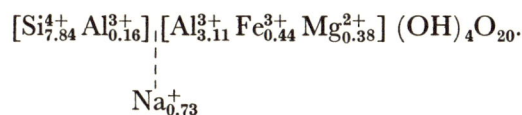

$$[Si^{4+}_{7.84} Al^{3+}_{0.16}]\,[Al^{3+}_{3.11}\,Fe^{3+}_{0.44}\,Mg^{2+}_{0.38}]\,(OH)_4 O_{20}.$$

$$Na^{+}_{0.73}$$

[137]

Hence the isomorphous substitution was primarily in the octahedral layer. This gave a cation exchange capacity of 0.91 meq g^{-1} and a surface charge density (σ) of 11.6 µC cm^{-2}.

The compression curves obtained using samples of lithium montmorillonite are shown in figure 3 in the form of lg P against V/m, where V = the volume of the clay dispersion and m the mass of clay contained in it. As explained in previous work the first compression on the clay dispersion rearranges the plates into an ordered parallel array (Barclay & Ottewill 1970; Callaghan & Ottewill 1974). Once this has occurred subsequent compression and decompression curves are very reproducible. Consequently, the results given in this work are the second compression runs.

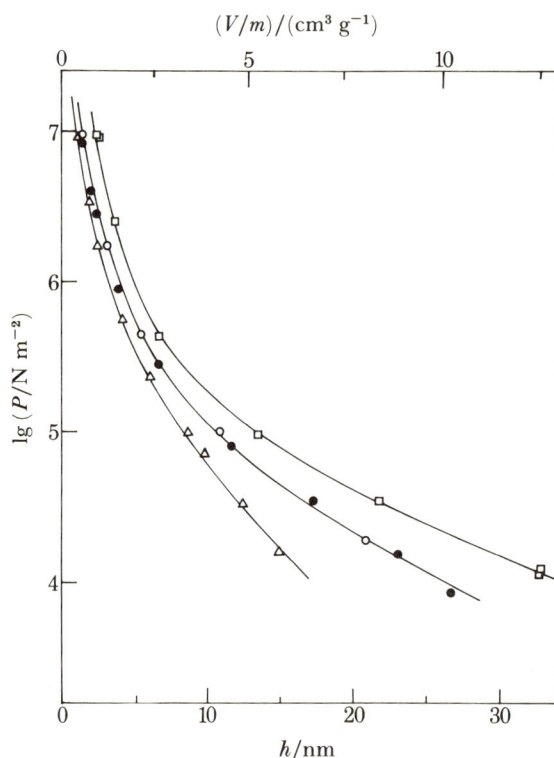

FIGURE 3. Lg P against V/m (and h) for lithium Wyoming bentonite in lithium chloride solutions. □—□, 10^{-4} mol dm^{-3}; ●—●, 10^{-3} mol dm^{-3}; ○—○, 10^{-3} mol dm^{-3} (centrifuged sample); △—△, 10^{-2} mol dm^{-3}.

For a material, such as a montmorillonite clay, essentially all the surface area is that due to the faces of the plates and for this situation the plate surface to surface separation, h, is given by the expression,

$$h = 2V/mA,\qquad(9)$$

where A = the specific surface area of the clay. For lithium montmorillonite this was found to be 763 m^2 g^{-1}. The use of this expression has been checked using both small angle X-ray diffraction (Barclay & Ottewill 1970) and small angle neutron diffraction (Cebula & Ottewill 1981) and good agreement between experiment and values of h calculated from (9) obtained. Moreover, neutron scattering studies on compressed lithium montmorillonite samples (Cebula et al. 1979) have confirmed that the plate arrangements are essentially parallel although some grain boundaries occur owing to the polydispersity in the face dimensions of the plates.

By the use of (9) it is possible to convert the values of V/m into values of h and the results plotted in the form, lg P against h are shown in figure 3 for three different electrolyte concentrations, 10^{-4}, 10^{-3} and 10^{-2} mol dm^{-3} lithium chloride. As in the case of the mica surfaces the effect of electrolyte concentration on the electrostatic forces is discernible and the long range nature of the interactions, particularly in 10^{-4} mol dm^{-3} lithium chloride is clearly apparent.

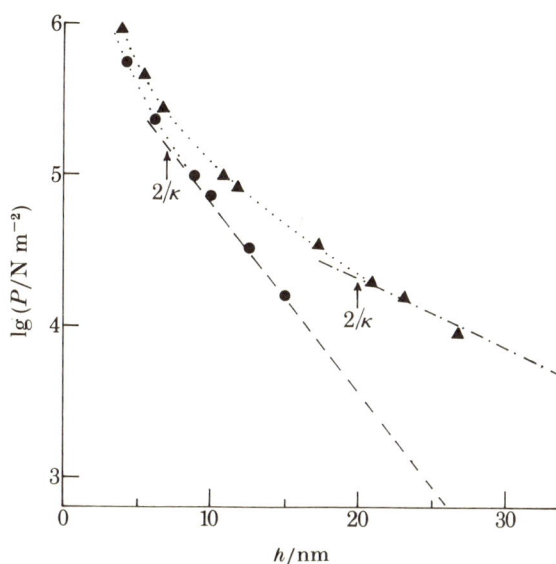

FIGURE 4. Lg P against h for Wyoming bentonite in lithium chloride solutions.
▲—▲, 10^{-3} mol dm^{-3}; ●—●, 10^{-2} mol dm^{-3}; – – –, calculated from simple electrostatic model.

In figure 4 the results obtained for the lithium bentonite are compared with the calculated curves of lg P_{el} against h for h values greater than $2/\kappa$ where (7) is valid. Good agreement is obtained between the calculated and experimental values of κ as shown by the dashed line. The estimated value of $|\psi_s|$ is of the order of 250 mV in 10^{-3} mol dm^{-3} lithium chloride solution and 150 mV in 10^{-2} mol dm^{-3} solution. These results strongly suggest that the lithium montmorillonite is well dispersed into single plates and that the h distances are essentially correct. Confirmation of the well dispersed state of the lithium montmorillonite was obtained from electron microscopy and from small angle neutron scattering studies (Cebula & Ottewill 1981; Cebula et al. 1980).

The compression results obtained on homoionic potassium Wyoming bentonite are given in figure 5 for electrolyte concentrations of 10^{-4}, 10^{-3} and 10^{-2} mol dm^{-3} potassium chloride. It is clear from these data that for the same interaction pressure the distance between the plates is apparently substantially decreased. It was also noted that there was a change in the optical characteristics of the samples in that they appeared somewhat whiter in colour. This suggests the formation of associated units. Electron microscopy also indicated that there appeared to be face–face association of the montmorillonite plates in the presence of potassium. However, the degree of association appears to be small and the evidence suggests that only a few units are involved in the stack. Greater binding of the potassium ions to the clay surface would be anticipated on the basis of the greater polarizability of the K$^+$ ion compared to the Li$^+$ ion.

[139]

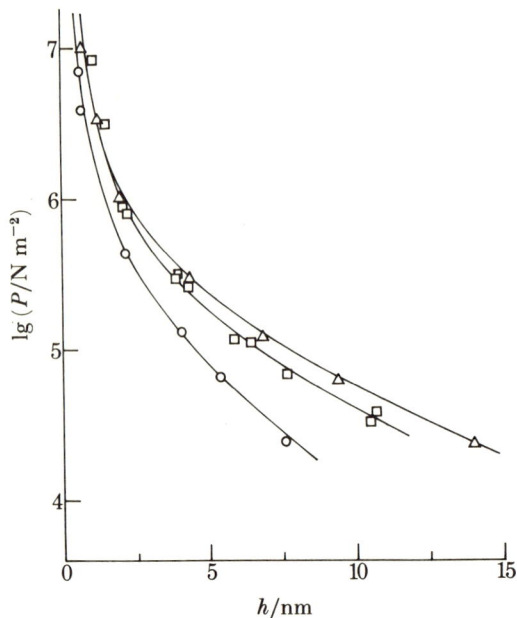

FIGURE 5. Lg P against h for Wyoming bentonite in potassium chloride solutions.
△—△, 10^{-4} mol dm^{-3}; □—□, 10^{-3} mol dm^{-3}; ○—○, 10^{-2} mol dm^{-3}.

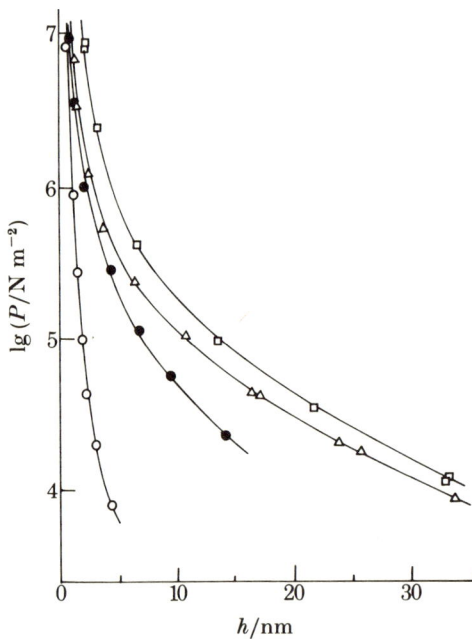

FIGURE 6. Lg P against h for Wyoming bentonite in 10^{-4} mol dm^{-3} salt solutions of various counter-ions. □—□, Li$^+$; △—△, Na$^+$; ●—●, K$^+$; ○—○, Cs$^+$.

As anticipated this effect becomes more pronounced in the presence of Cs^+ ions and figure 6 gives a comparison of the results obtained at an electrolyte concentration of 10^{-4} mol dm^{-3} with homoionic lithium, sodium, potassium and caesium Wyoming bentonites. Again the optical appearance of the caesium samples was quite different from that of the lithium samples in that they were much whiter in appearance. This again suggested the formation of stacks of montmorillonite plates in the presence of caesium ions in addition to the more pronounced binding of the Cs^+ ions to the clay surface.

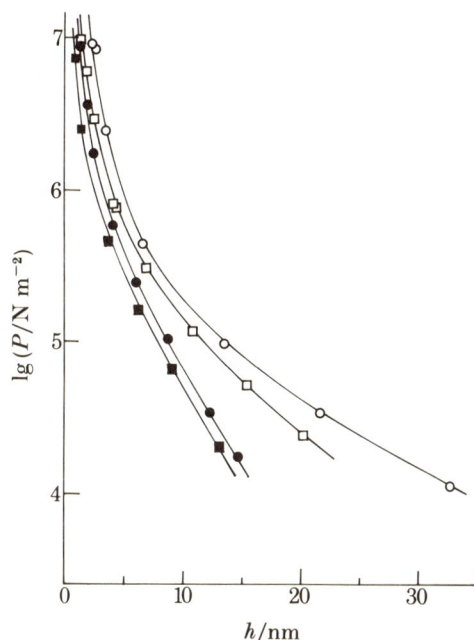

FIGURE 7. Lg P against h for clay dispersions in 10^{-4} mol dm^{-3} lithium chloride solutions. ○—○, Wyoming bentonite; □—□, Cypern montmorillonite. Filled symbols: results in 10^{-2} mol dm^{-3} lithium chloride solutions.

Cypern montmorillonite

The material used was found to have the composition

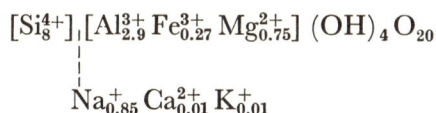

$$[Si_8^{4+}] [Al_{2.9}^{3+} Fe_{0.27}^{3+} Mg_{0.75}^{2+}] (OH)_4 O_{20}$$

$$Na_{0.85}^+ Ca_{0.01}^{2+} K_{0.01}^+$$

as originally obtained and hence had no isomorphous substitution in the tetrahedral layers. It was converted by the procedures described into the homoionic lithium form and the compressibility examined in 10^{-4} and 10^{-2} mol dm^{-3} lithium chloride solutions. The results obtained are presented in figure 7 as curves of lg P against h and compared with the results for Wyoming bentonite. At the same distances of plate separation the pressures exerted by the Cypern montmorillonite are significantly lower than those found using Wyoming bentonite. The surface charge density of the Cypern montmorillonite was found to be 10.3 μC cm^{-2} with a cation exchange capacity of 0.81 meq g^{-1} and hence was lower than that for the bentonite.

The results for the Cypern material can be compared with those of the bentonite at the values of $h > 2/\kappa$ in 10^{-2} mol dm^{-3} lithium chloride when it becomes clear that the slopes in

[141]

this region are essentially identical. The intercept obtained by extrapolation at $h = 0$ is how-ever lower for the Cypern material and using this value an estimate is obtained for the surface potential of 113 mV compared to 150 mV for the bentonite.

It is also noticeable that for the Cypern material the distance of closest approach of the plates at the highest pressures measured is 1.3 nm in 10^{-4} mol dm^{-3} lithium chloride whereas for the bentonite it is *ca.* 2.2 nm. Presumably this is also an effect of the lower surface charge density.

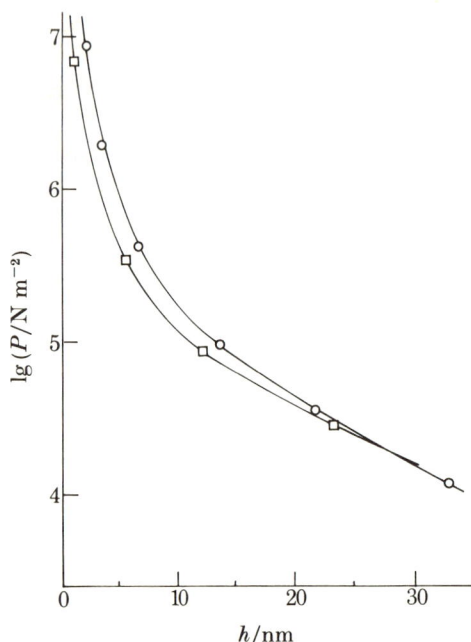

FIGURE 8. Lg P against h for clay dispersions in 10^{-4} mol dm^{-3} lithium chloride solutions.
○—○, Wyoming bentonite; □—□, beidellite.

Beidellite

The material used was found to have the composition,

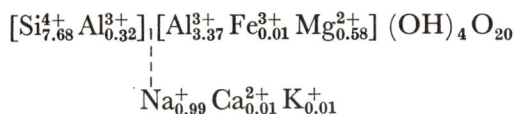

$$[Si^{4+}_{7.68} Al^{3+}_{0.32}]_1 [Al^{3+}_{3.37} Fe^{3+}_{0.01} Mg^{2+}_{0.58}] (OH)_4 O_{20}$$

$$Na^+_{0.99} Ca^{2+}_{0.01} K^+_{0.01}$$

in the original form. It was converted into the homoionic lithium form by the same procedure as that used for the other clays. However, for this material a substantial amount of isomor-phous substitution was present in the tetrahedral layer and it had the highest surface charge density of the clays examined, 13.5 μC cm^{-2}. The cation exchange capacity was 1.06 meq g^{-1}.

The curve of lg P against h obtained for the beidellite dispersed in 10^{-4} mol dm^{-3} lithium chloride is given in figure 8. The form of the curve is interesting in that when it is compared with the results for lithium bentonite there appears to be a cross-over at large separations. This, in principle, would indicate a lower surface potential at the beidellite–water interface than at the montmorillonite interface. The inference from this is that the deficit of charge in the tetrahedral layer causes a much stronger binding of Li$^+$ ions to the surface than occurs with montmorillonite.

In this context it is also significant that at the highest pressures examined the beidellite plates approach much more closely than the montmorillonite plates under comparable conditions; this also could be a consequence of tighter counter-ion binding.

5. Discussion

The direct measurements of electrostatic interaction forces between mica surfaces and between plates of lithium montmorillonite (bentonite) indicate that the dominant interaction forces are electrostatic in nature. By using the simple theory of electrostatic interaction embodied in (7) and (8) good agreement is obtained between experiment and theory for the condition $h > 2/\kappa$. The similarity of behaviour between mica surfaces and the lithium clays seems to provide strong evidence for the fact that the clay is dispersed as essentially single platelets. Small angle neutron scattering experiments by Cebula et al. (1980) on dilute dispersions of the same material also support this conclusion.

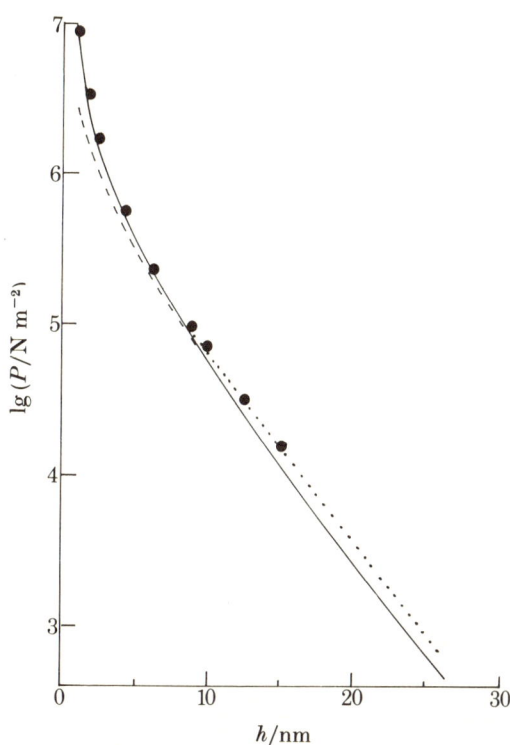

FIGURE 9. Lg P against h. ●—●, experimental data for lithium Wyoming bentonite in 10^{-2} mol dm^{-3} lithium chloride solution. −−−, constant potential model; ——, constant surface charge model: ····· approximate equation (7).

At the closer distances of approach of the lithium clays, when $h < 2/\kappa$, then the simple theory can no longer be expected to hold and the full theory given by (4) has to be used to give a comparison with the experimental data. Taking the lithium bentonite results in 10^{-2} mol dm^{-3} and using $\psi_s = 150$ mV, the value found for large distances from the simple theory, we obtain $z = 6$. Using the tabulations given by Verwey & Overbeek (1948) for interaction at constant potential this gave the curve of lg P against h shown as a dashed line in figure 9. Although the curve deviates from linearity at the closer distances and curves upwards, it falls below the experimental points.

[143]

However, since the electrostatic charge on clay surfaces arises as a consequence of isomorphous substitution in the lattice, it would appear physically more realistic to consider the interaction to occur at constant charge. This condition is satisfied by the equation

$$(2 \cosh z - 2 \cosh u)^{\frac{1}{2}} = \sigma \left(\frac{1}{2\epsilon_r \epsilon_0 n_0 k T} \right)^{\frac{1}{2}}, \tag{10}$$

where the right-hand side is a constant determined by the magnitude of the surface charge density. As $h \to \infty$ then $u \to 0$ and z has the same value as for interaction at constant potential. Hence taking $z = 6$ the constant charge curve of $\lg P$ against h was calculated for 10^{-2} mol dm^{-3} lithium chloride. The result is shown as the continuous line in figure 9. At the longer distances, as anticipated, it is identical to the curve obtained at constant potential. At the shorter distances, however, it curves upward more strongly and lies very close to the experimental results. The value obtained for σ was 11.8 μC cm^{-2} which was remarkably close to the value obtained from the cation exchange capacity. Although the agreement between experiment and theory is good it must be treated with caution. The theory, for example, does not take into account such factors as the finite size of the ions and it would be anticipated that this should be very important at close distances.

An interesting point is that at the longer distances the approximate theory, (7), appears to give a better fit to the experimental data than the more exact one. This may, however, be because the approximate theory slightly underestimates the surface potential.

A point of major interest is the fact that for counter-ions other than lithium the properties of the bulk clay–water mixtures deviate from that expected on the basis of electrostatic interaction with the deviation being the most pronounced in the case of the caesium clay. This is demonstrated clearly in the results presented in figure 6. It was also found (Middleton 1978) that the samples had different optical properties; the lithium and sodium bentonite dispersions were yellow whereas a dispersion of caesium bentonite was creamy white in appearance. The potassium bentonite was intermediate in colour between the lithium and caesium dispersions. The intensity of light scattered by the dispersions was greatest for the caesium dispersions. These results suggested some association between the plates. From small angle neutron scattering measurements, Cebula et al. (1980) using dilute dispersions of the same bentonite suggested that in potassium bentonite dispersions the average unit consisted of a pair of platelets with two molecular layers of water between them and in the case of caesium bentonite the associated unit consisted of three platelets interleaved with double layers of water. It is of interest from this point of view that the distances between the plates at the highest pressures (figure 6) are closely similar for the potassium and caesium clays and tend to a limiting value of ca. 0.7 nm which is close to the value found by Cebula et al. (1980). It would appear, therefore, that the changes in the $\lg P$ against h curves found by changing counter-ions can be explained by the formation of associated units. The explanation of this association phenomenon is not clear but is most likely connected with the order of the energy of ion-binding to the clay surface, in this case in decreasing binding order

$$Cs^+ > K^+ > Na^+ > Li^+.$$

Consistent with this explanation are the results obtained with Cypern montmorillonite and beidellite. In the former case comparison of the lithium form with lithium bentonite indicated that the pressures at the same distance of separation were lower for the Cypern material.

[144]

This seemed to be consistent with the lower surface charge density of 10.3 μC cm^{-2} which on figure 9 would shift the curves closer to the ordinate. The beidellite on the other hand had a much higher surface charge density and yet the pressure curve obtained with the lithium form crosses the curve for the lithium bentonite at longer distances and has a smaller slope which suggests a smaller surface potential than the beidellite. At higher pressures the plates move more closely together than those of lithium bentonite. A tentative explanation of these effects is that the isomorphous substitution in the tetrahedral layer in the case of the beidellite causes a much stronger ion binding of the Li$^+$ ion to the surface than occurs with the bentonite where the substitution is primarily in the octahedral layer.

One of the properties of clay–water systems which is important in their use is their ability to form gels or to exhibit elastic properties. If we use the bulk modulus as defined by

$$E = V\left(\frac{\mathrm{d}P}{\mathrm{d}V}\right)_T$$

it is immediately clear that for a *highly aligned* system of plates as apparently given by the lithium clays used in this work we can define the elastic modulus as

$$E = h\left(\frac{\mathrm{d}p}{\mathrm{d}H}\right)_T$$

and hence directly relate the modulus to the pressure against distance curves. As we pointed out in our previous work (Callaghan & Ottewill 1974) this suggests that the gel properties of well-dispersed montmorillonite plates should be directly related to the electrostatic interactions between them, an idea originally introduced by Norrish (1954). Evidence for the importance of this type of interaction in dispersions of sodium montmorillonite has also been found by Rand *et al.* (1980). We conclude therefore from our observations that the elastic modulus of lithium montmorillonite should be strongly dependent on lithium chloride concentration and that in the case of caesium montmorillonites, where plate–plate association occurs, the onset of elastic gel properties should occur at higher concentration of clay than required with lithium clay at the same homionic salt concentrations.

We wish to thank the Science and Engineering Research Council and English Clays Lovering Pochin and Company Limited for support of this work. We would also like to thank Dr W. B. Jepson for a number of valuable discussions.

References

Barclay, L. M. & Ottewill, R. H. 1970 *Spec. Disc. Faraday Soc.* **1**, 138–147.
Callaghan, I. C. & Ottewill, R. H. 1974 *Faraday Disc. Chem. Soc.* **57**, 110–118.
Cebula, D. J. & Ottewill, R. H. 1981 *Clays Clay Miner.* **29**, 73–75.
Cebula, D. J., Thomas, R. K., Middleton, S. R., Ottewill, R. H. & White, J. W. 1979 *Clays Clay Miner.* **27**, 39–52.
Cebula, D. J., Thomas, R. K. & White, J. W. 1980 *J. Chem. Soc. Faraday I* **76**, 314–321.
Greene, R. S. B., Murphy, P. J., Posner, A. M. & Quirk, J. P. 1974 *Clays Clay Miner.* **22**, 185–188.
Israelachvili, J. N. & Tabor, D. 1972 *Proc. R. Soc. Lond.* A **331**, 19–38.
Israelachvili, J. N. & Tabor, D. 1973 *Prog. Surface Membrane Sci.* **7**, 1–55.
Langmuir, I. 1938 *J. chem. Phys.* **6**, 873–896.
Lubetkin, S. D. & Ottewill, R. H. 1984 (In the press.)
Lyons, J. S., Furlong, D. N. & Healy, T. W. 1981 *Aust. J. Chem.*, **34**, 1177–1187.
Marshall, C. E. 1949 *The colloid chemistry of the silicate minerals*, vol. 1. New York: Academic Press.

Middleton, S. R. 1978 Ph.D. Thesis, University of Bristol.
Norrish, K. 1954 *Dis. Faraday Soc.* **18**, 120–134.
Olphen, H. van 1963 *An introduction to clay colloid chemistry.* London: Interscience.
Rand, B., Pekenc, E., Goodwin, J. W. & Smith, R. W. 1980 *J. Chem. Soc. Faraday* I, **76**, 225–235.
Tabor, D. & Winterton, R. H. S. 1969 *Proc. R. Soc. Lond.* A **312**, 435–450.
Tolansky, S. 1948 *Multiple beam interferometry of surfaces and films.* Oxford: University Press.
Verwey, E. J. W. & Overbeek, J. Th. G. 1948 *Theory of stability of lyophobic colloids.* Amsterdam: Elsevier.

Discussion

P. BARNES (*Department of Crystallography, Birkbeck College, Malet Street, London WC1E 7HX, U.K.*). The precision force–distance apparatus described by Professor Ottewill appears to resemble the design of Israelachvili *et al.* (Canberra, Australia) who has also studied mica–mica properties in various solutions. My memory of that system is that it depends for its success on having molecularly smooth crossed cylinders. This requirement would seem to restrict clay material usage to near perfect mica.

R. H. OTTEWILL. The direct measurement of interaction forces is restricted to using very smooth surfaces. The crossed-cylinder conformation that we used is similar to that originally used by Tabor & Winterton (1969) and in the subsequent studies by Israelachvili (for example, 1973 and 1978). The method of force measurement in our equipment, however, is different in design from that of Israelachvili and can be used for both kinetic measurements and constant force measurements. Our interest was essentially to compare direct measurements of force between single macroscopic surfaces with the pressures measured on assemblies of clay plates. The results obtained with lithium montmorillonite give us some confidence that the two can be related and that the compression technique can be used to obtain information about clay–water dispersions.

References

Israelachvili, J. N. 1978 *Faraday Disc. Chem. Soc.* **65**, 20–24.
Israelachvili, J. N. & Tabor, D. 1973 *Prog. Surface Membrane Sci.* **7**, 1–55.
Tabor, D. & Winterton, R. H. S. 1969 *Proc. R. Soc. Lond.* A **312**, 435–450.

P. NADEAU (*The Macaulay Institute for Soil Research, Craigiebuckler, Aberdeen AB9 2QJ, U.K.*). I would like to say that transmission electron microscopy and X-ray diffraction results obtained at the Macaulay Institute confirm the presence of elementary smectite layers in suspension. The achievement of complete dispersion of the clay is related not only to the exchangeable cation, but also to the use of dialysis to ensure quantitative retention of the finely dispersed material during final washings, and to the removal of incompletely dispersed aggregates by centrifugation. Another factor of importance is the concentration of clay material in the final suspension.

R. H. OTTEWILL. I agree with your comments. In our work, in the preparation of the homo-ionic clays, a number of dialysis stages were used, each followed by sedimentation to remove extraneous material. In the case of lithium montmorillonite, using cytochrome *c*-coated carbon grids as the substrate, single plates of clay were visible in transmission electron microscopy. In the case of caesium montmorillonite stacks of plates were common. It is certainly probable that the formation of tactoids is both clay concentration and electrolyte concentration dependent. This information should, in principle, be derivable from the compression curves given in the paper.

J. FRIPIAT (*C.N.R.S.*, *C.R.S.O.C.I.*, 1 *B rue de la Férollerie*, 45045, *Orleans, Cedex, France*). In his talk Professor Ottewill related the swelling pressure to the water/clay ratio. Then the water/clay ratio was transformed into an average distance between clay sheets by assuming the complete availability of the interlayer surface area (*ca.*800 m^2 g^{-1}) in agreement with Low & Oliphant (1982). It was shown that the swelling pressure can be predicted by the double layer theory for ratios higher than 10.

The procedure for calculating the average distance implies that infinite swelling occurs in dilute clay suspensions at low ionic strength.

This conception contradicts observations reported by Fripiat (this symposium) which lead to the conclusion that even under these extreme conditions, tactoids are formed and that these tactoids must be considered as 'textural units' from which the suspension is made. This conclusion came from the good agreement between heats of immersion and nuclear magnetic resonance relaxation time measurements which, in case of swelling silicates, shows that the surface available to water is about an order of magnitude less than the total (internal and external) surface area.

Ottewill has used, among others, Li and Na-Wyoming bentonites, whereas Fripiat studied Na and Ca-hectorites or Na-laponite but it seems rather improbable that the 10-fold difference can be explained by differences in clay compositions or morphology, or both.

There is thus an important disagreement between the approaches followed in these two contributions which deserves further study, particularly because they are important in many situations where the aggregation of clay particles is important (drilling muds, soil mechanics and so on).

Suppose, that only limited swelling occurs so that the intersheet distance never permits a fast exchange between interlayer water and bulk water. There may be an equilibrium between tactoids and individual sheets in the suspension but it should be strongly displaced in favour of the tactoids. As a first approximation let us assume that more than 80% of the sheets never expand for more than 2 nm (basal spacing about 3 nm). What would be the consequence in adopting such a model, which is strongly supported by two independent techniques (heats of immersion and n.m.r.) on the calculation of swelling pressures as carried out by Ottewill?

Reference

Low, P. F. & Oliphant, J. L. 1982 *J. Coll. Interf. Sci.* **89**, 366–373.

R. H. OTTEWILL. Thank you for your very interesting comments which are well taken.

In the case of lithium montmorillonite, thoroughly ion-exchanged over a substantial period of time, with frequent sedimentation stages to remove extraneous material, we are pretty certain that with Wyoming bentonite we do obtain a very well dispersed material. In the pressure measurements we also started with a very dilute dispersion, *ca.* 2% (by mass), and obtained alignment of the plates by successive compressions and decompressions. Determination of the interplate spacing by X-ray diffraction and neutron diffraction over a very wide concentration range gave distances that were within a small percentage of those calculated using the surface area of the clay as 763 m^2 g^{-1}. Moreover, as mentioned in my reply to Dr Nadeau transmission electron micrographs of lithium montmorillonite, using cytochrome c-coated carbon grids, showed the material to be essentially all single plates. In addition the compression measurements with lithium montmorillonite and their agreement with the theory of diffuse double layer interaction suggests that the plates are well dispersed. Clearly, however, there must be some 'grain boundaries' in the arrangement of the plates in the cell as a consequence of the polydispersity in the face areas of the plates. If tactoids do exist in aqueous dispersions of lithium montmorillonite I think it must be to a very small extent. It would be nice to check this by your n.m.r. technique but I think this would be difficult owing to the presence of isomorphously substituted Fe^{3+} in the lattice.

In the case of sodium, potassium and caesium montmorillonites we would certainly concur that tactoid formation occurs. In the case of caesium samples this is visible from the optical properties. The changes in the compression curves with the nature of the counter-ion can also be explained by face–face association with the counter-ions held between the plates. I have not attempted to put this on a quantitative basis but the question you have put has stimulated me to attempt this. The measurements with sodium montmorillonite would indicate a small extent of tactoid formation. This becomes more pronounced with potassium as the counter-ion and even more pronounced with caesium. This order of effect is confirmed by the small angle neutron scattering studies of Cebula *et al.* (1980) on dilute dispersions of the same materials.

I agree that there is an important point here in the difference between the methods, although the measurements have been made on different materials, which should be cleared up. I hope that in the near future we can pursue the matter further.

Reference

Cebula, D. J., Thomas, R. K. & White, J. W. 1980 *J. Chem. Soc. Faraday Trans.* I, **76**, 314–321.

Phil. Trans. R. Soc. Lond. A **311**, 369–373 (1984)   [369]
Printed in Great Britain

The importance of pore water chemistry on mechanical and engineering properties of clay soils

By I. T. Rosenqvist

Institutt for Geologi, Oslo University P.O. Box 1047, Oslo, Norway

The classical Coulomb–Terzaghi shear strength equation, $s = c + (\sigma - u) \tan \phi$, depends not only upon mechanical and mineralogical factors, but to a high degree upon the total chemistry of the system. Changes in chemistry owing to natural and anthropogenic factors may change the strength properties of a given clay soil at a given water content. The classical example of slow acting chemical changes is the development of quick clay properties. Modern infiltration of different electrolytes and detergents from leaking sewer systems may change shear strength and compressibility of the ground. These processes are different in clays of expanding and non-expanding minerals. The importance of the different cation species seems to depend upon valency and polarizability. Influence of anions is more complicated and depends upon interaction with edge charges and the crystalline properties of the clay minerals.

As we know clay does not represent any definite chemical or mineralogical compound and may vary considerably even within a limited geographical area. Different clays may have very different mechanical properties. The reason for calling similar types of geological material 'clay' in different languages seems to have been that they have some mechanical properties in common. These may be plasticity, thixotropy, etc., depending upon mineral/water ratios. However, these properties are dependent upon many factors.

Until X-ray studies, combined with various other methods of micromineralogy, made it clear how the mineral phases varied from clay to clay, there was little accurate idea of their nature. The shear strength properties of all soils have, however, some features in common. These had been recognized by Coulomb in 1776, when he introduced the shear strength equation:

$$s = c + \sigma \tan \phi. \tag{1}$$

This states that the total shear resistance of a soil (s) can be considered as the sum of cohesive resistance (c) and the frictional term, σ, being the normal stress upon the surface of slippage and ϕ the angle of internal friction.

Von Terzaghi (1925) established the fundamental basis for soil mechanics. He introduced the concept of *effective stresses*, showing that the total stresses of Coulomb gave misleading results. Terzaghi's equation may be given as follows:

$$s = c + (\sigma - \bar{u}) \tan \phi, \tag{2}$$

$\bar{u}$ being the pore water pressure, i.e. the pressure in the voids in the soil. $(\sigma - u)$ may be written as $\bar{\sigma}$, the effective stress on the shear plane. The next major step forward was taken by Hvorslev (1937). He confirmed in principle the law of Coulomb–Terzaghi, but for any given clay, cohesion was found to be a function of water content. He therefore extended Coulomb's equation:

$$s = \bar{c} + \bar{\sigma} \tan \phi \qquad (3)$$

in which $\bar{c}$ now depends on the water content whereas c in (1) may be independent of water content. This dependence can be expressed by the introduction of the *equivalent* consolidation pressure θ_c, with a constant H:

$$s = H\sigma_c + \bar{\sigma} \tan \phi. \qquad (4)$$

The significance of the result of Hvorslev's investigations is very profound. The two main results: dependency of the cohesion on water content with a given mineralogy and chemistry, and recognition of the angle of internal friction as a soil characteristic which determines the strength of even rather plastic and fine-grained clays were most important steps in the advancement of knowledge of the fundamental strength properties of soils.

Hvorslev's investigation was carried out on remoulded soils, and his results are valid only for such material. This has often been forgotten by many research workers dealing with clay soil properties. As seen from an electron micrograph of *normally consolidated clays*, these consist of aggregates of platy minerals touching each other by corner-to-plane contacts in more or less dense cardhouse structures. Any deformation of such a body will necessarily involve a mutual movement of the particles relative to each other and a corresponding flow of the water in interstitial voids. As chemical electrostatic forces will act between the individual minerals at the points of contact, the deformation process must necessarily be complex; indeed it may *be stated that the strength of clay materials is not fully understood.*

Terzaghi's concept of effective stresses is still fully valid. If we consider a soil sample, we conclude from a general mathematical principle on invariance that any scientifically correct failure criterion must include the three effective principal stresses σ_1, σ_2 and σ_3. Several attempts have been made to solve the general failure problem, but because of lack of experimental data and insufficient theoretical understanding, we are unable to treat the complete problem, even for isotropic soils, especially as these become anisotropic as a result of deformation. By using originally anisotropic bodies with changing anisotropy, a complete solution of the frictional term has been attempted, but in vain. That cohesion is a function of the stress history and not only of the actual set of stresses adds to the complexity of the problem.

In practice, however, it has been proved that σ_1 and σ_3 are of the greatest importance for the strength, and engineering practice is to work with failure criteria where only these two principal stresses are used. In most cases we choose to put $\sigma_2 = \sigma_3$. This is done in all tests of the triaxial compression type. Thus we ignore the influence of the intermediate principal stress σ_2.

According to the Hvorslev concept of failure, a clay body exposed to a stress ellipsoid with rotational symmetry will have a plane of failure with the angle $45° + \frac{1}{2}\phi_r$, where the subscript r is used to emphasize the rotational symmetry to the σ_3 direction. As shown above, the Hvorslev true or real conditions of failure are based upon effective stresses in a given soil material having a constant voids ratio (ratio of pore volume to solids volume) under variable external conditions. If, on the other hand, the voids ratio varies as a function of external stresses, we may determine a set of Mohr circles for the effective stresses at failure. In this way, the envelope curve will determine the effective failure conditions, and we obtain the values for c and ϕ called respectively *effective cohesion* and *effective angle of friction*.

To determine c_r, ϕ_r, $\bar{c}$ and $\bar{\phi}$, respectively, we have to know the pore water pressure at failure.

In certain cases it is possible to analyse the problem of failure approximately only by total stresses. This gives us c and ϕ at Coulomb's conditions of failure called *the apparent cohesion* and *the apparent angle of friction*.

Several hypotheses have been advanced as to the nature of cohesion and friction in clay soils. The fact that clays are colloidal materials indicates that electrochemical forces are involved, at least in the cohesional terms (see below), probably also in the frictional term.

As all the failure equations from Coulomb to Hvorslev deal only with the bulk properties, that is, the macroscopic stress ellipsoid and the pore water pressure, it is obvious that none of these equations can predict the attractive and repulsive forces acting between the charged mineral particles, nor do they evaluate the cementing action of precipitated mineral substances at all the points and areas of contact between the minerals. This stresses the importance of the title of Hvorslev's (1937) paper. His equation was given for completely remoulded material where any cementing action could be ignored. In all natural geological deposits, we have to assume that secondary cementation may exist.

Among other authors, Denisov & Reltov (1961) have demonstrated the influence of natural cementing bonds. Such a cementation may be included as a *nonreversible* cohesion term, whereas some of the electrochemical forces may be included in the stress conditions as a set of internal stresses adding to the set of external effective stresses. The latter especially, will depend upon chemical changes in the pore water and mineralogical changes of the minerals. Thus for example, the subpolar *quick clays* have obtained their peculiar properties from secondary chemical changes in flocculated and slightly overconsolidated clays, these chemical changes being of a nature which increased the zeta potential of the minerals. This was dealt with by Rosenqvist (1977).

The simplest chemical change is represented by the leaching of salt from marine deposits owing to the seepage of rain water. Another reaction which will cause a similar effect upon the zeta potential is represented by the action of organic complexing anions from decaying humus. In both cases, soils of high sensitivity may develop. By *sensitivity* we understand a permanent loss of strength by remoulding, observed in many clays. If a clay sample of a given natural strength is thoroughly remoulded and the strength immediately measured, we mostly observe a considerable drop in the shear strength value. By storage, the strength gradually increases, but in many soils, it will never arrive at the initial values unless the water content drops, either by desiccation or by syneresis. Such a stored clay will again by a subsequent remodelling arrive at the same low shear strength as the original sample. Thus there is a reversible loss of strength in remoulded samples (thixotropy) and a nonreversible loss of strength in natural soils. This loss of strength is called sensitivity, and the ratio between the initial strength and the remoulded values is called *sensitivity value*. In the typical quick clays, the sensitivity may have values of more than 100, whereas the thixotropy value may only be 2–5. Thus these clays will lose 99 % of the initial strength by remoulding, and only 2–5 % of the initial strength will be regained by storage.

These processes seem to be restricted to clays of non-swelling minerals (illites, chlorites, kaolinite). In soils of some swelling minerals such as smectites, the chemical changes that cause the high sensitivity of quick clays may cause a *decrease* in the sensitivity. This is probably due to the volume change of the mineral phase caused by swelling. In such cases we may find that changes that reduce the sensitivity value of illitic clays may increase the sensitivity value of smectitic sediments.

[151]

Ever since the early days of modern soil mechanics, much emphasis has been given to the problem of the water in clay soils. Only recently, modern techniques, using diffusion of isotope tracers combined with nuclear magnetic resonance and infra red spectroscopy, have shown that the greater part of the aqueous phase of clay soils has *properties that are not fundamentally different* from those of water. However, through a transition zone of more or less gradually increasing rigidity, this free water in the voids becomes firmly bound water molecules on the surface of the clay minerals. Thermodynamically, and probably also mechanically, these innermost *one to three* water layers seem to have properties like those of a solid body. As any failure in a clay sediment involves a shearing in this innermost water, it is necessary to have a better knowledge of the innermost water in order to understand the physical basis for any rheological model of clay.

Johnson (1970) developed a rheological model describing the flow behaviour of remoulded clay material at a finite rate of deformation. This model is called a *Coulomb-viscous model*. It has the form:

$$\tau = c + \sigma_n \tan \phi + \eta, \frac{d\epsilon}{dt}, \quad \text{i.e.,} \quad \tau > c + \sigma_n \tan \overline{\phi}, \tag{5}$$

where τ is internal shear stress, σ_n is internal normal stress, ϕ is the angle of internal friction, η is viscosity, and ϵ is rate of shear strain (velocity gradient). In this case, it has to be noted that the cohesional term is not identical to the cohesion at rest.

We observe in all clay water systems a certain relaxation time if the strain rate is changed. If the strain rate is decreasing, we first observe a decreasing resistance against deformation, and after some time, a slow return to the constant value. On the other hand, if suddenly the rate of deformation is increased, we observe a temporary increase in resistance against deformation, and then a slow return to normal value.

It seems that an explanation of such data needs a rheological model differing from most 'spring and dashpot models' previously presented. In most cases these do not satisfy the observed consolidation curves either. We may consider two clay mineral particles under undisturbed conditions. We assume that they more or less touch each other at a point or a small volume of contact, and we assume that surrounding that area we have a small volume of coherently organized oxygen and hydrogen. The material in this volume behaves more or less like a solid substance of a given breaking strength. When the shear stress in the contact area exceeds a certain limit, the coherent structure will be broken, and the resistance F against mutual displacement will decrease with increasing rate of strain, that is, in accordance with the formula

$$F = a \Big/ \left(1 + \frac{d\epsilon}{dt}\right), \tag{6}$$

where a is a constant. The material thus shows *negative strain hardening*. This is characteristic of crystals where a limited number of dislocations arise by strain, and where the dislocations may diffuse out of the structure after rest.

The influence of exchangeable cations upon the shear strength of a remoulded clay depends very much upon the type of minerals in the fine fraction. In systems of non-swelling clay minerals, such as illitic-chloritic clays in Scandinavia and Canada, we found that not only the liquid limit but also the shear strength of the same clay mineral, having the various alkali ions in the water phase, regularly increases in the series Li^+, Na^+, K^+, Cs^+ at any given water content.

The positive cations are expected to be present on the basal planes of the clay mineral flakes, and to be absent from the edges and corners of the flakes. It may therefore be suggested that the van der Waals attraction in corner-to-plane contact is a function of the first power and not of the second power of the polarizability of the counter ion.

It we have a system with remoulded expanding clay minerals belonging to the smectite group, we have a different relationship. In these cases the shear strength seems to increase through the series K^+, Rb^+ and Cs^+, but decreases from Li^+ and Na^+ to K^+, at the same water content. Thus, at any given water content a K^+ smectite clay is softer than an Na^+ smectite clay at the same water content. This may result from the prevention of lattice expansion by the larger alkali cations, whereas the lithium and sodium clays are much more expandable, which is well known and will not be dealt with further in this paper.

The action of anions upon the shear strength is much more complicated. The dispersing action of polymetaphosphates and several organic humic anions has been qualitatively discussed several times, but quantitative data are rare. The use of equivalent concentrations of sodium salts of fluorine, chlorine, bromine, iodine, chromate and sulphate give different values for viscosity of clay water semi-liquids. Wendelbo (1983) has described the influence of these ions upon clay material, in which he believed he had blocked the edge charges by polymeta-phosphate, and it seems that these comparatively small anions have a moderate influence upon the basal charge of the clay minerals.

The data for *critical coagulation concentration* in diluted suspensions of kaolinite and smectite show the mutual influence of some simple anions and polymetaphosphates upon the attractive and repulsive forces between the minerals. The Bingham yield value of more concentrated clay water suspensions follow similar relations for the simple anions, in the case of smectites and kaolinites.

In the case of gibbsite–water systems at pH values where the minerals are positively charged the c.e.c. depends to a marked degree upon the charge of the anions in the order $PO_4^{3-} < SO_4^{2-} < Cl^-$. These data indicate that there is negligible adsorption or ion pair formation of the investigated anions upon the basal surface of the layered silicates, and the influence of these relatively small ions upon the charge properties of clays is probably due to adsorption upon *the broken edges*. The softening of clays at constant water content by addition of phosphate and silicate ions from detergents may be explained at least qualitatively in this way.

REFERENCES

Bjerrum, L. 1954 Theoretical and experimental investigations on the shear strength of soils. Norwegian Geotechnical Institute Publication number 5.

Coulomb, C. A. 1776 Essai sur une application des règles de maximis et minimis a quelques problemes de statique, relatifs à l' architecture. *Mem. Div. Sav. Academie des Sciences.*

Denisov, N. Y. & Reltov, B. F. 1961 The influence of certain processes on the strength of soils. *Proc. int. Conf. Soil Mech. Found. Eng., 5th Paris.* pp. 75–78.

Hvorslev, M. J. 1937 Uber die Festigkeitseigenschaften gestörter bindiger Böden. *Ingeniørvidenskab. Skrifter A.* number 45.

Johnson, A. M. 1970 *Physical processes in genology.* San Francisco: Freeman, Cooper & Co.

Rosenqvist, I. T. 1977 A general theory for quick clay properties. *Proc. third meeting European Clay Groups*, Oslo, Norway: Nordic Society for Clay Research.

Terzaghi, K. 1925 Erdbaumechanik auf boden-physikalischer Grundlage. Leipzig: Frantz Deuticke.

Terzaghi, K. 1943 *Theoretical soil mechanics.* New York: John Wiley.

Wendelbo, R. 1983 Gjensidig påvirkning av leirmineralers overflater og uorganiske anioner, spesielt sulfat. (Surface interaction between clay minerals and inorganic cations, specially sulphates). MSc. thesis, Oslo University.

Phil. Trans. R. Soc. Lond. A **311**, 375–389 (1984)   [375]
Printed in Great Britain

The significance of clays in agriculture and soils

By A. C. D. Newman

Soils and Plant Nutrition Department, Rothamsted Experimental Station,
Harpenden, Hertfordshire AL5 2JQ, U.K.

In managing soils for agricultural production, soil texture or particle-size distribution, and the amount of clay present are very important. Soil structure depends very much on clay: soils with little clay have a simple structure, whereas soils with much clay have complex structures and multimodal pore size distributions. Their response to changes in water content is structurally quite different from that of sandy soils.

Clays have a large specific surface, often predominantly negatively charged, that retains nutrients against leaching and reacts with hydrogen and aluminium ions, while buffering the soil against extreme pH changes. The clay itself may be a source of plant nutrients when it degrades.

Despite these known effects it is still difficult to predict soil behaviour from clay mineralogy. This is partly because the complexities of real clays in soil have been ignored in correlative studies, and that their effects on soil bulk properties are not understood. Future research should stress both of these factors and with improvement of techniques for describing the many structural imperfections in clays, new possibilities exist for predicting those of their properties that are important in agriculture.

1. Introduction

As the increase in agricultural productivity since 1940 has shown, man's intervention can vastly improve soil fertility on all of the types of soil with which he has to work (Cooke 1967, 1977). The actual productivity does, however, vary widely which shows that there are aspects of productivity outside man's direct control. One intrinsic factor in yield variation is soil type, which is related to the soil's composition and its position in the landscape (Burnham 1974). The amount of clay present is a prime consideration in the composition of the soil and has a highly important effect on its properties.

A soil usually contains at least some clay, and its clay content strongly influences its management and productivity (Davies *et al.* 1972). Soils with very little clay can be just as difficult to manage, for different reasons, as soils that contain large amounts, and in broad terms loam soils containing 15–25 % clay with particle sizes of under 2 μm and a larger proportion of silt particles sized 2–60 μm are the most productive. Such soils seem to contain enough clay to provide an adequate surface for interaction with water and nutrients, and to have a friable structure beneficial for tillage and root growth. Soils with more than 30–35 % by mass weight of clay tend to take on the properties of the clay itself, with the implications that they waterlog more easily during periods of excess rainfall, stay wet longer, require greater draft in cultivation and form large aggregates (clods) that must be broken down to form a favourable seed bed. In short, they pose more management problems than loamy soils.

Despite these apparently unfavourable properties conferred on soils by an excess of clay, clay makes a vital contribution to soil fertility. In combination with organic matter and sesquioxides, clay contributes coherence and structural stability which enables the soil to resist the mechanically destructive effects of rain and wind. Because clays have a large specific surface that is predominantly negatively charged, they retain cationic nutrients like K^+ and NH_4^+, and also absorb toxic substances. Layer silicate clays may also have plant nutrients present in their structure, and K^+ and Mg^{2+} can be released to soil solution under appropriate conditions.

Thus there is ample evidence to show that the amount of clay in a soil has a very important influence on the properties of that soil. In the present discussion, it is more interesting to consider what significance clay mineral *type* has in soils and agriculture.

The first problem that this question poses is the adequate identification, let alone quantitative estimation, of the clays actually present in soils. Soil clays differ in many ways from typical or standard clay minerals, and the variants likely to be encountered are discussed first. In succeeding sections the chemical and physical effects that clays have on soil properties are considered and finally an attempt is made to evaluate how far clay mineralogy has a proven or likely influence on soil properties.

2. CLAY MINERALS IN SOILS

The clay mineralogy of a soil is the result of several factors interacting with the soil parent material: climate and consequent intensity of weathering; accumulation and losses of substance within the soil profile, often related to its position in the landscape; vegetation; and the duration of these influences. Only in certain combinations of circumstances do soil forming processes have a uniquely determining effect on the clay mineralogy. Examples where this does happen are the highly weathered soils of the tropics where sesquioxides and kaolinite are dominant clay minerals (Oxisols) and the black soils where shrink–swell processes are attributed to the dominance of smectite (Vertisols). Another unique group of soils are those formed on volcanic ash rocks where allophane (amorphous silico-aluminas) are dominant. In many other types mineralogy is not a dominant characteristic and the soils can contain illite, smectite, vermiculite, kaolinite, chlorite and mixed-layer minerals inherited from the parent material, but partly altered and supplemented by the soil-forming processes (Jackson 1964). In England and Wales, Avery & Bullock (1977) found that the mineralogy of clay soils depended quite markedly on inheritance from the parent material. Mineralogical variability within a single geological formation could be as great as interformational variability and tended to obscure the influence of soil-forming processes. Apart from the relation between vertic features (that is, features similar to but less intense than those in Vertisols) and smectite content, they concluded that in England and Wales mineralogy was not crucial in soil classification.

Therefore, except for the specific instances mentioned, a search for links between soil mineralogy and soil properties of agricultural importance is likely to be difficult. It will be unreasonable to expect strong relations if, for instance, the description of the minerals actually present in the soil is inadequate or incomplete. Further, as soil minerals often differ from those already known and described as 'type' minerals, it is necessary also to investigate the properties of these minerals relevant to bulk soil properties.

In general, soil clays are mixtures of several components, and each of these components may have a relatively low order of regularity in its structure. The task of uniquely identifying the components is therefore complex, and much effort may be required to give an adequate description of a soil clay. In favourable circumstances, it can be shown that soil kaolinites are *b*-axis disordered and that illites have stacking disorder and contain less K and more H_2O than true micas. Many smectite-like minerals contain mixtures of two or more types of layer, illite-smectite being the commonest, but kaolinite-smectite and chlorite-smectite also being found, and combinations of three layer types are sometimes suspected, but are difficult to identify. As clay minerals have a large specific surface accessible to soil solution, substances are sorbed on the surfaces and may fundamentally alter their properties. For instance the sorption of hydroxy-aluminium polymers on the negatively charged surface may decrease the effective charge density and make it pH-dependent (de Villiers & Jackson 1967). Soil organic matter is strongly associated with the clay fraction by coordinating to such surface-sorbed polyvalent metals (Edwards & Bremner 1967) and often can only be removed by total destruction. Polymers are sorbed by cooperative van der Waals forces and the sensitivity of infrared spectroscopy has shown likewise that complete dissociation from the clay surface is very difficult. Most soil clays contain a very significant proportion of fine clay under 0.2 μm, with a tendency for the clays to become more poorly crystalline as the particle size decreases. Such fine particle fractions have a large specific surface and may exert an influence on soil bulk properties, for example in the imbibition of water, out of proportion to their concentration in the soil.

3. CHEMICAL PROPERTIES OF CLAYS

3.1. *The buffering action of clays*

Free hydrogen ions with an activity greater than at about pH 5 react with clays, dissolving octahedral Al, Mg and Fe (Osthaus 1955). In time, H^+-saturated clays become partly Al-saturated (Eeckman & Laudelout 1961), because according to the mass action equations for cation exchange, the trivalent aluminium ions are preferentially sorbed on the clay surface, displacing H^+ ions (Chernov 1947; Coulter 1969a, b). In solution, Al^{3+} hydrolyses readily to the binuclear species $Al_2(OH)_2^{4+}$ and to polymeric cations like $[AlO_4Al_{12}(OH)_{24}12H_2O]_7^+$ (Aveston 1965) and possibly $Al_8(OH)_{20}^{4+}$ (Akitt & Farthing 1981). These are strongly sorbed on the interlamellar surfaces of smectites and vermiculites, forming hydroxyaluminium interlayers (Rich 1968). The stoichiometry of the reactions is quite well defined (Brown & Newman 1973) and similar equilibria have been shown to exist in soils (Richberg & Adams 1970).

Soil-forming processes usually cause an accumulation of hydrogen ions in the soil. Biological activity releases organic acids and HCO_3^- into the soil solution, and rainwater also contains acids and CO_2. When there is leaching through the soil, bases are carried down with the leachate and there is a net accumulation of hydrogen ions. Thus in soils not containing free calcium carbonate, colloids in the upper soil layers tend to retain aluminium compounds complexed with organic matter and sorbed on clay surfaces.

This process of acidification and the consequent formation of hydroxyaluminium coatings on the surface of soil colloids is fundamentally important in pedogenesis and in determining soil properties of agricultural significance (Jackson 1963). For instance, it means that if soil acidity is allowed to increase and fall below pH about 5, the point at which hydrogen ions begin to depolymerise the hydroxyaluminium complexes, free Al^{3+} is released into the soil solution. Even in small concentrations, Al^{3+} is toxic to many agricultural crops, and generally speaking, concentrations in the range 2–5 μg cm^{-3} inhibit the growth of sensitive crops like barley. As is well known from the early analyses by Voelcker of drainage water from the Broadbalk experiment at Rothamsted (Lawes *et al.* 1882), nitrogen fertilizers and ammonium salts in particular greatly accelerate the leaching of bases from soil, and the consequent accumulation of acidity must be neutralized by regular additions of calcium carbonate.

Clays that contain hydroxyaluminium (Al–OH) compounds have different properties from those of the parent clay, and (Al–OH) smectite neither swells nor dehydrates like the parent smectite (Kidder & Reed 1972). The Al–OH polymers form 'islands' between the individual aluminosilicate layers as well as on the surface, and these islands do not dehydroxylate below 600 K. This property, together with the acidity function of the residual water coordinated to the polynuclear cation, gives such clays industrially useful catalytic properties (Vaughan *et al.* 1979). It may be that the presence of such interlayers in soil clays plays an important part in the chemistry of soil organic matter. Because interlayer Al–OH polymers are relatively immobile and inhibit changes in the interlayer spacing they alter the field strength within the interlamellar space, with consequences for cation exchange equilibrium (Kozak & Huang 1971; Coulter 1969) and for the fixation of K^+ and NH_4^+ (Barshad & Kishk 1970; see §3.4).

In general the practical consequences of aluminium interlayering in soils have been little studied and one reason for this is that the extent and structural consequences of interlayering have proved very difficult to describe accurately. For example, aluminium interlayering in mixed-layer illite–smectite is likely in the clay fraction of many acid brown earth soils but has never been investigated in detail owing to the complexities that it introduces to the X-ray diffraction profiles.

3.2. *Cation exchange phenomena*

Clay minerals have net negative charge at soil pH-values and in natural conditions the cations countering this charge are Ca^{2+}, Mg^{2+}, K^+, Na^+, NH_4^+ and Al^{3+} with its hydrolysed derivatives. Soils with pH above 6 have Ca^{2+} as the predominant cation unless Na^+ has accumulated in the profile, but at more acid pH values, increasing amounts of Al balance the negative charge. Soils with kaolinite as the predominant clay mineral also contain iron oxides as important accessory minerals, and there has been much interest in whether the two minerals act independently. The pH-dependent charge characteristics of the combination is effectively the sum of the component parts, so that they appear to behave as a mixture. However, because soil kaolinites have a relatively large specific surface, the negative charge is considerably larger than usual with mineral kaolinites (Greenland 1975). Soils with 2:1 minerals as the dominant clays have large exchange capacities and pH-dependent charge contributes a much smaller fraction of the total than with kaolinitic soils.

In general, exchangeable cations can be regarded as accessible to plants and are considered to be the immediately available sources of nutrient cations, e.g. K^+ and Mg^{2+}. However, balance sheet calculations in plant growth experiments have often shown (Arnold & Close 1961; Feigenbaum & Hagin 1967; Talibudeen & Dey 1968) that more K can be extracted from soils by plants than is exchangeable, and this has led to the hypothesis that several categories of soil K exist, with transfer between them. Micas and illites are the commonest and most effective ultimate sources of K in soils and a large volume of research was done 10–20 years ago on the mechanisms of potassium release from these minerals (Norrish 1973). Even if definite categories of soil K exist, equilibrium between them may not be established, because solid-state diffusion plays an important part in limiting the rate at which K can be released from soil particles (Talibudeen *et al.* 1978). As these ultimate diffusion rates are so slow, it would not be surprising if particle size was an important factor in K-release, the smallest particles having the largest specific surface and releasing K faster than the larger, as was suggested by Arnold (1960). However, the smallest particle sizes in soil illites are very resistant to attack, the optimum size for exchange of K being 1–5 μm (Scott 1968). As noted earlier, many soil clays are mixed-layer minerals, illite-smectite combinations being especially common in the finer clay fractions, whereas the coarser fractions usually contain appreciable amounts of unstratified illite. Consequently it is predicted that K-release is more rapid from the coarser soil fractions than the fine. Thus, whole soil clays with particle sizes under 2 μm from an experimental field at Rothamsted were extracted with hot $BaCl_2$, which is a very effective reagent for removing potassium from micas (Avery *et al.* 1972). The products were then fractionated into the sizes less than 0.5 μm, 0.5–1.5 μm, and 1.5–2 μm, which were individually examined by X-ray diffraction and analysed chemically. Under laboratory conditions in which almost all the discrete illite in the coarse clay fraction was converted into vermiculite, the mixed-layer minerals in the finest fraction were scarcely altered at all (table 1) and the mixed-layer clay was more stable than the discrete illite. In the field situation, this may not necessarily be true, because the presence of organic acids and carbon dioxide may alter the course of the weathering reaction towards mineral dissolution, so that the fine particles may be more vulnerable to attack than the coarse.

TABLE 1. RELEASE OF K FROM CLAY FRACTIONS OF BARNFIELD SOIL,
ROTHAMSTED (AFTER AVERY *ET AL.* 1972)

(Mineralogy: before extraction: illite, mixed-layer; after extraction: vermiculite, mixed layer. The fine clay contains less illite and more mixed-layer mineral than the coarse clay.)

size fraction	1.5–2 μm	under 0.5 μm
K contents (%)		
before extraction with $BaCl_2$	2.36	1.70
after extraction with $BaCl_2$	1.22	1.55

Besides K, clays are also a potential source of Mg, which is essential not only for plant growth but also for the health of grazing animals (Bolton & Penny 1978). Most non-organic soil Mg is present in 2:1 layer silicate minerals, partly in the octahedral sheet, but also in the brucite sheets in chlorites. Much less research has been done on soil Mg than on soil K, but laboratory experiments show that the clay fractions of soils are able to release Mg to acid attack. Pot experiments have shown, however, that little Mg other than that exchangeable by ammonium acetate is actually extracted by ryegrass (Bolton 1973), and the rate of release is inadequate to supply the recommended amounts for plant and animal health.

Cation exchange equilibria studies on clays have generally used well-characterized minerals, but when the surface is modified by the introduction of Al–OH polymers, the selectivity relations may be changed. Kozak & Huang (1971) found that interlayering vermiculite with the Al–OH polymers decreased its selectivity for K^+ in K^+:Ca^{2+} exchange equilibria, although montmorillonite was little affected. This appears to be another instance in which the specific nature of soil clay surfaces must be taken into consideration.

[159]

3.3. *Soil clay–phosphate interactions*

The reaction that occurs between clays and phosphates depends both on the nature of the clay surface and the concentration of the phosphate. If the concentration of phosphate is high, new phases are formed: in molar solutions of sodium and potassium phosphates, variscite ($AlPO_4$ $2H_2O$) and strengite ($FePO_42H_2O$) are formed from kaolinite and greenalite (Kittrick & Jackson 1956), and taranakite group minerals are formed by the reaction of ammonium phosphate with halloysite and allophane (Wada 1959). Such reactions, however, have only transient significance in agriculture, and it is the reactions of soils with small concentrations of phosphate that are most important.

Most attention has been directed to the sorption of phosphate on surfaces of variable charge, in particular calcium carbonate and the hydrous oxides of iron and aluminium (for a review see White 1980). Iron hydrous oxides are the soil constituents most implicated in phosphate sorption. The reactions that occur when phosphate is sorbed on goethite (α-FeOOH) have been studied in detail by Parfitt *et al.* (1976) who proposed that phosphate formed a bridging ligand between two Fe atoms on the 100 face of the crystal. Phosphate sorption on goethite reaches a fairly well-defined maximum, and pH changes during the reaction suggest that release of OH^- ions from the surface partly compensate for the increase in negative charge carried by the phosphate ions. Whether a similar reaction occurs in the surface of other hydrous oxides of iron depends on the detailed structure and the distribution of OH ions on the surface. The sorption of phosphate on gibbsite (γ-Al(OH)$_3$) is less well defined and does not reach a maximum; this may indicate that recrystallization is occurring at the surface as the phosphate builds up (van Riemsdijk & Lyklema 1980), as has been shown to occur on the surface of $CaCO_3$ (Griffin & Jurinak 1974).

Relative to alumina, kaolinite sorbs much less phosphate per unit of surface, one phosphate anion occupying 240×10^{-20} m² at pH 4.4 and 740×10^{-20} m² at pH 7.6 for the sorption maximum indicated by application of the Langmuir equation to the sorption isotherm (Chen *et al.* 1973). It is believed that phosphate is sorbed only on the non-basal crystal faces and the decrease in sorption maximum with pH is in line with the decrease in positive edge charge (Schofield & Samson 1954). Few studies have been made of the sorption of phosphate by 2:1 clay minerals, but Hall & Baker (1971) investigated the effect of laboratory-synthesized Al–OH interlayers in montmorillonite and vermiculite on phosphate sorption. They found that increasing amounts of phosphate were fixed by montmorillonite (as measured by ^{32}P exchange) as the pH was raised, but that the opposite occurred with vermiculite. They suggested that this was because the Al–OH interlayers are more stable in vermiculite than in montmorillonite (Rich 1968), so that with the latter mineral, there was a greater possibility of forming insoluble aluminium phosphate phases. This study emphasizes the general principle that phosphate reactions in soil are strongly dependent on the presence of surface alumina (Veith & Sposito 1977) and that an adequate description of such surface-sorbed species in soil is needed before the full nature of phosphate interaction with soil clays can be understood.

3.4. *Nitrogen compounds and clays*

As mentioned in §3.2, ammonium cations are retained as counter ions at negatively charged clay surfaces but interaction between the interlayer cation and the higher charge density minerals like vermiculite and weathered mica residues promotes dehydration of the interlamellar region. When the cation has a low hydration energy like NH_4^+ or K^+, the dehydration develops to the point where the cation is enclosed within the twelve-fold coordination cavity of the basal oxygen surfaces and exchanges only with difficulty. This is the origin of 'ammonium fixation' in vermiculite (Allison *et al.* 1953) and only a small proportion of ammonium fixed in this way is isotopically exchangeable (Newman & Oliver 1966), although it can be exchanged by successive extraction in the same way as potassium from micas (Newman 1969). Small quantities of ammonium occur naturally in clays and shales (Yaalon & Feigin 1970) but frequently sedimentary clays contain organic nitrogen compounds, often amines, amino acids and bases (Dungworth *et al.* 1977).

The polarizing action of surface cations on water molecules in the inner hydration shell imparts proton-donating properties (Bronsted acidity) to the clay which is capable of reacting with nitrogen compounds and other bases. This is evident not only with ammonia sorbed on montmorillonite and vermiculite, which has been shown by infrared examination to be converted into NH_4^+ (Mortland *et al.* 1963), but also with other bases (Mortland 1970). Two factors influence the protonation: the degree of the polarizing action of the cation on coordinated water, and the strength of the base. Thus H^+-, Fe^{3+}- and Al^{3+}-clays will protonate urea and amides, but the Na^+ and Ca^{2+} forms do not. Water content of the clay appears to be important in the reactions: as the water content of the clay is decreased, the polarizing action of the cations is concentrated on a smaller number of water molecules, and this increases their proton-donating ability (Mortland & Raman 1968).

Nitrate is the principal plant nutrient form of nitrogen. Unlike phosphate, no specific chemisorption reactions with oxide surfaces are known, so the main effect of clay on nitrate in the soil is to exclude it from a volume of the soil solution adjacent to the clay surface (van Olphen 1977). The extent of this exclusion is still unknown (Wild & Cameron 1980) but in clay soils with very small pore sizes (see §4) it could have the effect of concentrating nitrate on the surfaces of soil aggregates where it would be more prone to leaching but also more accessible to plant roots.

3.5. *Pesticide–clay interactions*

A wide range of interactions occurs between organic pesticides and clay minerals, depending on the nature of the compounds. The extremes are exemplified by paraquat, which is cationic and strongly sorbed, and picloram, which is acidic and repelled from the clay surface. In an excellent review, Green (1974) emphasizes the wide range of behaviour by juxtaposing the sorption isotherms of Na- and Ca-montmorillonite for paraquat, monuron, prometone, lindane, picloram and an organophosphate insecticide. The shape of the isotherm gives some indication of the nature of the adsorption reaction. Paraquat, 1,1'-dimethyl 4,4'-bipyridinium dichloride, is strongly sorbed by coulombic forces enhanced by van der Waals interaction, so the isotherm is very steep initially but has a definite sorption limit. Prometone is also strongly adsorbed, presumably because its basic nature allows a proton donation reaction at the clay surface. Monuron is less strongly sorbed but, as noted in section 3.4, in a system of high acid activity there may be protonation of the urea base. Little interaction can be expected between chlorinated hydrocarbons such as lindane and clays, and the isotherm suggests that there is a constant partition between the solution and sorbent.

Thus in deciding what may be the principal interaction mechanism between clay and a pesticide, the main points to consider are the acidity of the clay and the chemical nature of the pesticide.

4. PHYSICAL PROPERTIES OF CLAY SOILS

In each of the aspects of soil physical behaviour considered in this section the influence of clay content and mineralogy is closely related to the ability of the clay component to take up and release water. After the work of Schofield (1935), the relation between soil water content and soil water suction (negative pressure in pore water) is often used to characterize soil–water interaction. The maximum quantity of water that can be stored in a soil is considered to be that retained when the soil profile in the field is fully wet and excess has drained away, often defined as the water content at -5 kPa. Plant-available water is usually defined as the difference between this water content and the water content at -1.5 MPa, most plants being unable to extract water held at greater suctions. At this suction most clay soils still contain a substantial amount of residual water, but as the main drying action on subsoils comes from the extraction of water by plant roots, it is rare for subsoils to dry to greater suctions.

TABLE 2. VOLUMETRIC WATER CONTENTS (PERCENTAGE) IN SOILS OF
CONTRASTING MINERALOGY

	soil mineralogy	water suction/MPa 0.01	1.0	difference
remoulded	a*	57.1	32.0	25.1
	b	49.9	31.2	18.7
	c	48.2	33.7	14.5
remoulded, dried and rewetted	a	46.4	32.1	14.3
	b	38.7	27.3	11.4
	c	46.2	32.8	13.4

* a: smectite; b; mixed-layer illite–smectite; c: illite.

The volumetric water contents at both of these limiting points increases linearly with clay content up to about 40%, but there is a wide scatter about the regression line. For instance, at 50% clay content, the water content at -1.5 MPa can vary between 22 and 37% (Hall et al. 1977) and although some of this scatter may be attributed to clay mineralogy, the structural arrangement of the particles may also contribute. If a London Clay soil is slurried with excess water and dried under gradually increasing suction, the water content decreases as shown by the solid line crossing figure 1 from right to left. If, after drying to 100 MPa suction, an attempt is made to rewet the soil by decreasing the suction, the drying curve is not retraced, and at each change in suction, the water content lags further behind that on the drying curve, giving the s-shaped full curve shown in figure 1. On further drying and rewetting cycles, the closed hysteresis loop shown is followed reproducibly. The natural soil, when dried, follows the intermediate (dashed) line. These results, which were obtained by Croney & Coleman (1954), appear to show that disruption of particle–particle associations in the clay (by remoulding or making the soil into a slurry) causes thicker water films to be interposed between the clay particles than would develop if the clay was allowed to swell freely but without the application of mechanical work.

The other soil physical property affected by loss of water is the soil volume, which contracts in response to the compression forces of capillary origin acting on the boundaries of the coherent soil aggregate. The general form of the relation between soil volume and water content is shown in figure 2. For remoulded clay soils, there is a region where the volume contraction

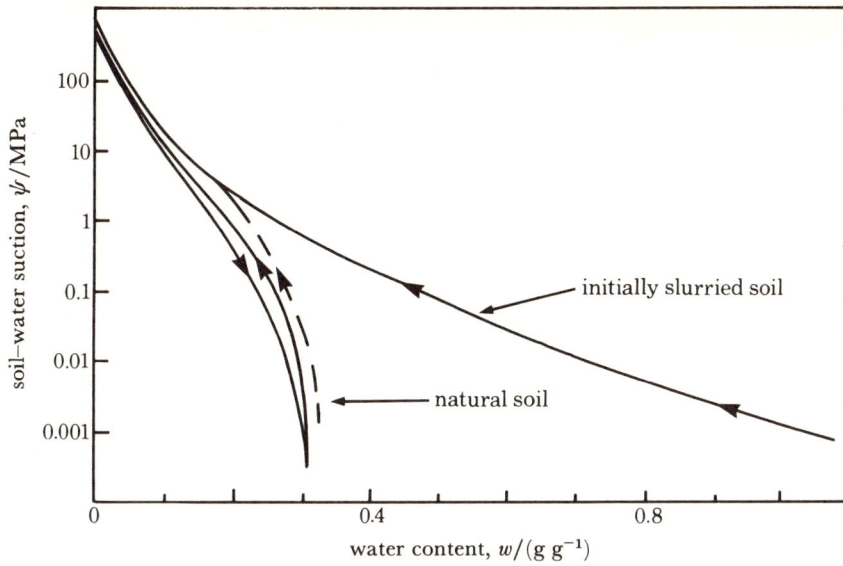

FIGURE 1. Matric suction–water content relation for London Clay soil.
(After Croney & Coleman 1954.)

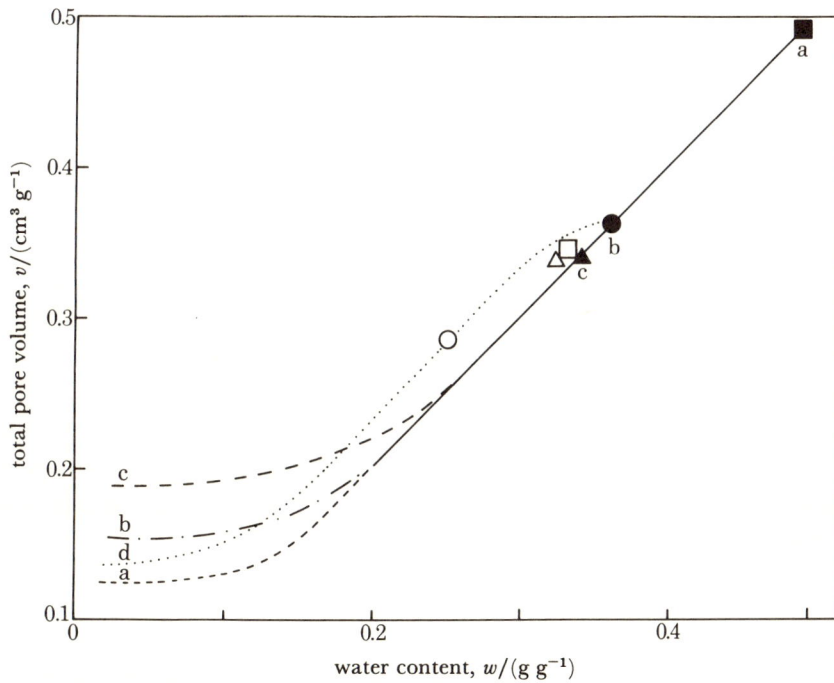

FIGURE 2. Shrinkage curves for remoulded soils containing (a) smectite, (b) illite–smectite (60:40), (c) illite. Curve (d) is the shrinkage curve for a natural unremoulded soil. The filled symbols (■, ●, ▲) represent soils freshly remoulded at -10 kPa, the open symbols (□, ○, △) soils dried to -100 MPa before wetting at -10 kPa.

(shrinkage) equals the volume of water removed by drying, this linear relation being known as 'normal shrinkage'. At some point on this line, which depends on soil composition, shrinkage starts to lag behind the change in water content, until no further shrinkage is possible and the 'shrinkage limit' is reached. At the point where shrinkage starts to lag, either fractures develop or air starts to enter the pore system of the soil, as the pore water pressure becomes sufficiently negative to draw liquid menisci through the soil capillaries. This condition is dependent on the size of the capillaries according to the Young–Laplace equation

$$-\Delta p = 2\sigma \cos \theta / r,$$

where Δp is the reduction in pressure across the menisci, σ is the surface tension of water, θ is the contact angle and r is the radius of the capillary. Fractures initiate and develop where the stress of shrinkage is concentrated at points of weakness on the boundaries.

An example of the shrinkage behaviour for a natural, unremoulded soil aggregate is given in figure 2. Natural aggregates show nearly normal shrinkage, the gradient of the linear section being very slightly less than unity for clay soils (Lawrence et al. 1979). As the amount of clay in a soil decreases, the length of the normal or nearly normal shrinkage line decreases until for soils with 25 % or less clay, shrinkage lags behind water content change over most of the range.

The effect of mineralogy on shrinkage behaviour is also shown in figure 2. All three soils contained 45–55 % clay, soil (a) having smectite as the dominant mineral, soil (b) containing mainly a smectite–illite mixed-layer mineral, and soil (c) containing predominantly illite. The soils were remoulded at − 10 kPa until no further water was taken up, or were remoulded at − 10 kPa, dried to − 100 MPa and then rewetted without remoulding at − 10 kPa. These two treatments of the soils are approximately equivalent to the initial state and the rewetted state in the experiment of Croney & Coleman (figure 1). The effect of remoulding and drying is mineral-dependent, with the smectite soil taking up a large quantity of additional water on remoulding at − 10 kPa, but the illite soil taking up very little extra, the mixed layer mineral being intermediate. Furthermore, the shrinkage limit is also dependent on mineralogy, with the illitic soil shrinking least and the smectitic soil most. However, contrary to expectation, the plant-available water differs little between the three soils (see §4.1).

The origin of these differences can be shown to lie in the pore structure of the soils. Mercury porosimetry is a useful technique for studying pore structure when used in combination with critical point drying to preserve the wet volume of soil (Greene-Kelly 1973; Lawrence 1977) and mercury penetration data for the three soils is shown in figures 3 and 4. The smectitic soil (figure 3a) on remoulding develops a large volume of pores in the size range 30–100 nm but remoulding the illitic soil develops a much smaller volume of pores with a wider pore size distribution; again the soil with mixed-layer mineralogy is intermediate. When dried to the shrinkage limit, however, the situations are reversed, the illitic soil containing a larger volume of coarser-sized pores whereas the smectitic soil contains a small volume of very fine pores (figure 4). These pore size distributions show that air-entry should occur at a higher water content in the illitic soil than in the smectitic, as was found in the shrinkage behaviour.

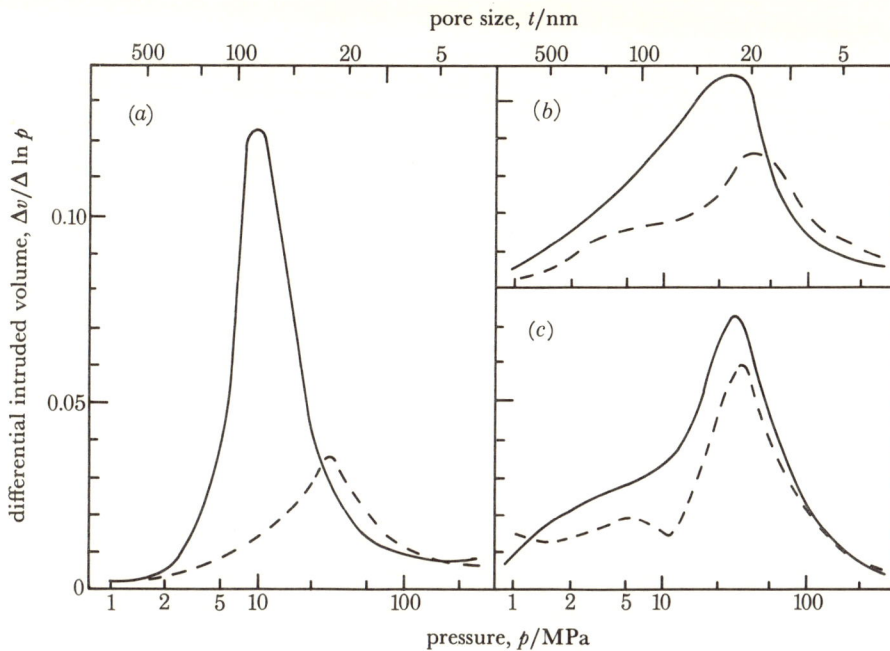

FIGURE 3. Mercury penetration of critical point-dried clay soils containing (a) smectite, (b) illite–smectite (60:40), (c) illite. Soil preparation: ———, remoulded at −10 kPa; – – – –, remoulded at −10 kPa, dried at −100 MPa, then rewetted at −10 kPa.

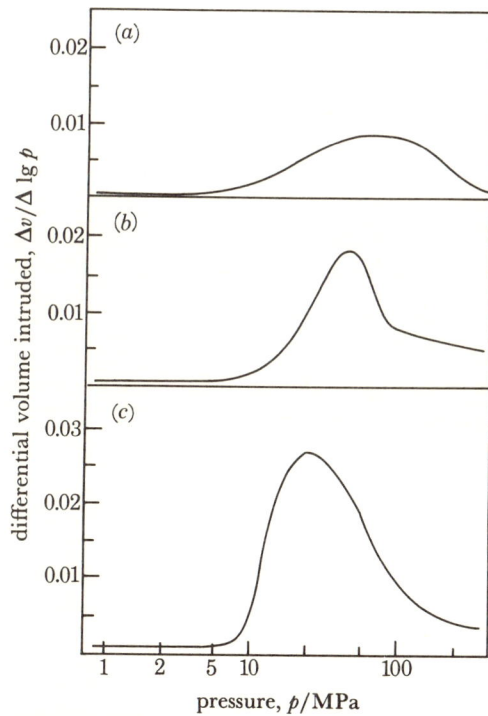

FIGURE 4. Mercury penetration of air-dried soils contains (a) smectite, (b) illite–smectite (60:40), (c) illite.

4.1. Plant-available water

There have been several investigations of correlations between conventionally defined plant available water and soil composition. Salter *et al.* (1966) used a limited range of coarse textured soils and derived a regression equation based on sand, silt and organic carbon contents. Using sieved and compacted topsoils, Archer & Smith (1972) found, with a smaller number of soils but of wider textural range, that available water was inversely related to dry bulk density (the weight of solid soil material in unit volume of whole soil in any specified condition). Reeve *et al.* (1973) confirmed this relation for some undisturbed soils and Hall *et al.* (1977) extended the correlation with samples from 261 soil profiles. The question that this work poses is whether dry bulk density at a defined pore water pressure is related to soil mineralogy when due allowance is made for the air-filled pore space within the soil sample. Clays that swell most should be capable of imbibing most water and contributing most to plant-available water. However, for a range of 63 non-remoulded clay soils Greene-Kelly (1974) found that both total and available water contents were poorly related to mineralogy. Equally, soils with widely different mineralogies may show little difference in water held between specific suctions (figure 2). The soil containing the mixed-layer smectite has relatively little plant available water presumably because microstructural arrangement of the clay particles affects the release of available water more than loss of interlamellar water from smectite layers (see §4.2).

4.2. Shrinkage and swelling phenomena

In clay soils the ability to swell and shrink in response to fluctuation in water addition and removal is vital to the active generation of the visible macrostructure, in which soil aggregates are bounded by fissures and channels that conduct excess water and allow air and roots to penetrate into the subsoil. Excess shrinkage and swelling, however, causes problems of structural stability in civil engineering and there has been much study of the relation of shrinkage and swelling to mineralogy. Smectite minerals in subsoils have been associated with such structural instability but Martin & Lambe (1957) suggested that mixed-layer smectite–illite was much less active in these processes than pure smectites. In the study of the relation of shrinkage and swelling to clay mineralogy referred to earlier (Greene-Kelly 1974) no distinction was made between smectite in the strict sense and smectite in mixed layer materials. It may be that a better correlation with mineralogy would have been found if this distinction had been made. Even so, the structural arrangement of the clay particles and the presence of sesquioxides and organic matter will also influence swelling.

The contribution of changes in interlamellar water content to shrinkage and swelling remains a debatable issue. Unless they are Na^+-saturated, smectites contain 3 interlamellar sheets of water when fully hydrated (1.9×10^{-9} m basal spacing), and it has proved difficult to define exactly at what water activity this dehydrates to the 2-sheet complex (1.5×10^{-9} m basal spacing). Present evidence suggests it is in the relative humidity range of 95–99 % (2–7 MPa negative pressure), which is outside the available water range and hence unlikely to be reached in subsoils (Ormerod & Newman 1983). This is supported by the data in table 2, in that the illitic soil contained as much water between 0.01 and 1 MPa as the smectitic soil.

4.3. *Soil and aggregate structure*

Soil structure may be briefly defined by the way a potentially continuous matrix of mineral particles is broken into a mosaic of subunits each with a quasi-independent existence. If the soil is coarse grained, the subunit is the individual mineral grain, but in coherent soil the subunits are compound and can themselves break up into smaller subunits at the submicroscopic level. Until recently, little has been known about these submicroscopic units except by inference from measurements like water retention and mercury porosimetry. However, the advent of the scanning electron microscope, and the preparation by ion-thinning of ultra-thin sections that can be examined in the transmission electron microscope (Bresson 1981) has been the stimulus for new ideas about soil microstructure. The present picture seems to be that soil structure is hierarchical, with a scale that ranges from the visible field structure with fissures 10–100 mm in size, down to pores of 1–10 nm diameter which represent the distances between clusters of units that diffract X-rays coherently, i.e. silicate layers which are oriented in a parallel way.

Within this hierarchy, the influence of clay mineralogy is likely to be considerable, but ancillary factors also contribute. Iron oxides affect the way that clay crystals aggregate (Blackmore 1973) and Al–OH polymers also induce strong aggregates to form (Kubota 1975; El-Swaify & Emerson 1975). The macroscopic strength of clays can be greatly modified by inter-action with Al–OH polymers (Foster & Gazzard 1975) and organic compounds and soil organic matter influence macroscopic soil structure through their interaction at the submicroscopic level (Greenland 1965). The macrostructure of soils that develops on drying can be modified by amendments like calcium sulphate, calcium carbonate (O'Callaghan & Loveday 1973) and dextran (Pagliai *et al.* 1980). In all of these studies, the clay and the amendment combine to produce new structural units.

These microstructural influences on the macrostructure of soils have great practical import-ance because macropores 'short-circuit' water flow in clay soils (Hoogmoed & Bouma 1980). The drawing of mole channels through clay subsoils is an important management technique in draining excess water, and Spoor *et al.* (1982) consider that swelling processes related to clay mineralogy are important in destabilizing mole drains.

4.4. *Soil tillage*

The mechanical effort needed to draw a plough through soil depends on several factors, one of which is the amount of clay in the soil (Haines & Keen 1925). Soil consistency and the change in liquidity or solidity of soils with water content are related to the amount and type of clay in the soil (Spoor & Muckle 1974): the preparation of a satisfactory seedbed depends on culti-vating at a soil water content at which the soil is friable, but not sticky (too wet) or cloddy (too dry). The range of water content over which the soil remains in a friable state is related to its mechanical composition, and also to the mineralogical and structural composition of the colloid fraction. Friability depends on the number of air–water interfaces within the soil aggregates, and this is likely to depend on the pore size distribution in the micropore regions.

5. THE ROLE OF SOIL CLAYS IN AGRICULTURE

Martin & Lambe (1957) in an investigation of soil composition and the engineering behaviour of soils, noted that although mineralogy was one essential factor, others, such as cementation and flocculation, and whether clays of different mineralogy were mixed together, were also vital.

A similar conclusion may also be reached from the present survey of soil properties with agricultural importance. Although clay mineralogy is clearly a primary factor, soil properties are likely to be influenced by the way that the surface of the clays is modified by relatively immobile contaminants and coatings. In particular, the modification of clay surfaces by cations and the acidic polynuclear products from the hydrolysis of Al and Fe affects cation exchange and enables anions to interact with clays negatively charged at the surface by isomorphous substitution. There is evidence also that the effect of mixed-layer minerals on soil properties is not quite what might have been anticipated, particularly in relation to their swelling properties.

It is clear that our present knowledge is still imperfect. The main need is to investigate the properties of the minerals, their mixtures and the surface modifications in the form that they occur in the soil. Normally we separate a clay fraction from the soil in order to identify it, but these procedures are destructive, particularly of the particle arrangement (fabric) in the soil, and also of the surface covering. Therefore, concurrent with the traditional procedures of clay mineralogy, we should be studying the properties of the clays *in situ* in the whole soil.

It must also be said that sometimes the identification and description of soil clays is over-simplified. A simple level of clay mineralogy does have its place in a preliminary study, and indeed should always precede a more detailed study. It is, however, unreasonable to expect detailed relations between soil mineralogy and soil properties unless the mineralogical description is accurate enough for the purpose intended. A scientifically exact description of soil clay mineralogy will always be complex: the aim is to simplify the aspects that are less important for the purpose in hand.

REFERENCES

Akitt, J. W. & Farthing, A. 1981 *J. chem. Soc. Dalton Proc.* 1617–1623.
Allison, F. E., Roller, E. M. & Doetsch, J. H. 1953 *Soil Sci.* **75**, 173–180.
Archer, J. R. & Smith, P. D. 1972 *J. Soil Sci.* **23**, 475–480.
Arnold, P. W. 1960 *Nature, Lond.* **187**, 436–437.
Arnold, P. W. & Close, B. M. 1961 *J. agric. Soc. Camb.* **57**, 295–304.
Avery, B. W. & Bullock, P. 1977 *Mineralogy of clayey soils in relation to soil classification.* Tech. Monogr. no. 10. Harpenden: Soil Survey of England and Wales.
Avery, B. W., Bullock, P., Catt, J. A., Newman, A. C. D., Rayner, J. H. & Weir, A. H. 1972 Rothamsted Experimental Station Report for 1971, 5–37.
Aveston, J. 1965 *J. chem. Soc.* 4438–4443.
Barshad, I. & Kishk, F. M. 1970 *Clays Clay Miner.* **18**, 127–137.
Blackmore, A. V. 1973 *Aust. J. Soil Res.* **11**, 75–82.
Bolton, J. 1973 *J. Sci. Fd Agric.* **24**, 727–738.
Bolton, J. & Penny, A. 1978 *J. agric. Sci., Camb.* **91**, 693–699.
Bresson, L. M. 1981 In *Submicroscopy of soils and weathered rocks* (ed. E. A. Bisdom), pp. 173–189. Wageningen: Pudoc.
Brown, G. & Newman, A. C. D. 1973 *J. Soil Sci.* **24**, 339–354.
Burnham, C. P. 1974 In *Soil type and land capability.* Tech. Monogr. no. 4 (ed. D. Mackney), pp. 1–12. Harpenden: Soil Survey of England and Wales.
Chen, Y. R., Butler, J. N. & Stumm, W. 1973 *J. Colloid Interface Sci.* **43**, 421–436.
Chernov, V. A. 1947 *On the nature of soil acidity* (in translation). Madison: Soil Science Society of America.
Cooke, G. W. 1967 *The control of soil fertility.* London: Crosby Lockwood.
Cooke, G. W. 1977 *Phil. Trans. R. Soc. Lond.* B **281**, 75–76.

Coulter, B. S. 1969a *J. Soil Sci.* **20**, 72–83.

Coulter, B. S. 1969b *Soils and fertilizers* **32**, 215–223.

Croney, D. & Coleman, J. D. 1954 *J. Soil Sci.* **5**, 75–84.

Davies, D. B., Eagle, D. J. & Finney, J. B. 1972 *Soil management.* Ipswich: Farming Press Limited.

de Villiers, J. M. & Jackson, M. L. 1967 *Proc. Soil Sci. Soc. Am.* **31**, 614–619.

Dungworth, G., Thijssen, M., Zuurveld, J., van der Velden, W. & Schwartz, A. 1977 *Chem. Geol.* **19**, 295–308.

Edwards, A. P. & Bremner, J. M. 1967 *J. Soil Sci.* **18**, 64–73.

Eeckman, J. P. & Laudelout, H. 1961 *Kolloidzeitschrift* **78**, 99–107.

El-Swaify, S. A. & Emerson, W. W. 1975 *Proc. Soil Sci. Soc. Am.* **39**, 1056–1063.

Feigenbaum, S. & Hagin, J. 1967 *J. Soil Sci.* **18**, 197–203.

Foster, R. H. & Gazzard, I. J. 1975 *Géotechnique* **25**, 513–525.

Green, R. E. 1974 In *Pesticides in soil and water* (ed. R. C. Dinauer), pp. 3–37. Madison, USA: Soil Science Society of America.

Greene-Kelly, R. 1973 *J. Soil Sci.* **24**, 277–283.

Greene-Kelly, R. 1974 *Geoderma* **11**, 243–257.

Greenland, D. J. 1975 *Clay Miner.* **10**, 407–416.

Griffin, R. A. & Jurinak, J. J. 1974 *Proc. Soil Sci. Soc. Am.* **38**, 75–79.

Haines, W. B. & Keen, B. A. 1925 *J. agric. Sci. Camb.* **15**, 395–406.

Hall, D. G. M., Reeve, M. J., Thomasson, A. J. & Wright, V. F. 1977 *Water retention, porosity and density of field soils.* Tech. Monogr. no. 9. Harpenden: Soil Survey of England and Wales.

Hall, J. K. & Baker, D. E. 1971 *Proc. Soil Sci. Soc. Am.* **35**, 876–881.

Hoogmoed, W. B. & Bouma, J. 1980 *J. Soil Sci. Soc. Am.* **44**, 458–461.

Jackson, M. L. 1963 *Proc. Soil Sci. Soc. Am.* **27**, 1–10.

Jackson, M. L. 1964 In *Chemistry of the soil* (ed. F. E. Bear), pp. 71–141. London: Chapman and Hall.

Kidder, G. & Reed, L. W. 1972 *Clays Clay Miner.* **20**, 13–20.

Kittrick, J. A. & Jackson, M. L. 1956 *J. Soil Sci.* **7**, 81–89.

Kozak, L. M. & Huang, P. M. 1971 *Clays Clay Miner.* **19**, 95–102.

Kubota, J. 1975 *Soil Sci. pl. Nutr.* **21**, 1–12.

Lawes, J. B., Gilbert, J. H. & Warington, R. 1882 *Jl R. agric. Soc.* **18**, 1–71.

Lawrence, G. P. 1977 *J. Soil Sci.* **28**, 527–540.

Lawrence, G. P., Payne, D. & Greenland, D. J. 1979 *J. Soil Sci.* **30**, 499–516.

Martin, R. T. & Lambe, T. W. 1957 *Clay Miner. Bull.* **3**, 137–150.

Mortland, M. M. 1970 *Adv. Agron.* **22**, 75–117.

Mortland, M. M., Fripiat, J. J., Chaussidon, J. & Uytterhoeven, J. 1963 *J. phys. Chem.* **67**, 248–258.

Mortland, M. M. & Raman, K. V. 1968 *Clays Clay Miner.* **16**, 393–398.

Newman, A. C. D. 1969 *J. Soil Sci.* **20**, 357–373.

Newman, A. C. D. & Oliver, S. 1966 *J. Soil Sci.* **17**, 159–174.

Norrish, K. 1973 *Proceedings of the International Clay Conference* 1972, *Madrid*, pp. 417–432. Madrid: Division de Aencias, C.S.I.C.

O'Callaghan, J. F. & Loveday, J. 1973 *J. Pattern Recog.* **5**, 83–98.

Ormerod, E. C. & Newman, A. C. D. 1983 *Clay Miner.* **18**, 289–299.

Osthaus, B. 1955 *Clays Clay Miner.* **4**, 301–321.

Pagliai, M., Guidi, G. & La Marca, M. 1980 *J. Soil Sci.* **31**, 493–504.

Parfitt, R. L., Russell, J. D. & Farmer, V. C. 1976 *J. Chem. Soc. Farad. Trans.* I **72**, 1082–1087.

Reeve, M. J., Smith, P. D. & Thomasson, A. J. 1973 *J. Soil Sci.* **24**, 355–367.

Rich, C. I. 1968 *Clays Clay Miner.* **16**, 15–30.

Richburg, J. S. & Adams, F. 1970 *Proc. Soil Sci. Soc. Am.* **34**, 728–734.

Salter, P. J., Berry, G. & Williams, J. B. 1966 *J. Soil Sci.* **17**, 93–98.

Schofield, R. K. 1935 *Trans. 3rd Int. Congr. Soil Sci.* **2**, 37–48.

Schofield, R. K. & Samson, H. R. 1954 *Discuss. Faraday Soc.* **18**, 135–145.

Scott, A. D. 1968 *Trans. 9th Int. Congr. Soil Sci.* **2**, 649–660.

Spoor, G. & Muckle, J. B. 1974 In *Soil type and land capability.* Tech. Monogr. no. 4 (ed. D. Mackney), pp. 125–134. Harpenden: Soil Survey of England and Wales.

Spoor, G., Leeds-Harrison, P. B. & Godwin, R. J. 1982 *J. Soil Sci.* **33**, 427–441.

Talibudeen, O., Beasley, J. D., Lane, P. & Rajendran, N. 1978 *J. Soil Sci.* **29**, 207–218.

Talibudeen, O. & Dey, S. K. 1968 *J. agric. Soc., Camb.* **71**, 405–411.

van Olphen, H. 1977 *An introduction to clay colloid chemistry* 2nd edn. New York: John Wiley.

van Riemsdijk, W. H. & Lyklema, J. 1980 *Colloids and Surfaces* **1**, 33–44.

Vaughan, D. E. W., Lussier, R. J. & Magee, J. S. 1979 U.S. Patent 4176090.

Veith, J. A. & Sposito, G. 1977 *J. Soil Sci. Soc. Am.* **41**, 870–876.

Wada, K. 1959 *Soil Sci.* **87**, 325–330.

White, R. E. 1980 In *Soils and agriculture* (ed. P. B. Tinker). Critical Reports on Applied Chemistry **2**, 71–114.

Wild, A. & Cameron, K. C. 1980 In *Soils and agriculture* (ed. P. B. Tinker). Critical Reports on Applied Chemistry **2**, 55–70.

Yaalon, D. J. & Feigin, A. 1970 *Israel J. Chem.* **8**, 425–434.

Phil. Trans. R. Soc. Lond. A **311**, 391–409 (1984)   [391]
Printed in Great Britain

Smectite clay minerals: properties and uses

By I. E. Odom

American Colloid Company, 5100 Suffield Court, Stokie, Illinois 60077, U.S.A.

[Plate 1]

The physicochemical properties of smectite clay minerals that determine their industrial utilization are reviewed. Smectite is the name used for a group of phyllo-silicate mineral species, the most important of which are montmorillonite, beidellite, nontronite, saponite and hectorite. These and several other less common species are differentiated by variations in chemical composition involving substitutions of Al for Si in tetrahedral cation sites and Al, Fe, Mg and Li in octahedral cation sites. Smectite clays have a variable net negative charge, which is balanced by Na, Ca, Mg and, or, H adsorbed externally on interlamellar surfaces. The structure, chemical composition, exchangeable ion type and small crystal size of smectite clays are responsible for several unique properties, including a large chemically active surface area, a high cation exchange capacity, interlamellar surfaces having unusual hydration characteristics, and sometimes the ability to modify strongly the flow behaviour of liquids.

In terms of major industrial and chemical uses, natural smectite clays can be divided into three categories, Na smectites, Ca–Mg smectites and Fuller's or acid earths. Large volumes of Na smectites and Na-exchanged Ca-Mg smectites and Fuller's earth are directly used in the foundry, oil well drilling, wine, and iron ore and feed pelletizing industries, and are also used in civil engineering to impede water movement. Significant volumes of Na smectites are used for various purposes in the manufacturing of many industrial, chemical and consumer products. Large quantities of Ca–Mg smectites are used directly in iron foundries, in agricultural industries and for filtering and decolorizing various types of oils. A significant fraction of the Ca–Mg smectites used for decolorizing has been acid treated. Large volumes of Fuller's or acid earths are commercially used for preparing animal litter trays and oil and grease absorbents, as carriers for insecticides, and for decolorizing of oils and fats.

Natural Na smectites occur in commercial quantities in only a few places, but Ca–Mg smectite and Fuller's earth deposits of considerable size occur on almost every continent.

Introduction

There are many industrial and chemical uses for smectite clays, all of which depend on one or more of the unique properties of this group of minerals. During the past two decades, the various properties of smectite clays have been investigated intensively, with the result that many new uses have been introduced, especially in the environmental and chemical areas. Even today, almost every issue of journals devoted to clay mineral science contains one or more papers describing some previously unknown or poorly understood property of smectite clays.

The vast range of uses and the scientific interest in the smectite clay minerals stem from their physicochemical properties, many of which are not displayed by any other known natural minerals. The unique physicochemical properties displayed by smectite clays are the result of (i) extremely small crystal size, (ii) variations in internal chemical composition, (iii) structural characteristics caused by chemical factors, (iv) large cation exchange capacity, (v) large surface area that is chemically active, (vi) variations in types of exchangeable ions and surface charge, and (vii) interactions with inorganic and organic liquids.

Smectite clays are formed and are most stable in surface and shallow subsurface environments. At burial depths of 2000–4000 m smectite clays undergo alteration to illite–smectite mixed layer clays, and at greater depths the mixed layer clay is altered to illite through K adsorption in interlayer positions or even to chlorite if Mg is abundant (Eberl, this symposium). Smectite clays are ubiquitous components of soils, especially in semi-arid and arid regions. Continental and oceanic sediments from Cretaceous to Recent frequently contain smectite clays. Although very pure smectite clays as old as Jurassic are known, most of the commercial deposits are Cretaceous or younger in age.

The widespread occurrence of smectite clay minerals is related to multiple modes of origin including precipitation from fluids containing the appropriate chemical elements, from the alteration of Si-, Al-, Fe- and Mg-bearing minerals, and especially from the alteration of volcanic glass. The greatest number of commercial smectite clay deposits have formed from the hydrolysis of extrusive volcanics, although certain types of intrusive rocks have also been altered to smectite clays in weathering environments near the surface. Because of the vast amount of extrusive volcanism that began throughout the world in the Cretaceous period, smectite clay deposits are known on every continent, except Antarctica.

Large amounts of smectite clays have been used for many decades, and by 1982 the annual worldwide use, not including China and the U.S.S.R., had grown to more than six million tons†. Smectite clays have been important in the industrial revolution and in petroleum energy resource development, as these clays have extensive use in preparing metal moulds and in drilling oil wells. It has even been proposed that the beginnings of life originated when organic molecules were photochemically synthesized on the surfaces of iron-rich smectite clays.

PROPERTIES OF SMECTITE CLAY MINERALS

The physicochemical properties of smectite clay minerals are described in considerable detail in numerous textbooks, reference books and professional papers. Several of the properties of smectite clays are further described in other papers in this volume. The following discussion focuses on only those properties of smectite clays that are most important in their industrial and chemical uses. As far as possible, interrelations among physicochemical properties are stressed because many uses of smectite clays are dependent on more than one property.

Crystal structure and chemical composition

The smectite clay minerals have a layer lattice structure similar to micas, but they differ from micas in that the bonds between layers are weakened because of internal chemical substitutions. Only in the smectite clays are interlayer cations exchangeable and interlayer surfaces and cations hydratable. The smectite clays consist of layers of negatively charged oxygen atoms within which several types of positively charged cations are fixed in specific positions. In a two-dimensional schematic diagram of the structure (figure 1), four layers of oxygen atoms can be seen that define upper and lower tetrahedral sheets containing tetravalent (Si) and sometimes trivalent cations (Al^{3+} and Fe^{3+}). The apexes of the tetrahedra point toward each other, and the oxygen atoms at the apexes form part of an octahedral sheet that may contain

† 1 ton ≈ 1016 kg.

trivalent cations, (Al, Fe), divalent cations (Fe, Mg), both divalent and trivalent cations, or divalent and monovalent (Li) cations. The presence of two tetrahedral sheets and one octahedral sheet is the basis for classifying the smectite structures as 2:1 phyllosilicates. This structural characteristic differentiates smectite clays from kaolinite clay structures containing one tetrahedral and one octahedral sheet and from chlorite clay structures that contain two tetrahedral and two octahedral sheets. Illite clay structures are similar to smectite clay structures, but in illite adjacent tetrahedral sheets are bonded by K^+ ions which are not exchangeable.

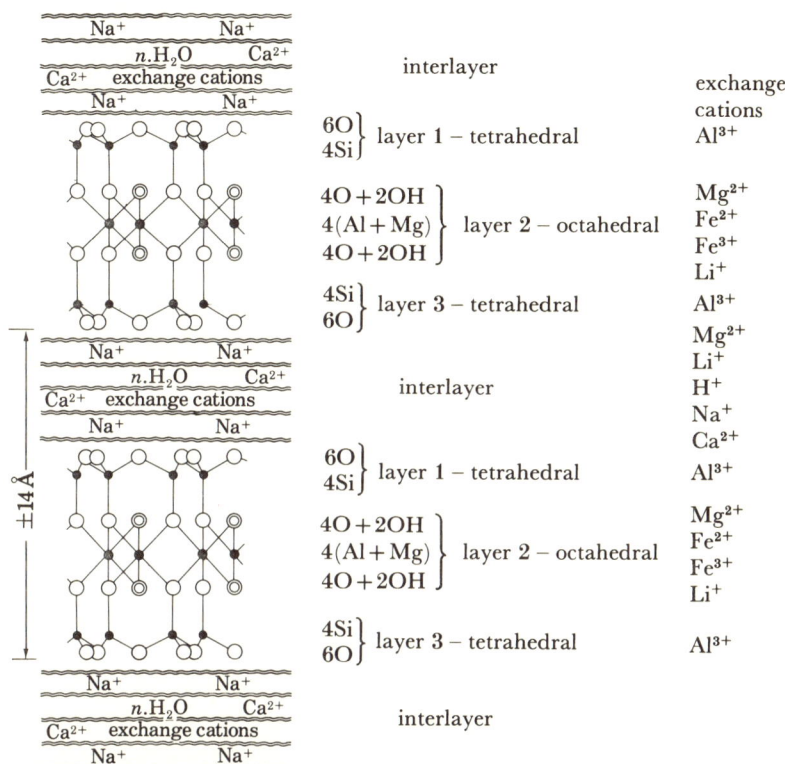

FIGURE 1. Structure of smectite clay minerals. The chemical composition shown is that of montmorillonite. Various ions that may be present in other smectite clay species are shown in column on right.

Smectite 2:1 structural units are separated by layers of loosely held hydrated cations. These cations are present to balance the negatively charged structure that results from internal substitutions of trivalent ions for silica in the tetrahedral sheet and especially from variations in trivalent and divalent ion content in the octahedral sheet. Interlayer surface and cation hydration between smectite structural units is a unique property of smectite clays.

The members of the smectite group of clays are distinguished by the species and location of cations in the tetrahedral and octahedral sites. Table 1 shows the currently accepted classification of smectite clay minerals and gives idealized structural formulas. This classification is based on both crystalline structure and chemical composition. In a half-unit cell containing ten oxygen and two hydroxyl ions, there are four tetrahedral cation sites and three octahedral cation sites. When two-thirds of the octahedral sites are occupied by trivalent cations, the structures are classified as dioctahedral, and when all three octahedral sites are occupied by divalent cations the structures are classified as trioctahedral. While wholly dioctahedral and trioctahedral structures exist, deviations in octahedral site occupations by divalent and tri-valent cations produce structures that may be intermediate between dioctahedral and triocta-hedral. Smectite clays have additional structural and chemical differences caused by limited substitutions of Al in tetrahedral sites.

A negative surface charge is characteristic of all smectite clay minerals and may range from 0.2 to $0.66e$ per S_4O_{10}. The net negative surface charge results from a structural positive charge deficiency due to the substitution of cations of lower charge for cations of higher charge, thus the net negative charge should theoretically equal the positive charge deficiency.

TABLE 1. CHEMICAL COMPOSITION OF MOST COMMON SMECTITE CLAY MINERALS

Modified from Brindley & Brown (1980)

dioctahedral smectites

montmorillonite	$M_y^+ nH_2O$ $(Al_{2y}Mg_y)$ $Si_4O_{10}(OH)_2$
beidellite	$M_x^+ nH_2O$ $Al_2(Si_{4-x}Al_x)$ $O_{10}(OH)_2$
nontronite	$M_x^+ nH_2O$ $Fe_2^{3+}(Si_{4-x}Al_x)$ $O_{10}(OH)_2$

trioctahedral smectites

saponite	$(M_{x-y}^+ nH_2O)$ $(Mg_{3-y}(AlFe)_y)$ $Si_{4-x}Al_x)$ $O_{10}OH_2$
hectorite	$(M_y^+ nH_2O)$ $(Mg_{3-y}Li_y Si_4O_{10}(OH)_2$

Dioctahedral montmorillonite has its structural charge originating from substitution of Mg^{2+} for Al^{3+} in the octahedral sheet whereas, in dioctahedral beidellite and trioctahedral saponite the surface charge comes from substitution of Al^{3+} for Si^{4+} in the tetrahedral sheet. Hectorite is also trioctahedral, but its charge stems from substitution of Li^+ for Mg^{2+} in the octahedral sheet. Nontronite is an analogue of beidellite that is iron rich.

The negative surface charge is balanced by cations intercalated between the structural units (interlayer sites). Typically, these cations are alkaline earth ions, Ca and Mg and, or, the alkali metal Na. These cations are always hydrated, as is the interlayer surface.

Other smectites having special structures or chemical characteristics include volchonskoite, sauconite, stevensite and lembergite. The basic properties of these smectites are briefly de-scribed by Brindley & Brown (1980). The structural and chemical characteristics of smectite clay minerals have been described in greater detail by others. Books by Grim (1968), Brindley & Brown (1980), and Weaver & Pollard (1973) include excellent discussions and many references to pertinent literature.

Crystal size and surface area

An excellent description of the crystal size and morphology of smectite clay crystals and aggregate types, based on high resolution electron microscopy, is given by Grim & Guven (1978). They show that smectite crystals may be as large as 2 μm and as small as 0.2 μm with an average size of about 0.5 μm. Individual crystal morphologies range from rhombic to hexagonal, lamellar to lath, and even to fibre shapes. Grim & Guven (1978) also describe a classification of aggregates including globular, mossy, lamellar, foliated, compact, and reticulated. The various aggregate types are determined by the habit or shape of individual crystallites and by their arrangement in the aggregate. Crystal shape and aggregate character-istics can have an important influence on physical and rheological properties. Even when dis-persed in water, smectite crystals tend to occur in aggregates rather than in single crystals.

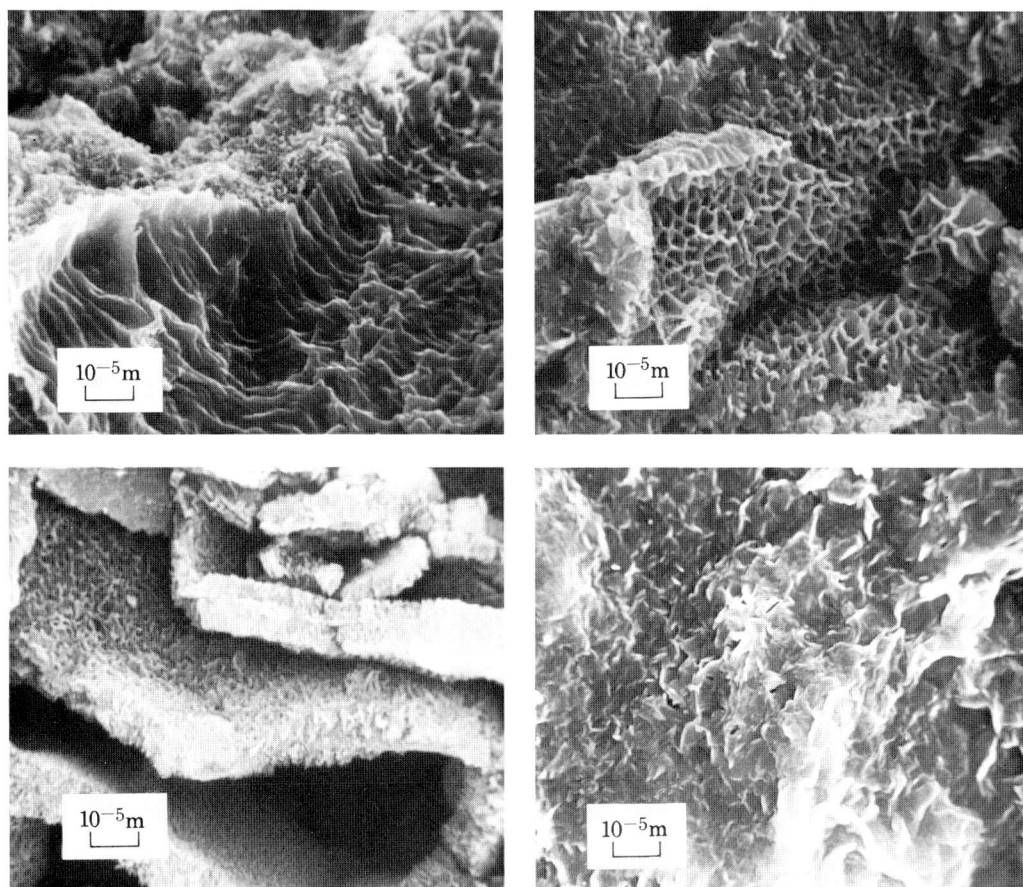

FIGURE 2. Electron micrographs showing the natural textures and crystal size variations of smectite clay minerals all of which formed from the hydrolysis of volcanic glass. The top two photographs are of Ca smectites; the bottom two photographs are Na smectites. Bar = 10 μm.

In the industrial use of smectites, it is important to recognize that because of aggregation the effective particle size and surface area of smectite clays are often considerably less than the actual particle size and surface area. This relation exists because during the growth of smectite clays by either transformation or neoformation crystals become interlocked and become difficult to separate except by a strong shearing force. Differences in the effective particle size of smectite clays are extremely important in the determination of properties such as ion exchange (Neal & Cooper 1983), viscosity and fluid loss. Natural smectites containing sodium as the predominant exchangeable ion yield the smallest effective crystal size and the largest surface area in clay–water systems because their natural crystal size tends to be small and swelling pressure tends to disrupt interlocked crystals. Smectite clays formed from volcanic ash may show considerable variation in primary texture (figure 2, plate 1), and this texture may persist to various degrees in clay–water systems.

Exchangeable ions and ion exchange

Perhaps the most unique property of smectite clays is the presence of exchangeable cations that are primarily adsorbed on interlamellar surfaces. The most common of these exchangeable ions are Ca, Mg, Na and H, but small amounts of exchangeable K and Li occur in some smectites. It is now generally realized that exchangeable Mg may be very abundant in some smectite clays, but as a rule Ca and Mg occur together and only in rare instances does the amount of exchangeable Mg exceed the amount of exchangeable Ca. As exchangeable ions, the chemical properties of Ca and Mg are similar, apparently because these ions have similar hydration characteristics. Among the major commercial smectite clay deposits, a far greater number contain Ca and Mg than contain Na or H as exchangeable ions.

The cation exchange capacity of relatively pure smectite clays ranges between 70 and 130 meq/100 g. According to Weaver & Pollard (1973), of the c.e.c. of smectite clays, about 80 % is due to charges resulting from structural substitution, and about 20 % is due to charges from broken bonds at the edges of crystals. Owing to natural chemical and physical factors and to methods of analysis, the abundance of natural exchangeable ions in even relatively pure smectite is usually less than the c.e.c. of the same smectite. If the abundance of exchangeable ions should exceed the c.e.c., then the presence of soluble salts should be suspected.

The exchangeable ions associated with smectite clays are easily and reversibly replaceable. Owing to its valency, Na is readily replaced by Ca and Mg, so that smectite clays tend to be depleted of Na when subjected to leaching conditions. For example, in local situations the Na originally present in South Dakota and Wyoming Cretaceous bentonites has partially been replaced by Ca and Mg in relatively recent times by groundwater leaching (figure 3). The Ca and Mg were derived from overlying shales or limestones, and the amount of Ca plus Mg exchange for Na increases as overburden thickness decreases. If present in sufficient concentration in groundwater, Na could easily replace Ca and Mg. Natural examples of replacement of Ca and Mg by Na are rare, but some examples are known in the western U.S. desert region where alkaline lake waters have percolated through smectite clays.

Exchangeable ions play a dominant role in the commercial use of smectite clays. Where Na is the predominant exchangeable ion, smectite clays, regardless of species, may have a high swelling capacity. Na tends to promote the development of many oriented water layers on interlamellar surfaces. The hydration associated with Na may produce swelling to the extent of complete dissociation of the individual smectite crystals, the result being a high degree of

[175]

dispersion and maximum development of colloidal-like properties, that is, high natural viscosity. On the other hand, smectites having similar internal chemistry but containing exchangeable Ca plus Mg even when fully hydrated show only a small degree of swelling. Differences in ion hydration characteristics are discussed more fully in a later section.

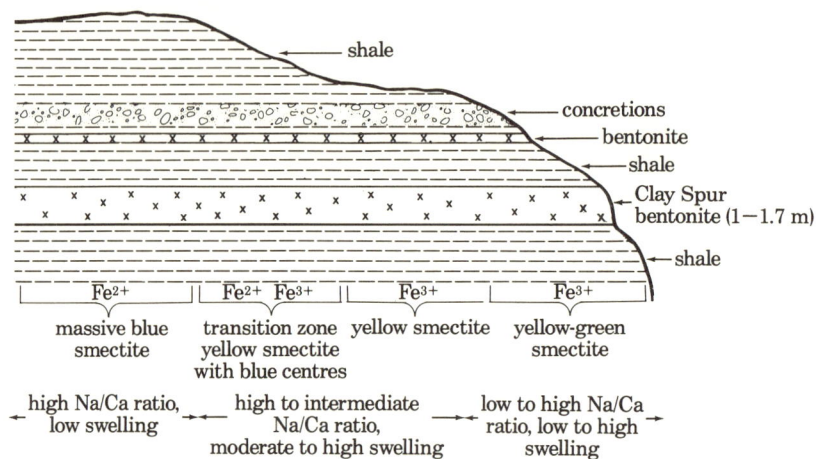

FIGURE 3. Diagrammatic cross-section of Clay Spur bentonite bed, Wyoming, showing variations in colour, iron oxidation, exchangeable ion ratio and swelling. The variations noted are caused by groundwater action and are related to overburden thickness and soluble ion content.

Exchangeable ion content of potentially commercial smectite deposits should be thoroughly studied or serious economic consequences could result. The exchangeable ions of smectite clays are usually determined initially by the chemical composition of the parent material if origin is by transformation, or by the concentration of ions in solution when formed by precipitation. The exchangeable ion content of smectite clays formed by transformation of volcanic glass that was desposited in a marine environment may be secondarily influenced by ions from pore solutions. These factors in addition to possible ion exchange from groundwater leaching can produce considerable variation of exchangeable ion content within a single smectite clay deposit.

Hydration and swelling

The hydration and dehydration of interlamellar surfaces are other unique properties of smectite clays, and these properties are of crucial importance in several industrial uses. Although hydration and dehydration occur regardless of the type of exchangeable cation present, the degree of hydration is dependent on the species of exchangeable ion, on the size and charge of the cations, and on the magnitude and location of the layer charge of the adjacent silicate sheets. Presently, it is generally accepted that the hydration of the interlamellar surfaces of smectite clays occurs in a series of steps when these clays are subjected to increasing relative humidity levels. At high humidity levels between one and four layers of water molecules may be formed, depending on the cation adsorbed. Most researchers believe that when only a few water layers are adsorbed, the water molecules (also organic liquids) are arranged in a highly ordered manner.

According to MacEwan & Wilson (1980), the hydration of smectites is controlled by three interrelated factors, (i) the magnitude of the negative charge on the clay mineral interlamellar surface; (ii) the exchangeable cations which occur between the interlamellar surfaces and balance the overall negative charge on the layer surface; and (iii) the interaction of the water molecules (or other neutral molecules) with the cations and with the clay mineral interlamellar surface.

The total net negative charge and the origin of this charge (octahedral or tetrahedral substitution) play an important role in the point localization of cations, especially multivalent species, on the layer surface. Charge resulting from tetrahedral Al substitution appears to cause ordering especially of multivalent cations on the silicate sheet. The tendency for exchangeable cation ordering is not found in octahedrally substituted smectites, since the net charge is more randomly distributed among the surface oxygens. This is a matter of great importance in certain areas of industrial use of smectites because it plays an important role in swelling behaviour.

The type of exchangeable cations and their tendency to hydrate has a strong influence on the arrangement of the water molecules and on the thickness to which oriented water layers could develop. Several investigators have demonstrated that multivalent ions tend to enhance the rigid structures of the first few adsorbed water layers, but as additional water layers are adsorbed these ions promote disorder or act to make the water more fluid. Monovalent ions such as Na and Li, however, tend to reinforce the rigid structure of water layers probably because they dissociate from the layer surface. The greater the degree of dissociation the less they disrupt the quasicrystalline water layers, thus these ions are, in effect, 'dissolved' in the water structure.

White & Pichler (1959) found that for air-dried smectite clays having Ca and Mg as adsorbed ions there is an initial rapid adsorption of water up to, or slightly in excess of, the liquid limit after which little further adsorption occurs, but when sodium and lithium are the adsorbed ions there is generally no sharp break in the adsorption rate and water may continue to be adsorbed far beyond the liquid limit. In naturally occurring smectites the highest water adsorption capacities are not always directly associated with the highest sodium ion content. The highest adsorption sometimes occurs when approximately one quarter to one fifth of the total exchangeable ions are Ca and Mg (figure 3). This relation suggests that Ca and Mg may strengthen the structure of water layers closest to the silicate layers, which then permits a greater number of rigid water layers 'reinforced' by Na ions to develop. In this situation, Na can diffuse a greater distance from the basal oxygen planes. The writer believes the primary reason some high-Na smectites have a low swelling capacity (figure 3) is that the crystallites are interlocked.

In summary, natural smectite clays range from strongly swelling to non-swelling depending on the smectite clay species and on the ratio of exchangeable Ca plus Mg present. The vast majority of naturally occurring smectite clays have Ca and Mg as the predominant exchangeable cations and are essentially non-swelling. Some Na and Li forms are particularly susceptible to swelling by water adsorption, and the layers may be separated by water layers many Ångstroms† thick, even to the point that dissociation of the layers may occur. A simple determination of hydration characteristics should be given priority in the evaluation of smectite clay deposits as so many industrial uses are dependent on this unique property.

Attempts to change Ca–Mg smectites to so-called Na smectites are a common practice in

† $1 \text{ Å} = 0.1 \text{ nm} = 10^{-10} \text{ m}$.

[177]

the clay industry, especially in Europe where naturally occurring Na smectite deposits are rare. The ion exchange is usually performed by mixing soda ash (sodium carbonate) with crude, moist clay using various mechanical methods. After the soda ash is added, it is a general practice to stockpile the clay for several days or weeks to permit maximum exchange to occur. In Europe, smectite clays are often 'activated' using several forms of extrusion. In addition to intimate mixing with the soda ash, extrusion effects considerable shear of the clay platelets making more interlamellar surfaces accessible for exchange. The main objectives of activation are to increase hydration and swelling, wet tensile strength and thermal resistance.

The author's investigations of activation indicate that the maximum improvement in hydration is usually attained when Na constitutes approximately one quarter of the natural exchangeable ions. Perhaps the presence of some natural sodium permits more complete exchange of Ca plus Mg when additional Na (soda ash) is added to the system.

Colloidal properties

When certain smectite clays are added in very small amounts to water, their chemical and hydration properties cause the crystals to separate and disperse. In addition, their electric potential causes the particles to repel each other, and because the crystals are so small they may remain suspended in the liquid indefinitely, so that a colloid state is attained. The same smectite clays when added to water in larger concentrations (5–6 %) may cause the liquid to become highly resistant to flow or viscous, and when the shear stress on the liquid is removed the smectite particles usually develop a rigid gel structure. The ability to impart high viscosity and to develop thixotropy are unique properties of naturally occurring montmorillonite, hectorite and some saponite clays having high amounts of exchangeable sodium. Ca–Mg smectites, on the other hand, neither give high viscosity values or display thixotropic behaviour even when the percentage solid is very high. The viscosity developed by Na smectites is believed to be due mainly to small particle size, large surface area, and high dispersibility, and secondarily to the electrical forces between the particles. Thixotropy results because the negatively charged basal surfaces and the positive charges present on crystal edges attract each other, and so forms an internal 'card house' structure. A thixotropic smectite clay system can be returned to a viscous system through shear or agitation. Both viscosity development and thixotropic behaviour are essential properties of smectite clays used for preparing drilling muds.

The viscosity properties of smectite clays within a single deposit or bed may be uniform or highly variable. Figure 4a illustrates the large natural apparent viscosity variations possible in a natural montmorillonite at various locations over a small pit in a bentonite bed in eastern Wyoming, U.S.A. In contrast, figure 4b shows that in a second pit located near Belle Fourche, South Dakota, the natural apparent viscosity of the Na montmorillonite is much lower and very uniform. These natural viscosity differences are primarily related to degree of dispersibility and secondarily to natural ion exchange. The montmorillonites yielding high viscosities (figure 4a) usually occur under moderately shallow overburden, have a yellow colour as Fe^{2+} iron has been oxidized to Fe^{3+} iron, and have some Ca and Mg exchangeable ions introduced by groundwater which have replaced Na ions. The montmorillonites having lower and more uniform viscosities (figure 4b), on the other hand, occur under more than 10 m of overburden, have a very high Na to Ca + Mg ratio, and a blue or green colour, indicating that little groundwater action or oxidation has occurred. The combined action of ion exchange and oxidation caused by groundwater has increased the hydration and dispersibility of the montmorillonite,

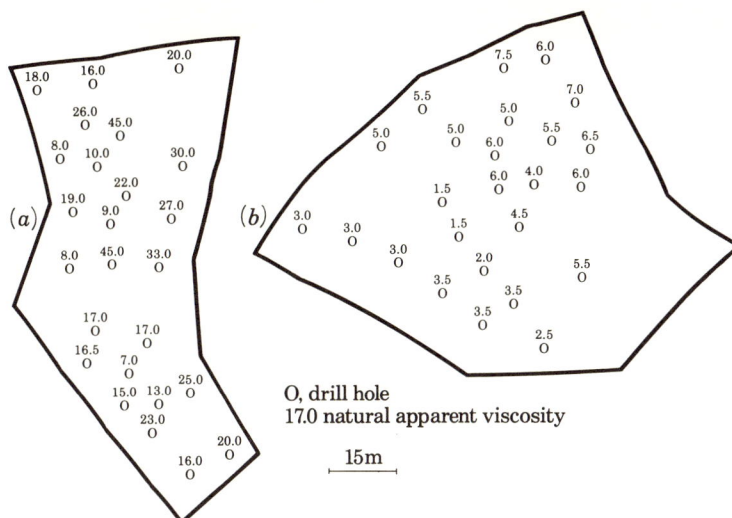

FIGURE 4. Variations in the natural viscosity of high Na smectite clays in different pits in South Dakota (a) and Wyoming (b). Numbers represent apparent viscosity in centipoises. Measurement made on a slurry containing 22.5 g in 350 ml of water using a direct-indicating viscometer.

TABLE 2. COLLOIDAL PROPERTIES OF SMECTITE CLAY SAMPLES FROM A SINGLE ACTIVE PIT, MEDITERRANEAN AREA, BEFORE AND AFTER SODA ASH ACTIVATION

| | crude smectite | | soda ash activated smectite | |
| | apparent viscosity | fluid loss | apparent viscosity | fluid loss |
sample number	cP	ml	cP	ml
1	2	31	26	15
2	2	21	39	13
3	2	40	7	21
4	2	30	10	16
5	2	31	13	18
6	2	26	5	20
7	2	20	54	14
8	2	27	16	19
9	2	23	19	24
10	2	36	24	14

Total Na content after activation is approximately 75 meq 100 g^{-1}. Measurement made on slurry containing 22.5 g in 350 ml of water using A.P.I. procedures.

leading to greatly increased natural viscosity. Table 2 shows the low natural viscosities of smectite clays from a bentonite deposit in the Mediterranean area in which approximately one quarter of the natural exchangeable ions are Na. Although these smectite clays initially have a low degree of dispersibility, treatment with sodium carbonate (table 2) greatly increases the dispersibility which in turn causes a dramatic improvement in the colloidal properties. In general, the influence of replacement by Na ions on the colloidal properties of Ca–Mg smectites is unpredictable and a careful evaluation of each deposit should be made.

Dehydration and rehydration

When smectite clays are heated two forms of water are lost, adsorbed and crystalline. The adsorbed water loss, most of which is interlayer water, occurs at low temperatures (100–200 °C). The amount of adsorbed water loss is contingent upon the nature of the adsorbed

[179]

ions, if pretreatment of the samples is the same, and also to a lesser degree on the structure of the smectite (Farmer & Russell 1971). When Na is the predominant adsorbed cation, adsorbed water is usually lost in a single stage, but when Ca and Mg are predominant the adsorbed water is sometimes lost in two stages (figure 5).

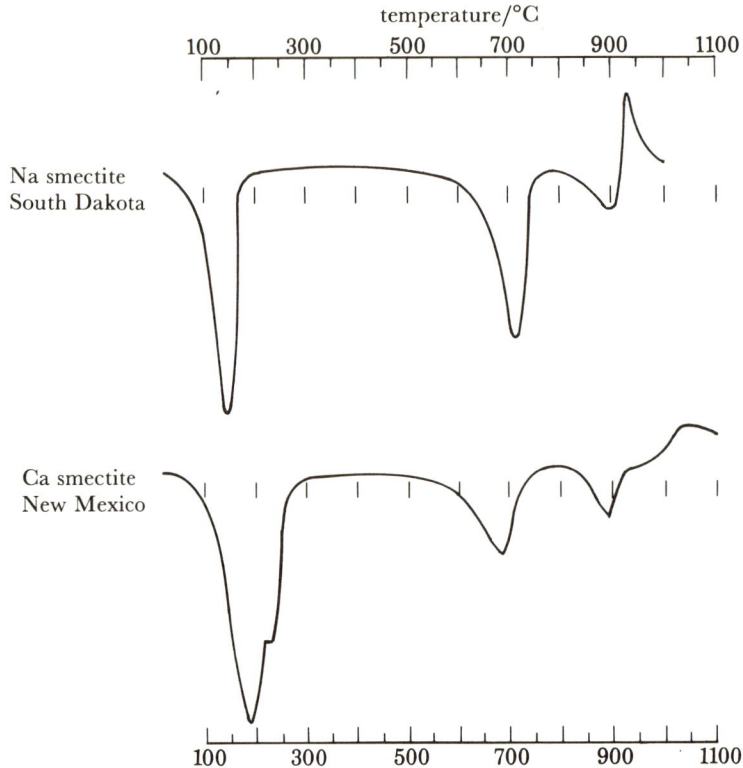

FIGURE 5. Differential thermal analysis patterns illustrating the single stage loss of adsorbed water that is typical of Na smectites and the two stage loss of adsorbed water that is often shown by Ca smectites.

Various smectites show wide variations of the temperature at which loss of OH or lattice water occurs. These variations are primarily related to crystalline structure, strength of cation-OH bonds, and chemical composition, since the OH loss involves a breakdown of the silicate structure. Among dioctahedral smectites, montmorillonites with low substitution of Fe and Mg for Al and having Na as the predominant exchangeable ion show loss of OH at the highest temperature, beginning gradually at 550 °C and ending at about 750 °C, with a peak at about 680–700 °C. Large replacement of Al by Fe or Mg causes a reduction in the temperature of the reaction, and for nontronite the peak temperature of OH loss is 500–600 °C. In magnesium-rich trioctahedral hectorite, however, the rapid loss of OH may not begin until 700 °C, possibly because the fluorine atoms that replace some hydroxyl groups are more tightly held within the lattice structure than the hydroxyl groups themselves. Other factors having some importance in the temperature of dehydroxylation of smectites are the intimate mixing of two smectites varying in composition and, or, structure (Grim & Kulbicki 1961), and possibly defect structures (Greene-Kelly 1957).

The rehydration characteristics of smectite clays after heating to various temperatures are not as well understood as the dehydration characteristics. If all of the adsorbed water is removed from smectite clays, some rehydration will occur, the amount depending on the relative humidity level and the properties of the clay. Smectites containing primarily exchangeable Ca and Mg usually rehydrate more rapidly, and at a lower humidity level, than do smectite clays containing exchangeable Na. This effect is due to the hydration properties of Ca and Mg ions. Also, the rate and amount of moisture a Ca–Mg smectite may adsorb seems to be greater if there is a moderate substitution of Mg for Al in the octahedral layer. The Ca–Mg smectite from Cheto, Arizona, has this chemical characteristic, and it rehydrates rapidly at low relative humidity levels (40%). Because of its rapid rehydration and textural properties, the Cheto smectite is used extensively as a desiccant.

When smectite clays are not heated much above 200 °C, most will rehydrate completely in the presence of liquid water, but generally there are some effects on certain properties such as swelling and, or, dispersion. Experience in drying bentonites commercially has shown that to retain primary properties it is unwise to remove the last traces of interlayer water, and in most industrial processing absorbed water content is seldom reduced below 8%.

In the presence of liquid water, smectite clays will generally rehydrate to some degree even after heating to temperatures as high as 500–700 °C. The degree of rehydration depends mainly on the degree of structural change owing to loss of OH lattice water. Naturally occurring Na montmorillonite having high thermal resistance (figure 6) is preferred for use in green sand moulding because the green compression strengths of the clay bonded sand moulds, which is a function of the rehydration of smectite, will be essentially the same even after the clay is heated repeatedly to temperatures of 600 °C.

Organic reactions

The absorptive and catalytic effects of smectite clays on organic compounds were long ago recognized as having fundamental importance in petroleum forming reactions, in soil forming processes, and in the chemical evolution of certain organic compounds. In the past 50 years an immense literature has developed on the properties of various smectite clay–organic interactions. This field of investigation has gained new vigour in the past decade as a result of the potential for using smectite clays as absorbents of various organic pollutants. Smectite clays have a large surface area for the adsorption and catalysis of organic molecules because of their small particle size and the large size and chemical activity of the oxygen layers forming the basal surfaces of the negatively charged 2:1 structure.

There are an almost endless number of organic compounds that interact in a variety of ways with smectite clays. The list is large because of the many types of bonding that are possible between the oxygen surfaces and the organic molecules (Mortland 1970). A simple example of a smectite clay–organic reaction that has great utility for the identification of smectite clays is the glycerol or ethylene glycol solvation technique. These organic compounds replace water layers on interlamellar surfaces and expand the interlayer distance to a nearly constant 17 Å spacing, which greatly aids identification by X-ray diffraction. In recent years, there has been considerable interest in organic molecules that are capable of carrying various metal ions as complexes, and that upon absorption may serve as pillars to prop open and increase the catalytic activity of smectite clay interlamellar surfaces (Pinnavaia 1983). These so-called pillared smectites are believed to have considerable promise for use as petroleum cracking catalysts. There is a growing use of smectite clays for absorbing various organic and inorganic contaminants from industrial waste water. In these processes a small amount of clay is added to the water to adsorb the contaminants, then other chemicals are added to flocculate the clay for easy removal from the system. Other uses of smectite clays related to their interaction with or action on organic compounds are discussed in later sections of this paper.

USES OF SMECTITE CLAYS

Smectite clays used in industry and in chemical and commercial products can be separated into three main types; Na smectites, Ca–Mg smectites and Fuller's or acid earths. The latter type may include materials that do not necessarily contain a large amount of smectite clays.

TABLE 3. PRIMARY INDUSTRIAL AND CHEMICAL USES OF SMECTITE CLAY MINERALS

large volumes	small volumes
foundry sand binder	medical, pharmaceutical, cosmetic
oil well drilling	paint
filtering, clarifying, decolorizing	building – brick, sewer pipe, roofing tile
pelletizing – iron ore	gypsum products
pelletizing – animal feed	radioactive waste disposal
waterproofing, water impedance	glazes
pet absorbent	pottery – ceramics
pesticide carrier	lubricants
oil and grease absorbent	fertilizers
	detergents
	mortar
	catalysts
	paper coating
	seed coating
	adhesives
	water purification
	other miscellaneous

Table 3 shows the primary uses of smectite clays. The uses are divided into two categories based on volume or tonnage used. Neither the large or small volume uses are necessarily ranked according to actual volume used, since annual consumption by various industries can change drastically depending on economic and political factors. However, over the past two decades the foundry, oil well drilling, and iron ore pelletizing industries have been the largest users of smectite clays, and these industries are likely to continue to be the largest users through the next several decades.

Foundry moulding sands

Several million tons of smectite clays are used annually in the metallurgical industry for preparing moulding sands. In Europe and Asia, foundry use is the main market for smectite clays, whereas in North America foundry use currently ranks second behind oil well use. In foundry usage smectites, in amounts varying from 5 to 10 %, are mixed with sand and water to make the sand plastic and cohesive so that it can be moulded around a pattern. After the pattern is moulded and removed the smectite clay must give the sand sufficient strength to maintain the shape of the cavity before, during and after the pouring of the hot metal into the mould. The properties of smectite clays that are crucial in foundry usage are green compression strength, dry compression strength, wet tensile strength, hot compression strength, flowability and durability. The methods for analysis of the above foundry properties are described in detail by Parkes (1971).

Green compression strength is the strength of the moulded sand–clay mixture at a specified water content. It is measured by determining the compressive force necessary to cause failure in a test specimen containing a fixed amount of tempering water that is compacted by ramming to a specific compactibility. An alternative method for preparing the test specimen is to add only the amount of tempering water that is necessary to obtain a specified compactability. The green strength of smectite clays varies with crystallinity, the type of exchangeable ions, the percentage of clay, and the percentage of tempering water. When evaluating the green strengths of smectite clays from a single deposit or different deposits, it is advisable to make all tests with a specific type of sand and at a fixed clay and tempering water content. Table 4 shows the green compressive strength of several smectite clays presently used in foundries containing various amounts of exchangeable Na, Ca and Mg determined at 2.5% tempering

TABLE 4. FOUNDRY AND OTHER PROPERTIES OF SMECTITE CLAYS AS RELATED TO EXCHANGEABLE ION CONTENT. SOME SAMPLES WERE ACTIVATED BY PRODUCER USING SODIUM CARBONATE. AMOUNT OF SODA ASH USED IN ACTIVATION IS NOT KNOWN

	green compression strength	dry compression strength	green compression durability (%)		liquid limit (%)
	pounds per square inch†	pounds per square inch†	500 °C	600 °C	
Na smectite, United States	10.9	104	100	100	670
Na smectite, Africa	9.3	125	100	100	465
Na–Ca–Mg smectite, Australia	8.6	85	100	42	338
Na–Ca–Mg smectite, Europe	11.3	84	75	31	225
Na–Ca–Mg smectite, Europe – activated	13.7	100	100	74	610
Ca–Mg smectite, Europe	13.6	54	11	0	200
Ca–Mg smectite, Europe – activated	14.6	53	78	44	430
Ca–Mg smectite, Europe	12.8	91	64	20	—
Ca–Mg smectite, Europe – activated	12.6	121	100	51	525
Ca–Mg smectite, Australia	11.1	58	70	16	200
Ca smectite, United States	13.8	56	50	16	200
Ca smectite, Europe	14.3	60	55	14	—
Ca smectite, Europe – activated	13.2	82	87	29	—

† 1 pound per square inch ≈ 6895 Pa.

water and 5% clay. In general well-crystallized smectites of high purity that contain primarily exchangeable Ca or Ca and Mg have somewhat higher green compression strength than natural Na smectites or Na exchange Ca–Mg smectites. The green strength differences between Na and Ca smectites are related to the influence that the various exchangeable ions have on the shear strength of a relatively small number of adsorbed water layers. Replacing Ca and Mg ions with Na usually decreases green strength.

Dry compressive strength is the force necessary to cause failure in a rammed specimen after all tempering water is removed by drying in an oven at 100 °C before cooling in a desiccator. Again, for routine comparative analysis of the dry strength of smectite clays, it is advisable to make all tests at a fixed moisture and clay content. Differing amounts of tempering water or clay content cause significant variations in dry compressive strengths as well as green compression strengths.

In smectite clay–sand systems containing 5 % clay and 2.5 % tempering water, the dry strength of Ca smectites is usually lower than that of Na smectites (table 3). If the tempering water is increased the dry strength of both Na and Ca smectites also increases. These variations in dry strength again illustrate the influence that the different common exchangeable ions have on the rigid structure of those absorbed water layers remaining after drying. It is possible that the higher initial dry strength of Na smectites is in part related to the dispersibility of the clay as Na smectites are more dispersible at lower moisture contents. Treating Ca smectites with Na salts tends to raise dry strength slightly or leaves it essentially unchanged (table 3).

When metal is poured into a smectite clay-bonded sand, the heat drives adsorbed water and water resulting from OH loss from the clay within a given distance from the metal–sand interface deeper into the mould. The excess water then condenses and concentrates in a narrow zone where the temperature is less than that required to remove absorbed water layers. The amount of water that condenses in this zone is in excess of the amount required for optimum green strength. Since green strength decreases with increasing moisture content, the mould becomes weak in the zone of moisture condensation or build-up. If the smectite clay in the mould contains an abundance of Na ions, the excess water is adsorbed on the clay surface in an organized or rigid structure, whereas if Ca–Mg are the predominant adsorbed cations the water remains in a liquid form. As a result of the ability of the Na ion to develop rigid water layers, the saturated zone becomes stronger and more resistant to tensile and compressional stresses that might cause a part of the mould wall to separate from the remainder of the mould (Patterson & Boenisch 1967). Natural Na smectites give high wet tensile strengths, whereas Ca–Mg smectites have very low wet tensile strengths. The wet tensile strength of many Ca smectites can be greatly improved by Na exchange.

In the preparation of moulds, the sand–clay mixture is compacted by a variety of mechanisms. In order to achieve the minimum porosity, the clay coated sand grains must 'flow', that is, the clay–water system acts as a lubricant reducing friction between the sand grains. Flowability is, of course, influenced by grain shape, clay, and moisture content, but for smectite clay–sand mixtures containing the same amount of clay and moisture Ca smectites generally have better flowability than Na smectites. The Na smectites tend to develop a greater number of water layers that have a rigid structure, whereas there is a sharp break between rigid and non-rigid water layers absorbed on Ca smectites.

When metal is poured into a mould, the clay–sand mixture inward a short distance from the metal–mould interface is heated to high temperatures, the maximum being dependent on the temperature of the metal. Smectite clays heated in the range 800–1100 °C begin to vitrify, and the compression strength of the clay-bonded sands through this temperature range shows a moderate to sharp increase, then a sharp decrease.

Natural Na smectites have much higher hot compression strengths than Ca smectites. These differences in hot compression strengths are primarily related to the influence different exchangeable ions have on the strength of the glass-like or mineral products formed during vitrification (when the mould is hot). Although the hot compression strength of Na-exchanged Ca smectites is often increased, the high strength characteristic of natural Na smectites is seldom attained.

Clay–bonded moulding sands are recycled many times in the mould making process. Each time metal is poured into a mould, a certain percentage of the clay is heated above the temperature where OH ions are removed from the clay structure. After loss of OH ions, the smectite loses its ability to develop green and dry compression strength properties. The OH loss from most smectite clays generally occurs in the range of 500 to 750 °C. As a general rule the higher the temperature of OH loss the greater the durability of the smectite clay. Although durability is primarily related to the temperature of OH loss, it is secondarily influenced by the type of exchangeable ions. An approximation of the durability of smectite clays can be obtained from differential thermal analysis (d.t.a.) (figure 6), but actual tests after heating and re-use of the smectite clays provide a more definitive measure of durability. As a general rule durability is greater for Na than for Ca smectite clays, and the durability of some Ca smectites can be improved by Na exchange.

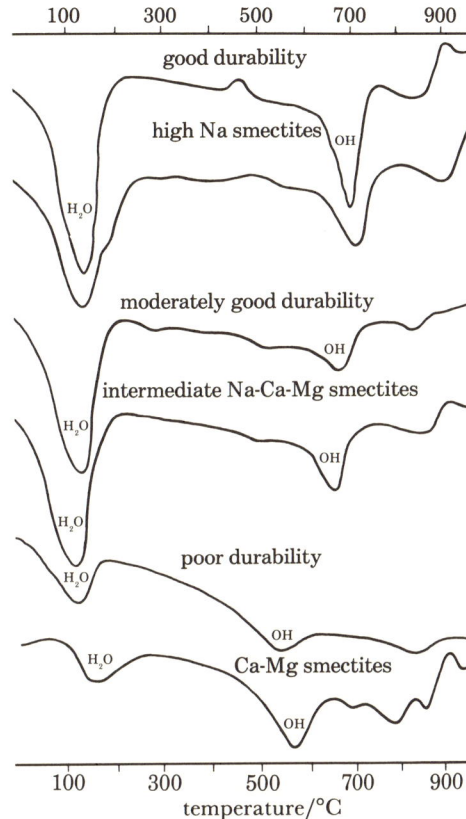

FIGURE 6. Differential thermal analysis patterns showing the relation of temperature of crystalline water (OH) loss to the thermal durability of some Na, Na–Ca–Mg, and Ca–Mg smectite clays.

The type of smectite clay used in each foundry depends on a variety of factors including geography of source, type of foundry practice, particular application, and price. Most foundries have their own specifications for green, dry and hot strengths, wet tensile strength, and liquid limit. Natural Na smectites have better durability and higher fusion temperatures, and are favoured for use in steel foundries. Na smectites, Ca smectites, and Na exchanged Ca smectites are suitable for preparing moulds for iron castings. In actual practice, many iron foundries specify a blend of specific types of Na and Ca smectites to achieve optimum green, dry, hot and wet tensile properties.

Drilling muds and related uses

The preparation of drilling muds currently constitutes the second largest end use of smectite clays, amounting to approximately 2.5 million tons in 1982. Smectite clays used for drilling muds must meet the American Petroleum Institute (A.P.I.) or the Oil Companies Materials Association (O.C.M.A.) standards set for grit, viscosity and related rheological properties, and fluid loss. Only certain natural Na and Na-exchanged Ca smectites have the potential for meeting the A.P.I. and, or, O.C.M.A. specifications. A.P.I. specifications require that 2000 pounds† of smectite clay (bentonite) yield at least 90 barrels‡ of drilling mud having a minimum apparent viscosity of 15 cP§, a plastic viscosity to yield point ratio not greater than 1:3, and a fluid loss of 15 ml or less. Although O.C.M.A. specifications are the same as the A.P.I., a higher concentration of smectite clay, 7.5 rather than 6.4%, is permitted to achieve the required properties and yield.

The worldwide occurrence of smectites that will meet all of the A.P.I. specifications is limited to a few geographic locations. However, if a Na or Na-exchanged Ca smectite clay meets the A.P.I. or O.C.M.A. fluid loss requirement, a small amount of a suitable polymer can be added during processing to develop the necessary minimum viscosity. Both fluid loss and viscosity of smectite clays can sometimes be favourably influenced by adding sodium carbonate and, or, certain types of polymers, but the cost of such additives is often prohibitive. There are many types of organic and inorganic additives used to improve or regulate the properties of drilling muds, the effectiveness of which depends on the surface, colloidal and ion exchange properties of the smectite clays used.

A relatively new and growing use of high-swelling Na smectites is to seal shallow exploration drill holes to prevent contamination of groundwater aquifers by surface or interformational leakage. The clay is compacted into large pellets that sink in the hole, after which hydration and swelling begin.

Na smectite clays specially formulated to resist high temperatures have recently been introduced for drilling deep wells and geothermal areas. These high temperature-stable smectites are not subject to the large increase in viscosity that elevated temperatures may induce in common well drilling grade smectites.

Pelletization

Smectite clays are extensively used as a binding agent in the pelletization of animal feeds, iron ore and other fine-grained solids. Smectite clays are added to animal feed in amounts of approximately 2% as a binder and to reduce die friction during the pelletization process. In addition, smectite clays are known to increase the nutritional benefits animals derive from the feed. Both Na and Ca smectites are used in feed pelletization.

The beneficiation of low grade iron ore (taconite) requires that the rock containing the ore be ground to a fine particle size so that impurities can be separated from the iron. The separation is usually done in a fluid medium. Pelletization of iron ore derived from taconite using smectite clays as a binder was introduced in the U.S.A. in the mid-1950s, and by the mid-1970s this was the largest single use of smectite clays. Both natural Na and Na-exchanged smectite clays are used for iron ore pelletization. Only the Na smectites promote good pellet

† 1 pound = 0.45 kg.
‡ 1 barrel = 0.159 m³.
§ 1 cP = 1 centipoise. 1 P = 10^{-1} Pa s.

development and give the pellets the required green and dry strengths and a high strength after calcination. Another important factor is that Na smectites are required to adsorb excess water that is not removed from the ore by vacuum filters. For this reason Na smectites having a high moisture adsorption capacity are preferred, and minimum moisture absorption capacity is often specified by iron ore pelletizing companies.

Civil engineering

The use of smectite clays in civil engineering applications first began in the mid-1950s, and these uses have grown substantially in the past three decades. In civil engineering smectite clays are used to grout cracks and fissures in rocks, to impede water or chemical waste movement through sand, gravel and permeable soils, to provide nonmechanical support of excavation and tunnel walls, to lubricate caissons and piles, to increase the plasticity of mortar, to make concrete pliable as well as impermeable, to waterproof concrete walls and floors that are below grade, and to lubricate cables and pipe for pulling through conduit.

The properties of smectite clays that are most significant in civil engineering uses are viscosity, thixotropy, impermeability and plasticity, thus smectites having a high swelling capacity and good dispersibility are normally required. Only natural Na smectites as well as certain Na exchanged Ca smectites have the necessary properties for engineering use.

Absorbents

Clay materials containing smectite clays and Fuller's earth or acid earth having absorbent properties are extensively used for pet litter, for absorbing oil and grease from floors, as a carrier for certain types of agricultural insecticides, and for packaging desiccant. In the preparation of pet litter, local customs or disposal methods may determine the type of smectite-containing clay material used. In the U.S.A., pet litter granules must be resistant to disaggregation in the dry state and resistant to slaking when wetted. Since high purity Na and Ca bentonites are not normally resistant to disaggregation or slaking, they are seldom used in the U.S.A. for pet litter. In some parts of Europe, however, granular Ca bentonites are used for pet litter because the favoured disposal method is to flush the litter into the sewage system, thus it is desirable for the litter to slake into fine particles when saturated.

Granular clay materials suitable for absorbing oil and grease from floors must not become slippery when wet. For this reason, only certain types of absorbent smectite clay materials are suitable for this use. Sometimes the slipperiness can be reduced and the absorbency increased by calcination. Smectite clays that have some hydrogen absorbed on surface exchange positions seem to be especially effective as carriers of agricultural insecticides, but high purity Na and Ca smectites are also used for this purpose. Ca smectite clays having high moisture absorption capacity at low relative humidity levels and high hardness are used for packaging desiccants.

Filtering, decolorizing and clarifying

More than 150 000 tons of smectite clays are used annually for the treatment of animal and mineral oils and greases, for decolorizing vegetable oils and for clarifying wine. Most of the smectites used for the filtering and decolorizing of oils are Ca smectite that has been acid treated to increase the decolorizing property. To qualify for these uses, a small amount of the acid-treated smectite must cause a large amount of colour reduction. Some natural acid or Fuller's earths are also used for filtering and decolorizing.

A moderately large amount of smectite clay is used to remove colloid impurities in wine. The colloidal organic impurities in wine carry a positive charge, and these particles are coagulated by mixing a small amount of a negatively charged smectite clay into the wine. Na smectites having a light colour and high dispersibility are preferred for use in the wine industry. Smectites are also used to clarify beer, vinegar and fruit juices.

Miscellaneous uses

There are so many miscellaneous small volume and speciality uses of smectite clays that only a few can be briefly mentioned. More than ever before, speciality uses of smectite clays are being investigated by numerous producing companies.

(i) Adhesives. Smectite clays having high dispersion and suspension characteristics are used in latex and asphaltic materials. Organic-clad smectites are used in some types of adhesives because of their gelling properties in certain organic liquids.

(ii) Atomic waste disposal. There has been a small use of Na smectites having high c.e.c. for absorbing radioactive isotopes of strontium and caesium and then fixing them against groundwater leaching by calcining the clay to the vitrification point. More recently, Na smectites have been used for sealing around buried canisters containing radioactive wastes.

(iii) Emulsifying suspending and stabilizing agents. Highly colloidal Na smectites and organic-clad smectites are used as suspending agents in liquid fertilizers and in many organic liquids.

(iv) Greases. Organic-clad smectites are widely used for stabilizing the gel properties of lubricating greases. The greases prepared with organic-clad smectites are said to have superior properties (Jordan 1950).

(v) Ink. Small quantities of organic-clad smectites are used in certain types of inks to control consistency, penetration and misting during the printing operation. The current extent of this use is not known. Smectites are also used to de-ink old newsprint through their ability to disperse and absorb ink pigments.

(vi) Medicines, pharmaceuticals and cosmetics. There is an increasing use of Na smectites in the preparation of medicines, pharmaceuticals and cosmetics. These uses generally require water washed Na smectites having a high brightness, and in some uses the smectite must also be high in magnesium.

(vii) Paint. Natural, Na-exchanged Ca and organic-clad smectites are used in both oil and water-based paints. The smectite clays act as suspending and thickening agents and the gel structure is said to improve brushability and spraying characteristics and to reduce pigment penetration into porous surfaces.

(viii) Catalysts. Certain types of Ca smectites were previously used for manufacturing petroleum cracking catalysts, but synthetic zeolites have largely replaced the clay catalysts. A small volume of Ca smectite is still used for manufacturing dimer acid.

(ix) Soaps. Because of the environmental problems caused by phosphates, there has been renewed interest in Na smectites as a component in soaps. Na smectites, in addition to their detergent action, have a beneficial water softening effect. The patent literature mentions a variety of cleaning and polishing compounds containing smectites.

(x) Ceramics. Smectite clays are not used in large volumes as ceramic raw materials because of their high water of plasticity. However, small amounts of low-iron smectites are added to ceramic bodies to impart strength and to give desired vitrification or colour properties. Smectites are sometimes added to kaolin slips to improve their suspension characteristics and to structural clay products to increase plasticity required for extrusion.

(xi) Anti-stick and drying agents. A moderately large tonnage of both Na and Ca smectites are used as absorbents to reduce moisture content and to prevent sticking of various types of grains and fertilizers. The potential use of granular Ca smectite as a substitute for natural gas for drying pecans has recently been reported.

(xii) Seed coating. Certain types of seeds are coated with Na smectite to improve germination and to increase their size to permit mechanical planting.

(xiii) Soil conditioners. There has been considerable research on the use of smectite clays having moderate to high water-holding capacity to retain moisture in porous and sandy soils. Because of costs, this use is limited to high profit crops. Ca smectites have been added to certain types of soils to improve tilth and to modify the pH.

(xiv) Water clarification. There are several commercial processes that use smectite clays for purification of water containing various types of industrial oils and organic contaminants. Na smectites are preferred for this use because of their dispersion and absorptive properties.

(xv) Mortar mixes. Specially formulated mortar mixes containing a small amount of sodium smectites are commercially available. The smectite clay improves the water retention, plasticity and general workability of the mortar. Microfine smectite clays are used for essentially the same purpose in the tape–joint compounds, plasters, putties and caulking compounds.

REFERENCES

Brindley, G. W. & Brown, G. 1980 *Crystal structures of clay minerals and their X-ray identification.* London: Mineralogical Society.
Farmer, V. C. & Russell, J. D. 1971 Interlayer complexes in layer silicates. The structure of water in lamellar ionic solutions. *Trans. Faraday Soc.* **67**, 2737–2749.
Greene-Kelly, R. 1957 The montmorillonite minerals (smectites). In *The differential thermal investigation of clays.* Ch. 5. London: Mineralogical Society.
Grim, R. E. 1968 *Clay mineralogy*, 2nd edn. New York: McGraw-Hill.
Grim, R. E. & Kulbicki, G. 1961 Montmorillonite: high temperature reactions and classification. *Am. Miner.* **56**, 1329–1369.
Grim, R. E. & Guven, N. 1978 *Bentonites: geology, mineralogy and uses.* New York, Amsterdam: Elsevier.
Jordan, J. W. 1950 Lubricants. U.S. Patent 2531440.
MacEwan, D. M. C. & Wilson, M. J. 1980 Interlayer and intercalation complexes of clay minerals. In *Crystal structures of clay minerals and their X-ray identification.* Ch. 3. London: Mineralogical Society.
Mortland, M. M. 1970 Clay–organic complexes and interactions. *Adv. Agron.* **23**, 75–117.
Neal, C. & Cooper, D. M. 1983 Extended version of Gouy–Chapman electrostatic theory as applied to the exchange behavior of clay in natural waters. *Clays Clay Miner.* **31**, 367–386.
Parkes, W. B. 1971 *Clay-bonded-foundry sand.* London: Applied Sciences Publishers Ltd.
Patterson, W. & Boenisch, D. 1967 *British Foundrymen* **55**, 438. Also 1958 *Giesserei* **45**, 465.
Pinnavaia, T. 1983 Intercalated clay catalysts. Abs. 32nd Annual Clay Minerals Conference, Buffalo, New York.
Weaver, C. E. & Pollard, L. D. 1973 *The chemistry of clay minerals.* New York, Amsterdam: Elsevier.
White, W. A. & Pichler, E. 1959 Water sorption characteristics of clay minerals. Illinois State Geological Survey, circular 266.

Phil. Trans. R. Soc. Lond. A **311**, 411–432 (1984)   [411]

Printed in Great Britain

Kaolins: their properties and uses

By W. B. JEPSON

Central Laboratories, English China Clays p.l.c., St Austell, Cornwall PL25 4DJ, U.K.

The main commercially important kaolin resources are the primary deposits in Cornwall and the sedimentary deposits in South Carolina and Georgia in the U.S.A. Annual world production is about 18 Mt $(1\,t = 10^3\,kg)$. In 1981 the U.K. production was 2.6×10^6 t of which about 77 % represented paper filler and paper coating grades. About 15% of the U.K. production is used in ceramics.

The kaolin extracted from the commercial deposits contains kaolinite as the major component together with ancilliary minerals. The kaolinite particles take the form of pseudo-hexagonal platelets with widths that vary from 10 to 0.1 μm and less.

Surface chemical properties have been widely researched. There is no uniformity of view on the origin of the cation exchange capacity. Aqueous suspensions of kaolin are flocculated at low pH. This can be understood in terms of the positively charged edges and the negatively charged faces of the platelets. The deflocculation of kaolin under alkaline conditions is important, and is assisted by adding a polyanion.

The Kubelka–Munk equations have provided a useful basis for interpreting the optical properties of kaolins in terms of an absorption coefficient (k) and a scattering coefficient (s). The parameter k is a bulk property and is interpreted in terms of light absorption by ancilliary minerals, by coatings of hydrated iron oxides on the kaolinite particles and by transition metals in the kaolinite structure. The parameter s depends on the size distribution of the platelets and the manner in which they are packed together.

The rheological properties of deflocculated suspensions of kaolin in water are important in modern paper coating applications. The relation between the bulk physical properties of kaolin and rheology is complex. Viscosity at low rates of shear increases with increasing specific surface area of the kaolin. Viscosity at high rates of shear depends on the shapes of the particles and their packing.

The extraction and production of kaolins are described for both primary and sedimentary deposits. The main use of kaolin is as a paper filler and a paper coating pigment. The various filler and coating grades are indicated, and effect of kaolin properties on coated sheet properties is discussed. Kaolin is a major component in most ceramic bodies. Examples of formulations are given. The manufacture of sanitaryware and tableware is discussed in terms of kaolin properties. Kaolin is used as an extender in aqueous based paints and as a filler in rubber and synthetic polymers.

1. BRIEF OUTLINE OF HISTORY

By the 16th century, Chinese porcelain was so highly regarded in Europe that many attempts were made to produce it from local materials. In the early 18th century a Jesuit missionary, Father Francis d'Entrecolles, working in China, observed the processing of the raw materials and the manufacture of the several porcelains. He wrote to his superior in Paris giving details (Pounds 1948) and supplied specimens of the two constituents, petunse and kaolin.

The samples were passed to a French scientist, Baron de Reaumur, who demonstrated that two types of porcelain could be produced, a soft paste and a hard paste, depending on the firing conditions. The hard paste porcelain was similar to that produced in China. Howe (1914) gives an excellent description of the methods used by the Chinese to produce their raw materials. A sample of what is thought to be the original petunse provided by d'Entrecolles is held by the

[191]

British Museum (BM 60022). Keller *et al.* (1980) describe kaolin from the locality of the original Kauling mine.

The word kaolin is used as an alternative name for the clay material known as china clay. The description china clay is retained in the United Kingdom and the description kaolin in most other countries including the United States of America.

William Cookworthy is credited with the discovery of the china clay and china stone deposits in Cornwall. He recognized their properties and identified them with the materials described by d'Entrecolles. Cookworthy took out a British patent and started a small porcelain factory in Plymouth. The history of the growth of the pottery industry in Staffordshire is well documented (Barton 1966; Wedgwood & Wedgwood 1980). Earthenware and bone china bodies were developed and hard porcelain manufacture declined.

The early 19th century saw the introduction of china clay as a filler in the manufacture of paper. It was also used in the textile industry as a constituent of size to add 'body' to cotton goods and to whiten them. The first coated papers were produced around 1900 to meet the demands from improved methods of printing.

2. TERMINOLOGY AND CRYSTAL STRUCTURE

Kaolin is tentatively defined by the International Committee on Correlation of Age and Genesis of Kaolin as 'an earthy rock characterized by a significant content of kaolin minerals' (Keller 1978). The kaolin minerals include all the dioctahedral 1:1 layer aluminosilicates (Bailey 1980). Of the many theoretically possible structures, kaolinite is the only known 1-layer form; dickite and nacrite are the only known 2-layer forms.

A platelet of kaolinite is made up of layers about 7.2Å† thick that are continuous in the a and b directions (excluding twinning) and stacked in a given order along the c-axis. In reality there are varying degrees of disorder in the layer stacking. Brindley (1980) has reviewed the various experimental and interpretational approaches and emphasized the important papers by Plancon & Tchoubar (see, for example, Plancon & Tchoubar 1977) whose models include stacking faults by linear displacements, rotation of layers and displacement of Al cation vacancies. The number of layers per domain was calculated from the breadth of the 001 X-ray reflection. Values varied from 25 layers for a French kaolinite to 75 layers for a Georgia kaolinite. The layer stacking imperfections are seen by electron microscopy as folds, voids or other inclusions (Tchoubar *et al.* 1982).

Material extracted from commercial deposits contains kaolinite as the major component together with ancilliary minerals that vary with the type and quality of the deposit. The extracted material is known and well understood as kaolin or china clay; usage is followed below although this conflicts with the geological definition. There is no known large scale exploitation of either dickite or nacrite. This account excludes certain industrial minerals which contain kaolinite as a major component. These include ball clays (Watson 1982a), flint clays (Keller & Fitzpatrick 1981) and the refractory clays (Dickson 1982). Halloysite is also excluded.

---

† $1 \text{ Å} = 10^{-10} \text{ m} = 10^{-1} \text{ nm}$.

3. Deposits

The most important kaolin resources are in Cornwall in SW England and in S Carolina and Georgia in the United States of America. Other important resources are found in the Amazon basin in Brazil, in Brittany and in the Aquitaine basin in France, in Bavaria and Bohemia, in Spain and in parts of eastern Europe, particularly Russia. There are unexploited deposits in South East Asia including Australia; it remains to be seen whether certain rheological problems can be overcome to permit their use in paper coating (O'Neill 1983).

The deposits in SW England are described by Bristow (1968, 1977) and the deposits in Georgia are described by Patterson & Murray (1975) and Hurst *et al.* (1979). Keller (1977a) has described the fabric of kaolin deposits throughout the world.

Kaolin deposits are classified (Keller 1978; Bristow 1980) into primary and secondary or sedimentary deposits. Those in SW England are primary deposits, probably formed by a combination of weathering and hydrothermal alteration of granite (Bristow 1977). On a mass basis, the commercial product represents about 15% of the whole, with the balance largely present as quartz, feldspars and unaltered granite. Those in S Carolina–Georgia are sedimentary deposits; they have been transported from their place of origin and the recovery of commercial product can exceed 90%. The kaolinitic sands are closely related to sedimentary kaolins but the kaolinite content is much lower, about 10–15%. Some of the Spanish and German deposits are of this type.

4. Commercial aspects

The total annual world production of kaolin is about 18 Mt (Watson 1982b). In 1981 the U.K. production was 2.6 Mt, of which 2.2 Mt were exported principally to Europe. Markets in Europe for paper grades of kaolin have been reviewed (Anon 1982). About 5.2 Mt of refined kaolin were produced in the U.S.A. in 1981. Some 90% of this was produced in Georgia. Harben (1979) reviews production in Georgia and Clark (1982) reviews kaolin markets in the U.S.A.

The main use of kaolin is as a filler and as a coating pigment in the production of paper. Some 78% of the U.K. production is used in this sector. About 15% of the U.K. production is used in the production of ceramic articles, mainly in sanitaryware, tableware and wall tiles. The balance of 7% is used in a variety of lesser applications of which those in paints, plastics and rubber predominate.

5. Properties

(a) Shape and size

The shape and size distribution of kaolin is important in its commercial uses, particularly in paper filling and paper coating. It affects the mechanical, optical and printing properties of the finished sheet of paper. Particle size is expressed as e.s.d. (equivalent spherical diameter) (Allen 1975) even though the kaolin particles are in the form of platelets and not spheres. Particle size distribution is normally measured by sedimentation from a deflocculated suspension in water. Measurements below 0.15 μm are best made with a centrifuge. Commercial instruments have been developed for routine measurements (Sennett *et al.* 1974; Allen 1975). The cumulative mass per cent below 2 μm value is often used as a control in production.

[193]

27-2

Transmission electron microscopy of a fraction of a kaolin under 5 μm e.s.d. prepared from a Cornish primary deposit shows it to be largely composed of pseudo-hexagonal plates. In this fraction the plate widths would vary from an upper limit of about 8 μm down to at least 0.1 μm. The aspect ratio, which is the ratio of plate width to thickness, would vary through the size distribution from about 10:1 at the coarse end to about 50:1 at the fine end (Noble & Golley 1970, unpublished work).

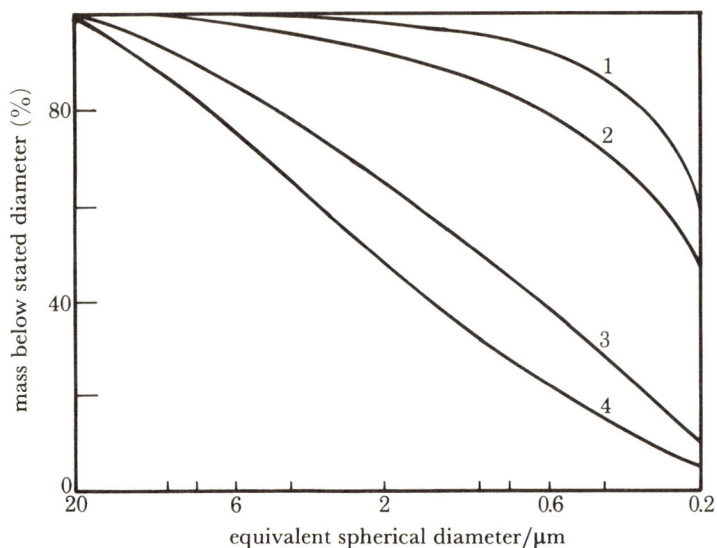

FIGURE 1. Cumulative particle size distribution curves of kaolins. Curves (1) and (2), sedimentary kaolins from East Georgia; curve (3), sedimentary kaolin from Central Georgia; curve (4), primary kaolin from Cornwall.

Kaolins refined from the central Georgia sedimentary deposits show a number of important differences (Conley 1966). In an under 5 μm e.s.d. fraction from a 'coarse' deposit the particles are much closer to hexagonal in shape. The plate widths again extend to at least 0.1 μm but the particles are now much smaller. The aspect ratios lie between 6:1 and 10:1 and vary little through the size range. If a fraction exceeding 5 μm e.s.d. is examined by scanning electron microscopy, many exfoliated booklets of kaolinite are seen; they are up to 50 μm thick and up to 20 μm across (Morris et al. 1965; Olivier & Sennett 1973). They are alternatively described as vermicules or stacks. They are porous and numerous slit-like voids can be seen between the kaolinite platelets making up each booklet. Keller (1977b) reviews their formation and origin. In commercial production the fraction over 5 μm e.s.d. is used either as a component of paper filling grades or is broken down mechanically (Whitley & Iannicelli 1972; Murray 1976) to produce grades called 'delaminated kaolins'. Booklets are also found in the kaolin deposits of SW England. They seem to break down readily during normal production and are not a significant component of the commercial products.

Examples of cumulative size distributions of kaolins refined from deposits in Cornwall and Georgia are shown in figure 1. The two kaolins from the East Georgia deposits (1 and 2) are finer than the kaolin from Central Georgia (3) which in turn is finer than that from Cornwall (4).

If a kaolin between narrow size limits is prepared by repeated sedimentation and then examined by transmission electron microscopy, the range of plate widths can be compared with the equivalent spherical diameter. For the primary kaolins, the centre of the plate width distribution exceeds the e.s.d. of the fraction by a factor of between 2 and 3. These measurements are tedious and time consuming. Encouraging progress is now being made in the application of electrically induced birefringence methods to the sizing of anisometric particles in the colloidal size range (Oakley & Jennings 1982). By assuming a thin disc model and a log-normal size distribution, a number-diameter distribution can be calculated. These methods are likely to find wide applicability once comparative measurements have been made on a range of samples and any limitations have been identified.

The microstructural features of kaolinite particles are receiving attention. Deposition of gold from the vapour phase has been used to decorate the faces; it reveals a multiplicity of micro steps (Thompson *et al.* 1981). Edges of the platelets, when examined at a magnification of about 1 million, show a frayed appearance. These findings are significant when judging the worth of calculations based on model particles to interpret the colloidal properties of kaolin suspensions in water.

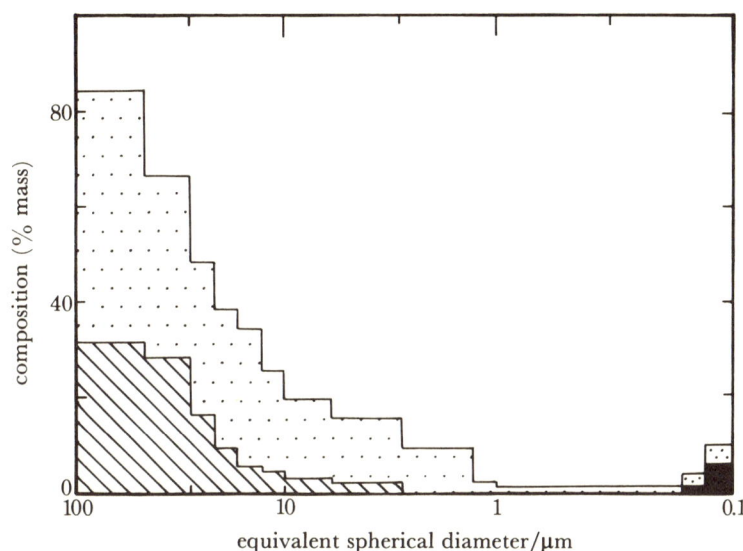

FIGURE 2. The variation of mineralogy of a Cornish kaolin with particle size. Oblique lines: feldspar and quartz; dots: mica; solid region: smectites; non shaded region: kaolinite.

(b) Elemental and mineralogical composition

The mineralogy of a primary kaolin varies with particle size. In the experiment shown in figure 2, a fraction under 50 μm was prepared from matrix and divided into 15 fractions by sedimentation from a deflocculated suspension. The mineralogical composition of each fraction was then determined by X-ray powder diffraction. The main ancilliary minerals are quartz, micas and feldspars. Small amounts of smectite are sometimes found in the finest fractions. The kaolinite content increases with decreasing particle size and this is the basis of much of the commercial production.

The mineralogical and partial elemental compositions of two refined kaolins, approximating to a commercial filler grade and a commercial coating grade, are shown in table 1. The finer fraction has the greater kaolinite content and the smaller mica content. The analytical TiO_2 content of each fraction is small. Anatase is sometimes found in kaolins derived from primary deposits overseas. The amount depends on the parent granite but is generally below 0.5% by mass.

[195]

The mineralogical compositions of the Georgia kaolins vary much less with particle size. The crude kaolin contains small amounts of mica, quartz, anatase and other ancilliary minerals amounting to perhaps 10% by mass (Hurst *et al.* 1979; Van Olphen & Fripiat 1979). Indicative mineralogies and partial elemental compositions of two Georgia kaolins are shown in table 2. Anatase is a major impurity but mica is present in only small amounts.

Kaolinite prepared from a geode source gave a chemical analysis close to the theoretical value and has been suggested as a reference kaolinite mineral (Keller *et al.* 1966). Research samples are best selected and prepared under laboratory conditions from matrix. Large samples may contain a mixture of polymorphic types (Keller & Haenni 1978).

TABLE 1. ANALYSES OF KAOLINS PREPARED FROM CORNISH MATRIX

	composition (% mass)	
	under 20 μm	under 10 μm
kaolinite	83	90
mica	14	8
feldspar	2	1
Fe_2O_3	0.9	0.6
TiO_2	0.05	0.06
K_2O	1.9	1.1

TABLE 2. ANALYSES OF TWO KAOLINS (UNDER 10 μm) PREPARED FROM GEORGIA MATRIXES

	composition (% mass)	
	central Georgia	east Georgia
kaolinite	> 95	> 95
mica	≈ 1	≈ 2
Fe_2O_3	0.25–0.60	0.8–1.5
TiO_2	1.0–2.0	1.5–3.0
K_2O	≈ 0.1	0.2

The identification and measurement of the ancilliary minerals in kaolin by X-ray powder diffraction together with quantitative elemental analysis by x.r.f. are essential prerequisites to any investigation. A major advance took place in the early 1970s when transmission electron microscopes equipped with an analytical facility became commercially available. The presence of iron in the kaolinite structure, previously inferred from lengthy Mossbauer experiments with magnetically extracted samples (Malden & Meads 1967), was confirmed (Jepson & Rowse 1975). There was a wide variation in Fe:Si atom ratios between individual kaolinite particles extracted from a small matrix sample. In the Georgia kaolins almost all of the Fe found by elemental analysis is contained in the kaolinite structure. Mossbauer and electron spin resonance spectroscopy have provided valuable information. Most of the iron in the kaolinite structure is present as Fe(III) isomorphously replacing Al(III) in octahedral sites (Angel & Hall 1972; Jefferson *et al.* 1975; Meads & Malden 1975; Hall 1980; Komusinski *et al.* 1981).

The association of mica with kaolinite has been inferred from analytical electron microscopy (Jepson & Rowse 1975). Ultramicrotomy and high resolution electron microscopy (Lee *et al.* 1975) showed it to be present as occlusions within the kaolinite layers in the samples examined.

Most of the titanium found by elemental analysis of refined Georgia kaolins is present as anatase with only minor amounts in the kaolinite structure (Dolcater *et al.* 1970; Jepson & Rowse 1975; Sayin & Jackson 1975; Rengasamy 1976). Small amounts of rutile are sometimes found. Transmission electron microscopic examination of refined Georgia kaolins (under 5 μm) shows the anatase to be present as particles about 0.1 μm across. Probe analysis shows some combined iron, hence the term 'titanoferrous impurities'.

(c) Surface chemistry

The crystal structure of kaolinite suggests that the octahedral basal surface of a platelet should be gibbsite-like and should present close packed hydroxyl groups; the tetrahedral basal surface of the platelet should be silica-like and present siloxane bonds. At the edges of a platelet, the lattice is disrupted and a broken bond surface should be exposed.

(i) Contamination of surfaces

Surfaces of kaolinite platelets found in nature deviate from the above. They are contaminated (Jefferson et al. 1975; Angel & Vincent 1978) with hydrated iron oxides which may be present either as particles, several 100 Å across, or as an amorphous coating combined with other elements and perhaps occupy localized areas. The hydrated iron oxides are removed during commercial production by leaching with sodium dithionite (Conley & Lloyd 1970). The surfaces are also contaminated with organic material and values up to 0.05%, expressed as carbon, are not unusual (Ferris & Jepson 1972). Commercial kaolins may contain residual organic compounds added as flocculants or deflocculants during processing.

(ii) Cation and anion exchange

The cation and anion exchange properties of kaolinite have been widely researched. Present day views have developed from the work of Schofield (1949), Schofield & Samson (1953) and Van Olphen (1951, 1964). If a well chosen and carefully prepared sample of kaolin is repeatedly washed with 1 M NaCl at pH 7, for example, then certain cations, the exchange cations, predominantly Ca^{2+} and Mg^{2+} ions, are displaced. If the sample is allowed to equilibrate with 0.01 M NaCl at pH 7 it is found that some Na^+ ions are associated with the kaolin. If the sample is rinsed with water most of the Na exchange ions are retained by the kaolin. The amount is typically about 20 μmol Na per gram of sample which is equivalent to about 1 Na atom per 100 Å^2 of surface using the BET (Brunauer, Emmett and Teller) nitrogen area. Figure 3 shows that Na^+ ion adsorption increases with increasing pH (Ferris & Jepson 1975). The adsorption around pH 3 is sometimes larger than zero (figure 3), and seems to depend on sample and preparative technique. Kaolins also show a small anion adsorption that decreases with increasing pH (figure 3).

There are three different views on the origin of the cation adsorption. The first is that isomorphous replacement of Al for Si in the kaolinite structure produces an excess of negative electrical charge. Only 1 Si atom in 400 would need to be replaced by 1 Al to give a cation adsorption of 20 μmol Na per gram. It has not been possible to demonstrate this deviation from stoichiometry either by wet chemical analysis on a bulk sample or by probe analysis of individual kaolinite particles (Jepson & Rowse 1975). Work by Weiss (1959) and Weiss & Russow (1963) suggested that the isomorphous substitution was confined to the silica tetrahedral layer. Comments by Uytterhoeven (1963) and by Bolt (1963) are pertinent. The second view attributes either some or all of the cation adsorption to the presence of small amounts of expanding minerals. Lim et al. (1980) examined 7 Georgia kaolins and apportioned the cation adsorptions, which varied from 30 to 90 μmol Cs per gram, to contributions from 2:1 layer minerals and from isomorphous substitution of Al for Si in the kaolinite structure. The external surface cation adsorption ranged from zero to 10 μmol Cs per gram. Talibudeen & Goulding (1983) measured differential heats of K^+–Ca^{2+} exchange on a number of kaolins and inferred the presence of vermiculite, micaceous and smectite layers in association with the kaolinite. The third view (Ferris & Jepson 1975) is based on studies of two kaolins prepared under laboratory conditions. It suggests that the kaolinite particles are coated with a gel-like material, rich in silica, which dominates the cation adsorption. This view does not exclude isomorphous substitution.

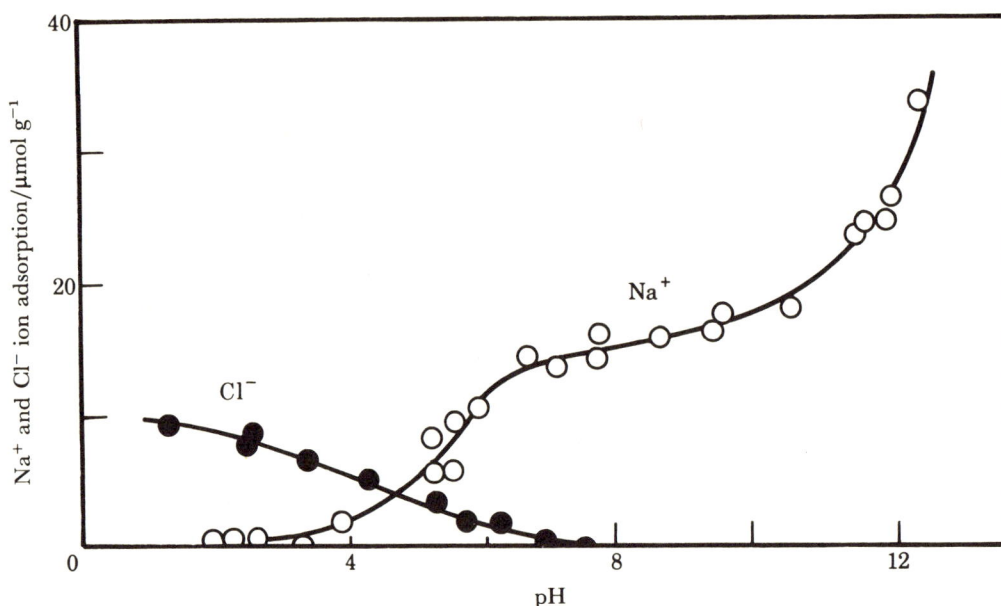

FIGURE 3. Change of Na^+ ion and Cl^- ion adsorption on kaolin with pH. Electrolyte: 6×10^{-3} mole dm^{-3} NaCl. After Ferris & Jepson (1975). Reproduced with permission from *J. Colloid Interface Sci.*

The anion adsorption and some of the cation adsorption are attributed to adsorption at the particle edges (see § 5(a)). The model proposed by Schofield & Samson (1953) which incorporates earlier suggestions by Van Olphen (1951) is still accepted. Under acidic conditions the edges are positively charged through protonation of hydroxyl groups attached to Al atoms exposed at the platelet edges. Above about pH 7, by analogy with gibbsite, protonation becomes negligible. As the pH is increased so the exposed edge silanol groups start to dissociate and the cation adsorption increases. The paper by Van Olphen (1964) on the flocculation of clay suspensions gives a deeper insight.

Rand & Melton (1977) have made rheological measurements using aqueous suspensions of kaolins in the homoionic Na-form and inferred that the platelet edges have an apparent isoelectric point of 7.3. In other work the zero point of charge of the platelet edges has been estimated by combining separate data for silica and for alumina; Williams & Williams (1978) suggest a value of pH 7.2.

The following simple description emerges. Each platelet carries a negative charge on one or both of its faces which will be of constant charge (isomorphous replacement or adsorbed 2:1 layer mineral models) or of constant potential (gel coating model). The negative charge on the faces increases with increasing pH (gel coating model). The particle edges are positively charged

[198]

at low pH, pass through an isoelectric point and become negatively charged at high pH. The model is supported by experiments with radioactively tagged cations followed by autoradiography at the electron microscope level (Fordham 1973).

(iii) *Colloid chemistry*

Kaolin suspensions have the interesting and important property of being dispersed (deflocculated) under alkaline conditions (say pH 9) and being coagulated (flocculated) under acidic conditions (say pH 4). This is readily understood in terms of the simplified account given above of the electrical charges carried by the kaolinite platelets. Stability calculations have been made by using double layer theory for different H^+ ion (pH 4–10) and electrolyte concentrations (10^{-2} to 10^{-4} mol dm^{-3}). Idealized geometry is used and electrical potentials are assigned to the edge and face surfaces (Flegmann *et al.* 1969; Williams & Williams 1982). Face-to-face interaction showed a high primary maximum in all environments. At pH 6 there is no energy barrier to either edge–edge or edge–face interactions. At pH 4 there is a slight barrier to edge–edge interactions which decreases with increasing salt concentration. At pH 10 there are moderate energy barriers to both edge–edge and edge–face interactions. These findings are in broad accord with practical observations.

Electron microscopy, freeze drying and critical point drying methods have been used to examine the spatial relations between the particles of kaolinite in floccules and in filter cakes. Results by O'Brien (1971, 1975), by Lanier & Jones (1979) and by Lockhart (1980) are in broad agreement. The flocculated kaolin fabric consists of abundant stepped clusters of plates in face–face contact with some zones of edge–face oriented plates randomly arranged in a porous 3-dimensional network. Lockhart (1980) uses the term 'microporous honeycomb'.

The flocculation of kaolin is important in commercial production. On lowering the pH of a deflocculated suspension, the platelets interact to form aggregates that grow into flocs by collision and by 'sweeping' up individual platelets. After flocculating and thickening, the kaolin is separated from most of the aqueous phase by filtration. The permeability of a kaolin depends quite markedly on its size distribution and on the pH of the suspension. The permeability of a Cornish coating grade (see table 1) increases by a factor of about 2.5 on decreasing the pH of the suspension from 8 to 4. In commercial production, the kaolin slurries are often filtered at elevated pressures to lower the final water content of the filtercake. In terms of mechanism the cake consolidates until its strength increases to the new applied pressure.

The deflocculation of kaolin under alkaline conditions is important in both the production and use of kaolin, and is explained by the stability calculations described above. Dispersion or separation of the particles is assisted by stirring to impart some shear which dislodges smaller particles from larger particles. It is also helped by adding a polyanion which adsorbs on the platelet surfaces and increases their negative charge.

Intermediate situations exist and are important in the casting of shapes in the ceramic industry (Bailey 1976). Casting rate and cast properties depend on achieving and maintaining the intermediate condition of 'under deflocculation'.

(iv) *Further discussion of kaolinite surfaces*

Experimental work with kaolin normally involves some pretreatment to remove adsorbed impurities. This may include leaching with sodium dithionite, contacting with hydrogen peroxide solution followed by washing with dilute hydrochloric acid (Schofield 1949) until the concentration of aluminium in solution reaches some small and arbitrary limit. Such an approach ignores the solubility of kaolinite in water and fails to recognize that the dissolved aluminium and silicon will eventually attain equilibrium values. Small concentrations of aluminium and silicon in solution do not of themselves guarantee that dissolution and readsorption processes are not taking place. Bolland *et al.* (1976), in work that merits detailed study, attempted to reduce dissolution and contamination with soluble Al species by conducting their experiments around 0 °C.

The above general approaches contain the implicit assumption that the pretreatments reveal the ideal kaolinite surfaces and that these surfaces persist during the subsequent experiments. This has never been convincingly demonstrated.

An alternative approach (Ferris & Jepson 1975) is to prepare the sought size fraction with a hydrocyclone, and hence add no chemical other than water. The experimental objective then becomes that of characterizing the surface of the naturally occurring material. There are prospects that the surface compositions of kaolinite platelets will be determined by using the newer techniques of electron spectroscopy and some progress is being made (McBride 1976, 1978; Koppelman & Dilland 1977; Dilland & Koppelman 1982).

TABLE 3. OPTICAL PROPERTIES OF TWO COATING GRADES OF KAOLIN

	kaolin C	kaolin G
$B(\%)$	86.8	86.8
$k/(\mathrm{cm^2\,g^{-1}})$	23.8	28.5
$s/(\mathrm{cm^2\,g^{-1}})$	2400	2850

(d) Absorption and scattering of light

The optical properties of kaolins are important in most of the commercial uses. They are usually interpreted in terms of an absorption coefficient (k) and a scattering coefficient (s) by using the Kubelka-Munk (K.–M.) theory (Kerker 1969; Starr & Young 1975).

The parameter k is a property of the kaolin itself and can be modified by chemical treatment or by beneficiation to remove coloured impurities. It can be determined by measuring the diffuse reflectance of light of a particular wavelength, generally 457 nm, from a series of paper sheets containing different loadings of kaolin.

The parameter s depends on the difference in refractive index between the kaolin and the medium, generally air. It also depends on the shape and size distribution of the particles and, in particular, on their state of aggregation. Conceptually it is useful to regard the pores in, for example, a compacted powder or a coating on paper as the light scattering units.

If a sample of flocculated kalin is powdered and compacted into a disc in a reproducible manner and the diffuse reflectance is measured at 457 nm, the powder brightness B (Windle & Gate 1968) is obtained. The value is expressed on an absolute basis as a percentage and is measured relative to a subsidiary standard which has been calibrated in a laboratory approved by the International Standards Organization.

Having already determined k the value of s can be calculated by using the K.–M. equations. Table 3 shows values for a coating grade (C) prepared from a Cornish deposit and a premium coating grade (G) prepared from a blended Georgia crude by refining and flotation. It is seen that sample G has a higher value of k and hence greater light absorption. It is however finer than sample C and when compacted shows more light scatter as evidenced by the greater value of s. It is interesting that in this example the greater value of k is offset by the greater value of s so that the powder brightnesses are equal (N. A. Climpson & J. H. Taylor 1972, unpublished work).

If a series of kaolins with different size distributions is examined, the calculated values of s can be related to some measure of average particle size. One common choice is the d_{50} value which is the e.s.d. corresponding to the 50 % value on the cumulative size distribution curve. It is found that s increases from about 1500 cm^2 g^{-1} for d_{50} equal to 3 μm up to a maximum of about 2900 cm^2 g^{-1} for d_{50} equal to 0.5 μm. It then decreases as d_{50} decreases further. With k set equal to 23.8 cm^2 g^{-1} the above values of s would give powder brightnesses of 83.7 and 88.0.

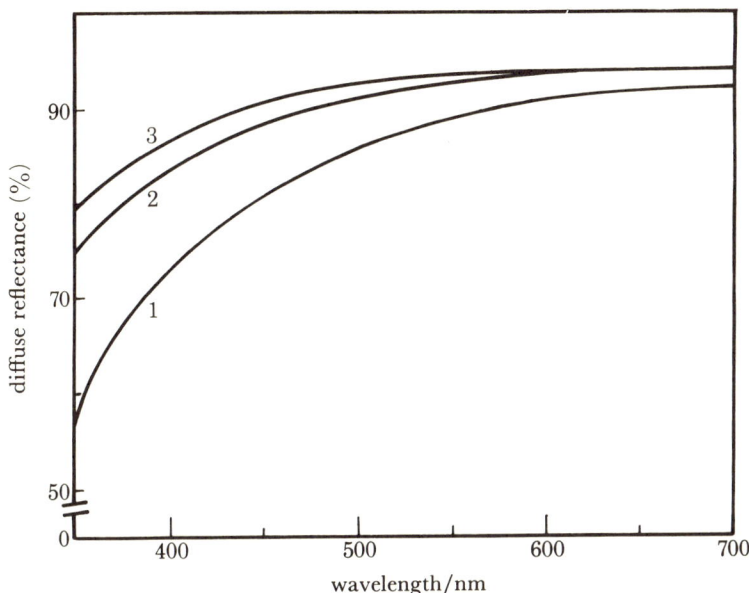

FIGURE 4. The change in the diffuse reflectance of light from several kaolin compacts plotted against wavelength. Each sample had been refined to under 10 μm. The ordinate is in absolute units. Curve (1), sedimentary kaolin from Georgia; curve (2), primary kaolin from Cornwall; curve (3), primary kaolin from Cornwall after leaching with sodium dithionite.

The optical properties of papers have been widely researched with both real and model systems. Ultra-microtomed sections have been used to make direct measurements of the size and distribution of the scattering elements (Climpson & Taylor 1976; Lepoutre 1976; Borch & Lepoutre 1978; Starr & Young 1978; Alince & Lepoutre 1980a, b). The scattering coefficient of a coating of kaolin and adhesive on paper is well below the value for the power compact because the particles are laid down from a deflocculated suspension. They adopt a preferred orientation broadly parallel to the base sheet and are more closely packed; some of the pore volume is occupied by adhesive. The values of s for the samples C and G in table 3 when coated on paper would each be about 800 cm^2 g^{-1}.

Figure 4 shows the change of diffuse reflectance with wavelength. The kaolins were prepared from flocculated suspensions. Reflectance decreases with decreasing wavelength and falls markedly in the ultraviolet region despite the increasing contribution there from scattering.

Ancilliary minerals (micas, tourmaline, titanoferrous impurities) and coatings of hydrated iron oxides on the kaolinite particles contribute substantially to the light absorption which arises from charge transfer (c.t.) transitions involving metal–ligand and metal–metal transfers (Weyl 1959; Robbins & Strens 1972; Karickhoff & Bailey 1973; Faye et al. 1974).

If the sample shown as curve (2) in figure 4 is leached with sodium dithionite and then filtered, dried and recompacted, the reflectance follows curve (3). This is because hydrated iron oxide coatings are dissolved. The analytical Fe_2O_3 content only fell from 0.54 to 0.52% by mass. It seems that quite large amounts of iron can be tolerated provided they are in the kaolinite structure. Some coating grades analyse about 2% Fe_2O_3. Provided the iron atoms are well separated from each other the c.t. reinforcement is limited.

Figure 4 also shows the reflectance curve (1) of a Georgia kaolin. When the sample was beneficiated by flotation and the reflectance redetermined, the experimental curve was close to curve (2). The analytical Fe_2O_3 content of 0.90% showed little change but the analytical TiO_2 content fell from 1.70 to 0.60%. This illustrates the major effect of titanoferrous impurities on light absorption.

The naturally occurring organic material adsorbed on the surfaces of the kaolinite particles also contributes to light absorption and oxidative treatments are sometimes used in processing to improve powder brightness. Ozone (Allegrini *et al.* 1970; Cecil & Jacobs 1971; Malden 1972), hydrogen peroxide (Bundy & Berberich 1969), potassium permanganate (Duke 1967) and sodium hypochlorite (Iannicelli *et al.* 1972) have been used.

(e) *Rheology*

When a kaolin is stirred under appropriate conditions into an aqueous solution of tetrasodium pyrophosphate (TSPP) that contains a small sodium hydroxide addition, a solids content of about 70% by mass (or 47% by volume) can be reached. The optimum TSPP addition is about 0.3% and the optimum pH is about 8. The suspension can be easily poured when freshly made and has a viscosity of about 500 m Pa s at a shear rate of 30 s^{-1}.

The deflocculation of kaolin with TSPP and the higher polyphosphates (Michaels 1958; Lyons 1964; Millman 1964) is due to Ca^{2+} and Mg^{2+} exchangeable cations being displaced from the kaolin surface by Na^+ ions and forming complexes with the polyphosphate ions in solution. Additionally, polyphosphate ions are strongly adsorbed on the kaolin surface, probably at exposed Al atoms on the platelet edges and at steps on the platelet surfaces. The negative charges built up together with the mechanical shear applied during mixing causes the particles to separate. The system is unstable and the polyphosphates slowly hydrolyse to orthophosphate. Complex aluminium phosphates also form (Dennison & Toms 1967).

Some kaolin slurries that have been prepared by using a polyphosphate will form a gel structure within a few hours if left undisturbed (Dennison & Toms 1967; Mallett & Craig 1977) which can give discharge problems when the kaolin is transported as a deflocculated slurry. Gelation is retarded or stopped by adding sodium polyacrylate along with the polyphosphate when the kaolin is dispersed in water. The polyacrylate may adsorb in the form of a looped configuration, which gives steric stabilization and prevents bridging of the particles (Ottewill 1978).

In paper coating a mixture of the deflocculated kaolin and an adhesive is applied to one or both sides of the paper web and dried. The rheological properties of this mixture or coating colour are very important in modern coating processes. Rates of shear can reach 10^6 s^{-1} along the line of contact of the blade with the paper in blade coating. Dilatency or shear thickening can develop to the extent that the coating colour assumes a rigid form as long as the high rate of shear is applied. If this were to occur in commercial practice, coat weight and smoothness would be adversely affected and printing properties would be degraded. The rheology of kaolin–water mixtures is also important during normal operation. An excessively dilatent suspension can give screening problems.

The relation between bulk kaolin properties and coating colour rheology is complex (Beazley 1972). The important variables seem to be specific surface area, particle shape and particle packing. Viscosity at low rates of shear increases as the specific surface area increases. The shear rate at which shear thickening becomes significant depends on the aspect ratios of the particles and their packing. Anisometry seems important. In general, the platier the particles then the lower the rate of shear at which shear thickening is first observed. Thus a suspension of kaolin that contains many thin particles with large diameters has a greater viscosity than a suspension of the same solids that contain 'blocky' particles with a range of sizes. In a well packed assembly, voids between the spheres of rotation of the larger particles are occupied by smaller particles and this is continued through the size distribution.

Expanding 2:1 layer minerals have an adverse effect on rheology. They are sometimes present in amounts of up to several per cent in kaolins. Processes have been developed to remove or modify chemically the expanding minerals; for example, they can be concentrated in the finest fraction of the size distribution and then removed in a continuous centrifuge (Lyons 1950; Brociner 1971). Alternatively they can be converted to a non-swelling form by adding a solution that contains positively charged aluminium–hydroxy species to a suspension of the kaolin in water and then by raising the pH to about 7. Coagulation occurs and the smectite layers become attached to each other and to the larger kaolinite platelets (McCartney & Yeo 1965; Adams *et al.* 1977; Kunkle & Kollmar 1977, 1978).

In ceramic production which involves plastic forming, particles of high aspect ratio play an important role in providing cohesion and strength. For such applications 2:1 expanding minerals at small concentrations are beneficial and the kaolin matrix is selected accordingly.

6. Commercial production

(a) From primary deposits

The extraction and production of refined kaolins from primary deposits are described (Menadue 1979). Powerful jets of water are directed at the working faces in the pit to break up the kaolinized granite or matrix. The washing streams gravitate to centrifugal gravel pumps which lift the suspension to spiral classifiers and then to hydrocyclones which produce an overflow containing particles mainly below 53 μm in size. The underflow contains quartz sand as the major component.

The product, which has a pH of about 4.5 and is naturally flocculated, is stored in tanks and allowed to thicken. It is then pumped to a refining plant where a deflocculant is added. Subsequent refining exploits the increase of kaolinite content with decreasing particle size as shown in figure 2. The slurry passes to refining tanks operated in series to give a product with about 12% over 10 μm and 45% under 2 μm and essentially 100% by mass under 20 μm (table 1).

The fraction under 20 μm is suitable as a feed for a paper filling grade or a ceramic clay, although in practice kaolin slurries from several pits are blended together in a central plant. Paper coating grades are further refined in solid bowl centrifuges which typically give a product with about 99.5% under 10 μm and 75% under 2 μm (table 1). The oversize fraction may be further beneficiated by flotation. The refined clay from either stage is flocculated by adding sulphuric acid. Paper grades are generally leached with sodium dithionite and most of the ferrous salts brought into solution are removed during filtration.

The slurry is allowed to thicken and is then filter pressed. In the production of a paper coating grade, a mixture of filter cake and dried clay is passed through a pug mill (Brociner 1970) to improve certain properties, particularly rheology. Product forms include powder, lump containing about 10 % water and deflocculated slurry at 65 % solids and above. Filler clay product from a tube press (Gwilliam 1971) contains only about 17 % water and can be shipped directly without thermal drying.

(b) From sedimentary deposits

The sedimentary deposits include sedimentary kaolins and kaolinitic sands. The extraction and production of refined kaolins from sedimentary deposits are described (Anon 1955; Murray 1963, 1976; Coope 1979; Harben 1979; Hurst et al. 1979). The simplest and cheapest production route uses the airfloat process. The matrix is crushed, dried and ground. It then passes into an air stream of constant velocity. Grit and coarse kaolin particles remain behind. The product is used as a low cost filler where brightness and viscosity are not important.

Most of the kaolins used in the paper industry are produced by wet refining. The kaolin matrix is mined by drag line or front-end loader, crushed and made down with added deflocculant as a suspension in water. The matrix may be slurried at the mine and transported as slurry by pipeline to a central plant. It is degritted, screened and then classified through continuous centrifuges to the sought particle size distribution. The coarse middle Georgia crudes, as mined, have a particle size of 60–70 % under 2 μm and are classified to 80–90 % under 2 μm for the coating grades. The east Georgia crudes have a particle size of 90 % under 2 μm as mined and do not require the same classification.

Subsequent processing involves flocculation, leaching with sodium dithionite followed by dewatering with either parallel plate filter presses or rotary vacuum filters. The kaolin may be thermally dried after extrusion in the form of noodles or it may be deflocculated and the slurry then spray dried for bulk shipment. There is an increasing emphasis on shipment as a deflocculated slurry.

7. DEVELOPMENTS IN SEPARATION

In the production routes outlined above, the kaolin is beneficiated by size separation or refining. Micromineral separation methods have been introduced to remove some of the unwanted minerals from the refined kaolin. In particular, the introduction of flotation and of high extraction magnetic filtration (HEMF) has been of great commercial significance. It has been possible to extend reserves, to reduce overall production costs and to produce new grades. One major application of HEMF has been the removal of titanoferrous and micaceous impurities from Georgia kaolins.

(a) Magnetic separation

High extraction magnetic filtration was developed between 1963 and 1968 and put into full commercial production by the Kaolin Industry of Georgia between 1973 and 1975 (Kolm 1975; Kolm et al. 1975; Iannicelli 1976; Dawson 1977). A slurry of deflocculated kaolin is passed through a canister packed with ferritic steel wool and surrounded with water cooled copper coils through which electric current is passed. The magnetic field in empty space is 2T.

Particles with the greatest volume magnetic susceptibilities and hence analytical Fe_2O_3 contents are retained on the filter. The supply of electric current is halted at intervals and the steel wool is flushed with water to remove the retained impurities.

(b) Discrimination between surfaces

(i) Froth flotation

Froth flotation (Greene *et al.* 1961; Greene & Duke 1962; Cundy 1968) is used to remove titanoferrous impurities from appropriate Georgia kaolin feeds. The analytical TiO_2 content is reduced to about 0.6 % and the shade is improved for the reasons discussed in § 5(d). About 500 Mt of high brightness coating grades are produced each year by flotation.

Froth flotation has also been applied to the processing of primary kaolins (Clark 1971). The oversize fraction from the refining stage has hitherto been of limited use as a filler grade because of its relatively high feldspar and quartz content (see § 6(a) and figure 2). In the flotation process, the kaolinite and mica particles are separated and represent a valued starting material for several commercial grades.

(ii) Selective flocculation

In selective or differential flocculation (Yarar & Kitchener 1970; Kitchener 1972) a small quantity of a polymeric flocculant is added to an aqueous suspension of kaolin. The titano-ferrous impurities are selectively flocculated to form aggregates and then flocs (Maynard 1974). The method has been used commercially to beneficiate kaolins refined from East Georgia crudes.

(iii) Selective coagulation

When a kaolin is deflocculated with a substantial excess of sodium hexametaphosphate and the suspension is left undisturbed, some of the titanoferrous impurities coagulate and settle out preferentially (Maynard *et al.* 1969). The method was operated commercially for some years before being replaced by selective flocculation.

A later example of selective coagulation is described by Nott & Price (1978a, b). A suspension of freshly prepared magnetite is added to the deflocculated kaolin slurry. The magnetite particles and the titanoferrous impurities mutually coagulate and are removed by magnetic separation. It seems that almost all of the titanoferrous impurities can be removed under laboratory conditions. Cook & Cobb (1982) describe later developments.

8. USE IN PAPER

Kaolin is used as a filler and as a coating pigment in the production of paper. Major sectors in terms of kaolin consumption are the supercalendered (SC) papers used in the production of large circulation magazines and the light weight coated (LWC) papers used in the production of mail order catalogues. In the SC papers the filler may represent up to 25 % of the total sub-stance. A LWC paper composed of a base sheet of 38 g m^{-2} with a coating of 8.5 g m^{-2} on each side contains about 33 % kaolin. Coat weights (pigment plus binder) may reach 35 g m^{-2} for the high quality art papers.

(a) As a filler

In papers made from chemical pulp the filler is added to provide opacity and to reduce the cost of the furnish. In the SC papers made from mechanical pulp, the filler again reduces furnish costs and more significantly improves paper gloss and printability.

Excellent accounts of papermaking are given by Casey (1960), Britt (1970) and Macdonald & Franklin (1970) while detailed aspects are described in the specialist publications of the Technical Association of the Pulp and Paper Industry in the U.S.A. (see, for example, Britt 1975; Davison 1982; Tanaka *et al.* 1982).

Filler grades are controlled in terms of particle size, powder brightness and abrasion index. In the U.S.A. they range from the air-floated grades to the wet refined grades including those grades produced mainly for coating applications.

The optical properties of a paper sheet are interpreted (Robinson 1976) by using the Kubelka–Munk equations (§ 5(d)). The absorption coefficient k is the algebraic sum of the values for the fibre and filler and is easily interpreted. The interpretation of s is more complex (Bown 1981, 1983) because the filler prevents some of the external fibrillation collapsing on to the fibre surfaces and disrupts the fibre–fibre bonding. Because fibres are held apart, the filler reduces the sheet strength and increases the bulk of the fibre network. The strength of a paper sheet, whether burst or tensile, decreases with decreasing filler particle size at a given loading (Beazley & Petereit 1975).

(b) As a coating pigment

Paper and board are coated with a mixture of pigment and adhesive to cover a relatively rough substrate. The adhesive (or binder) in the coating colour bonds the pigment particles to themselves and to the base sheet when the coating is dried. The appearance of the paper is improved and a smooth and receptive surface is provided for ink transfer in printing. The coated paper is generally calendered to improve its smoothness and gloss. Kaolin is the major pigment used in coating followed by calcium carbonate. Titania and calcined kaolins (Jones *et al.* 1982) are sometimes added to improve opacity. A mixed coating pigment system of kaolin and calcium carbonate may be used to reduce costs. Dean (1979) gives an excellent introduction to the aqueous coating of paper and board.

European coating grades are controlled in terms of particle size, powder brightness and rheology. There are many unspecified properties that have become established through usage and that contribute to the required consistent performance. In the U.S.A. over 10 coating grades are produced by the different companies. They range in particle size from 73–95 % below 2 μm which reflects the wide variation in size distribution of the available curves (see figure 1). They are classified in three groups: regular; delaminated (see §5(a)); and high brightness (see §6(b)).

The two major factors that determine the properties of a coated paper are the shape and size distribution of the pigment and the mass of the coat applied. Gloss, brightness, opacity and smoothness all generally improve with increasing coat mass and with decreasing particle size. A very finely sized kaolin can give problems. For example, the amount of adhesive needed to bond the platelets to themselves and to the base sheet increases with increasing fineness. The ultrafine grades impart excellent gloss but may give coatings deficient in opacity because the voids formed between the platelets are below the size required for optimum light scattering.

The ultrafine grades can also give a 'closed' coating which leads to difficulties when the paper is printed by offset lithography.

One recent development has been the deliberate narrowing of the size distribution of a platey Cornish kaolin to give an improved coating grade for LWC papers produced for rotogravure printing (Brociner & Beazley 1980). For this application the coating must be sufficiently compressible to permit contact with the inked cells during impression.

The smoothness of a coated paper is of importance and there is increasing interest in the mechanism of imbibition of coating colour and aqueous phase by the base sheet and the immobilization of solids that results (Baumeister & Weigl 1980; Baumeister & Kraft 1981; Kent 1983). Macroscopic roughness is determined by the shrinkage of the coating and the rate of the capillary imbibition and affects printing properties. Microscopic roughness is determined by the dimensions and orientations of the pigment particles and influences specular reflectance and hence gloss (Oittinen 1981).

9. USE IN CERAMICS

Kaolin is a major component in most ceramic bodies. It is used in the production of sanitaryware, tableware, fine china, electrical porcelain, wall tiles and floor tiles. Introductory accounts are given by Rado (1969) and by Worrall (1982) while Konta (1980) discusses the properties

TABLE 4. ILLUSTRATIVE CERAMIC BODY FORMULATIONS

| | composition (% mass) | | feldspathic | | firing temp/°C | |
product	kaolin	ball clay	flux	silica	biscuit	glost
hard porcelain	50–55	0	15–25	20–30	900	1400
soft porcelain	40	10	20–30	20–30	1230	1100
bone china†	25	0	25	0	1270	1120
vitreous sanitaryware‡	28	24	18	30	1200	—
earthenware	25	25	10–20	30–40	1170	1060
lime earthenware white tiles§	25	25	0	40	1080	980

† Also contains 50 % by mass bone ash.
‡ Nepheline syenite is the normally preferred flux.
§ Also contains 10 % by mass calcium carbonate.

of ceramic raw materials. Certain grades of china clay are calcined to about 1400 °C, ground and graded to give products used in the manufacture of kiln furniture and of moulds for use in investment casting.

Illustrative formulations are given in table 4, together with firing temperatures. The sanitaryware and tableware sectors are the most important in terms of kaolin consumption.

(a) Kaolin requirements

Kaolins with consistent properties are required in the production of tableware and sanitaryware (Bailey 1965). Historically, single pit clays were used. In the mid-1950s standardized ceramic grades were introduced: kaolins from four or five pits are refined and blended to give the sought properties. The two papers by Clark (1957, 1963) that discuss this change and the need for consistent properties make interesting reading.

Kaolins for sanitaryware are controlled in terms of analytical Fe_2O_3 and K_2O contents, particle size distribution, strength and rheological properties.

Kaolins for tableware are again controlled in terms of particle size, chemical composition and strength. Casting properties are also controlled because some shapes are still made by casting. The plastic properties of the kaolin are particularly important in bone china and hard porcelain formulations which contain little or no ball clay. There are other unspecified properties which have become established through usage.

(b) Use in vitreous sanitaryware

Shingler (1979) has given an excellent review of production methods. In simple terms the ceramic shape or cast is produced by pouring the deflocculated slip (see table 4) at a total solid content of about 72 % by mass into a plaster mould. A filtercake builds up on the plaster surface. When the required thickness is reached, the mould is inverted and the excess slip is allowed to drain. The cast is allowed to dewater in the mould. It is then removed and its surface is cleaned by sponging and holes are cut in the shape for waterways and for fittings. The cast is then dried, sprayed with glaze and fired.

The rheological properties of the casting slip are closely controlled. The deflocculant addition is carefully adjusted to be less than the amount required for optimum deflocculation (Bailey 1976). Recent technical developments have included battery casting (Gemmell 1981) and pressure casting (Ahlgren 1976), which have been successful in reducing production costs.

(c) Use in tableware

Examples of formulations and firing schedules are given in table 4. The glost temperature is that of the second firing after the glaze has been applied. Hard porcelain is made by one company in the U.K. and is the normal high quality product on the Continent, particularly in W Germany. Soft porcelain is not as strong or as translucent as hard porcelain. It is the general quality of much of the continental tableware. The body is more plastic and easily formed than hard porcelain but generally has an inferior colour. Bone china (Franklin & Forrester 1975; Rado 1981) is almost exclusively made in the U.K.; it is strong and translucent. Earthenware is produced in a wide range of qualities. It is not as strong as vitreous ware and has a lower chip resistance. The high content of plastic clay makes for easy fabrication. Opaque and transparent glazes are used.

Most tableware is produced by shaping a plastic mass in or on a plaster mould. Round items, including the majority of plates and saucers, are made in this way. Cups are made in a similar fashion with the handles being added later. Items with complicated shapes are cast from slip. The articles are dried before the subsequent stages of firing, glazing and decorating. These techniques require the constituents to be mixed, usually without the addition of deflocculants, at a relatively low solids content. The slurry is then filtered in a plate press, after which the cakes are fed to a deairing pug which extrudes a large diameter slug from which plastic body is passed to the 'making' machines.

A number of European companies now produce flatware directly from spray-dried granulate (Hahn 1978) which is metered into a press. The method has the advantage of complete automation of the 'making' machine and eliminates the need to dry the ware. A further way of eliminating the filterpress is to 'dry mix' the body constituents (Nilsson 1976). The components which must be free from contaminants, are blended and then made down to a well-controlled moisture content; production then follows the conventional route.

In translucent ceramics (hard porcelain, soft porcelain and bone china), light absorption depends on the concentrations of transition metal ions and which are largely derived from the kaolin. The Kubelka–Munk equations (Kerker 1969) have been applied (Hogg 1979). Values of s are interpreted in terms of light scattering from pores.

10. USE AS AN EXTENDER

Hydrous and calcined grades of kaolin are used as extenders in the paint and polymer industries. Properties of importance are fineness, aspect ratio and whiteness.

In paints, the requirements include some of those sought in the grades used in paper coating. Optical properties of brightness and opacity in the finished paint are important, as are physical properties such as scrub resistance and crazing resistance.

In plastics and rubbers, various other parameters may be important. Surface finish may be improved by fine well dispersed fillers, electrical properties enhanced by calcined clays, and rigidity, and other physical properties such as impact resistance may be significantly modified.

I am indebted to colleagues in the Central Laboratories of English China Clays p.l.c. for their help in preparing this review.

REFERENCES

Adams, R. W., Bidwell, J. I. & Jepson, W. B. 1977 British patent 1 481 118.
Ahlgren, R. 1976 *Trans. J. Br. ceram. Soc.* **75**, xviii–xix.
Alince, B. & Lepoutre, P. 1980*a* *J. Colloid Interface Sci.* **76**, 182–187.
Alince, B. & Lepoutre, P. 1980*b* *J. Colloid Interface Sci.* **76**, 439–444.
Allegrini, A. P., Jacobs, D. A. & Mercade, V. V. 1970 U.S. patent 3 503 499.
Allen, T. 1975 *Particle size measurement*, 2nd edn. London: Chapman & Hall.
Angel, B. R. & Hall, P. L. 1972 In *Proc. Int. Clay Conference, 1972, Madrid* (ed. J. M. Serratosa), pp. 47–60. Madrid: Division de Ciencias C.S.I.C.
Angel, B. R. & Vincent, W. E. J. 1978 *Clays Clay Miner.* **26**, 263–272.
Anon 1955 *Kaolin clays and their industrial uses*, 2nd edn. New York: J. M. Huber Corporation.
Anon 1982 *Ind. Miner.* no. 182, 41–43.
Bailey, R. T. 1965 *Interceram* **3**, 210–212.
Bailey, R. T. 1976 *Trans. J. Br. ceram. Soc.* **75**, 123–126.
Bailey, S. W. 1980 In *Crystal structures of clay minerals and their X-ray identification* (ed. G. W. Brindley & G. Brown), monograph no. 5, pp. 1–123. London: Mineralogical Society.
Barton, R. M. 1966 *A History of the Cornish china-clay industry*. Truro: D. Bradford Barton Ltd.
Baumeister, M. & Kraft, K. 1981 *TAPPI* **64**, 85–89.
Baumeister, M. & Weigl, J. 1980 *Wbl. PapFabr.* **108**, 145–151.
Beazley, K. M. 1972 *J. Colloid Interface Sci.* **41**, 105–115.
Beazley, K. M. & Petereit, H. 1975 *Wbl. PapFabr.* **4**, 143–147.
Bolland, M. D. A., Posner, A. M. & Quirk, J. P. 1976 *Aust. J. Soil Res.* **14**, 197–216.
Bolt, G. H. 1963 In *Proc. Int. Clay Conference* (ed. I. Th. Rosenquist & P. Graff-Petersen), p. 268. Oxford: Pergamon Press.
Borch, J. & Lepoutre, P. 1978 *TAPPI* **61**, 45–48.
Bown, R. 1981 *Wbl. PapFabr.* **109**, 263–266.
Bown, R. 1983 *Wbl. PapFabr* **111**, 737–740.
Brindley, G. W. 1980 In *Crystal structures of clay minerals and their X-ray identification* (ed. G. W. Brindley & G. Brown), monograph no. 5, pp. 125–195. London: Mineralogical Society.
Bristow, C. M. 1968 In *Int. Geological Congress. Report of the Twenty-Third Session Czechoslovakia, 1968. Proc. Symposium I, Kaolin Deposits of the World: A – Europe* (ed. M. Malkovsky & J. Vachtl), pp. 275–288. Prague: Academia.
Bristow, C. M. 1977 In *Proc. 8th Int. Symposium and Meeting on Alunite* (ed. E. Galen), pp. 1–19. Madrid–Rome: Servicio de Ministerio de Industrie y Energia.
Bristow, C. M. 1980 *Acta miner. petrogr.* XIV/1980. (Supplementum *Proc. 10th Kaolin Symposium in Budapest 3 September*, 1979, I.G.C.P. Project no. 23, pp. 19–25.)
Britt, K. W. 1970 *Handbook of pulp and paper technology*, 2nd edn. New York: Van Nostrand Reinhold.

Britt, K. W. 1975 Retention of fine solids during paper manufacture. CA Report no. 57. Atlanta, U.S.A.: Technical Association of the Pulp & Paper Industry.

Brociner, R. E. 1970 British patent 1194866.

Brociner, R. E. 1971 British Patent 1221929.

Brociner, R. E. & Beazley, K. M. 1980 *TAPPI* **63**, 55–58.

Bundy, W. M. & Berberich, J. P. 1969 U.S. patent 3477809.

Casey, J. P. 1960 *Pulp and paper*, 2nd edn, vol. II Papermaking. New York: Interscience Publishers Inc.

Cecil, T. A. & Jacobs, D. A. 1971 U.S. patent 3616900.

Clark, D. A. 1982 *Ind. Miner.* no. 182, 27–28.

Clark, N. O. 1957 *Trans. J. Br. ceram. Soc.* **56**, 389–401.

Clark, N. O. 1963 *J. Br. ceram. Soc.* **1**, 262–268.

Clark, N. O. 1971 British patent 1224877.

Climpson, N. A. & Taylor, J. H. 1976 *TAPPI* **59**, 89–92.

Conley, R. F. 1966 *Clays Clay Miner.* **14**, 317–330.

Conley, R. F. & Lloyd, Mary K. 1970 *Ind. Engng Chem. Process Des. Develop.* **9**, 595–601.

Cook, J. A. & Cobb, G. L. 1982 U.S. patent 4343694.

Coope, B. M. 1979 *Ind. Miner.* no. 136, 31–49.

Cundy, C. K. 1968 British patent 1103585.

Davison, R. W. 1982 In *1982 Papermakers Conference*, pp. 153–164. Atlanta, U.S.A.: TAPPI Press.

Dawson, A. M. 1977 *Ind. Miner.* no. 121, p. 16.

Dean, T. W. R. (ed.) 1979 *An operators guide to aqueous coating for paper and board*. London: Technical Division British Paper and Board Industry Federation.

Dennison, S. R. & Toms, G. L. 1967 *TAPPI* **50**, 502–508.

Dickson, T. 1982 *Ind. Miner.* no. 175, 93–99.

Dilland, J. G. & Koppelman, M. H. 1982 *J. Colloid Interface Sci.* **87**, 46–55.

Dolcater, D. L., Syers, J. K. & Jackson, M. L. 1970 *Clays Clay Miner.* **18**, 71–79.

Duke, J. B. 1967 U.S. patent 3353668.

Faye, G. H., Manning, P. G., Gosselin, J. R. & Tremblay, R. J. 1974 *Can. Mineralogist* **12**, 370–380.

Ferris, A. P. & Jepson, W. B. 1972 *Analyst, Lond.* **97**, 940–950.

Ferris, A. P. & Jepson, W. B. 1975 *J. Colloid Interface Sci.* **51**, 245–259.

Flegmann, A. W., Goodwin, J. W. & Ottewill, R. H. 1969 In *Clays and Other Colloidal Systems*, no. 13. Proc. Br. ceram. Soc. pp. 31–45. Stoke-on-Trent: British Ceramic Society.

Fordham, A. W. 1973 *Clays Clay Miner.* **21**, 175–184.

Franklin, C. E. L. & Forrester, A. J. 1975 *Trans. J. Br. ceram. Soc.* **74**, 141–145.

Gemmell, A. 1981 *Trans. J. Br. ceram. Soc.* **80**, 48–49.

Greene, E. W. & Duke, J. B. 1962 *Min. Engng, N.Y.* **14**, 51–55.

Greene, E. W., Duke, J. B. & Hunter, J. L. 1961 U.S. patent 2990958.

Gwilliam, R. D. 1971 *Filtration Separation*, **8**, 173–182.

Hahn, C. 1978 *Ber. dt. keram. Ges.* **55**, 121–122.

Hall, P. L. 1980 *Clay Miner.* **15**, 321–335.

Harben, P. 1979 *Ind. Miner.* no. 147, 23–35.

Hogg, C. S. 1979 In *Mineralogy of ceramics* (ed. D. Taylor & P. S. Rogers), pp. 23–35. Shelton, Stoke-on-Trent: British Ceramic Society.

Howe, J. A. 1914 *A handbook to the collection of kaolin, china-clay and china-stone, in the Museum of Practical Geology, Jermyn Street, London, S.W.* London: H. M. Stationery Office.

Hurst, V. J., (ed.) 1979 Field conference on kaolin, and Fuller's Earth. Bloomington, Indiana, U.S.A.: the Clay Minerals Society.

Iannicelli, J. 1976 *Clays Clay Miner.* **24**, 64–68.

Iannicelli, J., Kunkle, A. C. & Maynard, R. N. 1972 U.S. patent 3661515.

Jefferson, D. A., Tricker, M. J. & Winterbottom, A. P. 1975 *Clays Clay Miner.* **23**, 355–360.

Jepson, W. B. & Rowse, J. B. 1975 *Clays Clay Miner.* **23**, 310–317.

Jones, J. P. E., Cook, H. D. & Hollingsworth, R. L. 1982 *Pulp Pap.* **56**, 128–132.

Karickhoff, S. W. & Bailey, G. W. 1973 *Clays Clay Miner.* **21**, 59–70.

Keller, W. D. 1977*a* *Clays Clay Miner.* **25**, 347–364.

Keller, W. D. 1977*b* *Clays Clay Miner.* **25**, 311–345.

Keller, W. D. 1978 *Clays Clay Miner.* **26**, 1–20.

Keller, W. D., Cheng, H., Johns, W. D. & Meng, C. S. 1980 *Clays Clay Miner.* **28**, 97–104.

Keller, W. D. & Fitzpatrick, Wm. D. 1981 In *1980 Proc. Fourth Industrial Minerals Int. Congress, Atlanta* (ed. B. M. Coope), pp. 19–28. London: Metal Bulletin p.l.c.

Keller, W. D. & Haenni, R. P 1978 *Clays Clay Miner.* **26**, 384–396.

Keller, W. D., Pickett, E. E. & Reesman, A. L. 1966 In *Proc. Int. Clay Conference, Jerusalem, Israel*, vol. 1 (ed. L. Heller & A. Weiss), pp. 75–85. Israel Program for Scientific Translations, Jerusalem.

Kent, H. J. 1983 *Wbl. PapFabr.* **111**, 223–228.

Kerker, M. 1969 *The scattering of light*. New York: Academic Press.

Kitchener, J. A. 1972 *Br. Polym. J.* **4**, 217–229.

Kolm, H. H. 1975 *IEEE Trans.* **Mag-11** (5), 1567–1569.

Kolm, H. H., Oberteuffer, L. & Kelland, D. 1975 *Scient. Am.* **233**, 47–54.

Komusinski, J., Stoch, L. & Dubiel, S. M. 1981 *Clays Clay Miner.* **29**, 23–30.

Konta, J. 1980 Properties of ceramic raw materials. *Ceramic monographs-handbook of ceramics*, pp. 1–22. Verlag Schmid GmbH Freiburg i. Brg.

Koppelman, M. H. & Dilland, J. G. 1977 *Clays Clay Miner.* **25**, 457–462.

Kunkle, A. C. & Kollmar, C. E. 1977 U.S. patent 4 030 941.

Kunkle, A. C. & Kollmar, C. E. 1978 U.S. patent 4 105 466.

Lanier, W. P. & Jones, D. L. 1979 In *Scanning electron microscopy*, part I, pp. 525–529. U.S.A.: Scanning Electron Microscopy, Inc.

Lee, S. Y., Jackson, M. L. & Brown, J. L. 1975 *Clays Clay Miner.* **23**, 125–129.

Lepoutre, P. 1976 *TAPPI* **59**, 70–75.

Lim, C. H., Jackson, M. L., Koons, R. D. & Helmke, P. A. 1980 *Clays Clay Miner.* **28**, 223–229.

Lockhart, N. C. 1980 *J. Colloid Interface Sci.* **74**, 520–529.

Lyons, J. W. 1964 *J. Colloid Sci.* **19**, 399–412.

Lyons, S. C. 1950 U.S. patent 2 524 816.

McBride, M. B. 1976 *Clays Clay Miner.* **24**, 88–92.

McBride, M. B. 1978 *Clays Clay Miner.* **26**, 101–106.

McCartney, E. R. & Yeo, L. M. 1965 *J. Aust. ceram. Soc.* **1**, 13–17.

Macdonald, R. G. & Franklin, J. N. 1970 *Pulp and paper manufacture*, vols I–III. New York: McGraw-Hill.

Malden, P. J. 1972 U.S. patent 3 635 744.

Malden, P. J. & Meads, R. E. 1967 *Nature, Lond.* **215**, 844–846.

Mallett, A. S. & Craig, R. L. 1977 *TAPPI* **60**, 101–104.

Maynard, R. N. 1974 U.S. patent 3 857 781.

Maynard, R. N., Millman, N. & Iannicelli, J. 1969 *Clays Clay Miner.* **17**, 59–62.

Meads, R. E. & Malden, P. J. 1975 *Clay Miner.* **10**, 313–345.

Menadue, K. J. 1979 *Inst. Mining Metall.* A44–A50.

Michaels, A. S. 1958 *Ind. Engng Chem.* **50**, 951–958.

Millman, N. 1964 *TAPPI* **47**, 168–174.

Morris, H. H., Sennett, P. & Drexel, R. J. 1965 *TAPPI* **48**, 92A–99A.

Murray, H. H. 1963 *Georgia Minerals Newsletter* **16**, 3–11.

Murray, H. H. 1976 In *Paper coating pigments* (ed. R. W. Hagemeyer), pp. 69–109. Tappi Monograph Series no. 38, Atlanta, U.S.A.

Nilsson, H.-O. 1976 *Trans. J. Br. ceram. Soc.* **75**, 143.

Nott, A. J. & Price, W. M. 1978a U.S. patent 4 087 004.

Nott, A. J. & Price, W. M. 1978b U.S. patent 4 125 460.

Oakley, D. M. & Jennings, B. R. 1982 *Clay Miner.* **17**, 313–325.

O'Brien, N. R. 1971 *Clays Clay Miner.* **19**, 353–359.

O'Brien, N. R. 1975 *Bull. Inst. chem. Res., Kyoto University* **53**, 71–76.

Oittinen, P. 1983 In *BPBIF 7th Fundamental Research Symposium, Cambridge, 1981* (ed. J. Brander). pp. 635–654. London: Mech. Eng. Pub. Ltd.

Olivier, J. P. & Sennett, P. 1973 *Clays Clay Miner.* **21**, 403–412.

O'Neill, B. M. 1983 *Appita* **37**, 41–48.

Ottewill, R. H. 1978 In *Physical chemistry of pigments in paper coating* (ed. C. L. Garey). TAPPI Monograph no. 39, pp. 117–171. Atlanta: TAPPI Press, Inc.

Patterson, S. H. & Murray, H. H. 1975 In *Industrial minerals and rocks (non-metallics other than fuels)*, 4th edn, pp. 519–585. New York: American Institute of Mining, Metallurgical & Petroleum Engineers, Inc.

Plançon, A. & Tchoubar, C. 1977 *Clays Clay Miner.* **25**, 436–450.

Pounds, N. J. G. 1948 *The discovery of china clay, The economic history review*, 2nd series, vol. I, 1948–49, pp. 20–33. London: A & C. Black (pub. for the Economic History Society).

Rado, P. 1969 *An introduction to the technology of pottery*. London: Pergamon Press.

Rado, P. 1981 Bone china. *Ceramic monographs – handbook of ceramics*, pp. 1–10. Verlag Schmid GmbH Freiberg i. Brg.

Rand, B. & Melton, I. R. 1977 *J. Colloid Interface Sci.* **60**, 308–320.

Rengasamy, P. 1976 *Clays Clay Miner.* **24**, 265–266.

Robbins, D. W. & Strens, R. G. J. 1972 *Mineralog. Mag.* **38**, 551–563.

Robinson, J. V. 1976 *TAPPI* **59**, 77–82.

Sayin, M. & Jackson, M. L. 1975 *Clays Clay Miner.* **23**, 437–443.

Schofield, R. K. 1949 *J. Soil Sci.* **1**, 1–8.

Schofield, R. K. & Samson, H. R. 1953 *Clay Min. Bull.* **2**, 45–51.

Sennett, P., Olivier, J. P. & Hickin, G. K. 1974 *TAPPI* **57**, 92–95.

432 W. B. JEPSON

Shingler, T. 1979 Vitreous china sanitaryware, monograph 2.3. *Ceramic monographs – handbook of ceramics*, pp. 1–24. Verlag Schmid GmbH Freiburg i. Brg.
Starr, R. E. & Young, R. H. 1975 *TAPPI* **58**, 74–78.
Starr, R. E. & Young, R. H. 1978 *TAPPI* **61**, 78–80.
Talibudeen, O. & Goulding, K. W. T. 1983 *Clays Clay Miner.* **31**, 137–142.
Tanaka, H., Luner, P. & Cote, W. 1982 In *1982 Paper makers Conference*, pp. 129–144. Atlanta, U.S.A.: TAPPI Press.
Tchoubar, C., Plancon, A., Benbrahim, J., Clinard, C. & Sow, C. 1982 *Bull. Mineral.* **105**, 477–491.
Thompson, D. W., Macmillan, J. J. & Wyatt, D. A. 1981 *J. Colloid Interface Sci.* **82**, 362–372.
Uytterhoeven, J. 1963 In *Proc. Int. Clay Conference* (ed. I. Th. Rosenquist & P. Graff-Petersen), p. 267. Oxford: Pergamon Press.
Van Olphen, H. 1951 *Discuss. Faraday Soc.* **11**, 82–84.
Van Olphen, H. 1964 *J. Colloid Sci.* **19**, 313–322.
Van Olphen, H. 1977 *An introduction to clay colloid chemistry*, 2nd edn. New York: John Wiley.
Van Olphen, H. & Fripiat, J. J. (eds) 1979 *Data book for clay materials and other non-metallic minerals.* Oxford: Pergamon Press.
Watson, I. 1982*a* *Ind. Miner.* no. 179, 23–39.
Watson, I. 1982*b* *Ind. Miner.* no. 176, 17–26, 29–39.
Wedgwood, B. & Wedgwood, H. 1980 *The Wedgwood circle* 1730–1897. London: Studio Vista.
Weiss, A. 1959 *Z. anorg. allg. Chem.* **299**, 92–120.
Weiss, A. & Russow, J. 1963 In *Proc. Int. Clay Conference* (ed. I. Th. Rosenquist & P. Graff-Petersen), pp. 203–213. Oxford: Pergamon Press.
Weyl, W. A. 1959 *Coloured Glasses*. Originally published in 1951 by the Society of Glass Technology, Sheffield. Reprinted by Dawson's of Pall Mall, London 1959.
Whitley, J. B. & Iannicelli, J. 1972 U.S. patent 3 667 689.
Williams, D. J. A. & Williams, K. P. 1978 *J. Colloid Interface Sci.* **65**, 79–87.
Williams, D. J. A. & Williams, K. P. 1982 *Trans. J. Br. ceram. Soc.* **81**, 78–83.
Windle, W. & Gate, L. F. 1968 *TAPPI* **51**, 545–551.
Worrall, W. E. 1982 *Ceramic Raw Materials*, 2nd edn. Oxford: Pergamon Press.
Yarar, B. & Kitchener, J. A. 1970 *Trans. Instn Min. Metall.* **79**, C23–C33.

SBS
£33.50